# Embedded Design Using Programmable Gate Arrays

# Embedded Design Using Programmable Gate Arrays

**Dennis Silage**
**Electrical and Computer Engineering**
**Temple University**

Published by
Bookstand Publishing
Gilroy, CA 95020
2200_4



ISBN 978-1-58909-486-4

Printed in the United States of America

# Acknowledgements

This text would not have been possible without the support of the Xilinx University Program (XUP, *www.xilinx.com/univ*) and their commitment to assist Faculty to prepare the future workforce in the face of a sea change in electrotechnology. Chris Sepulveda and Jeff Weintraub of XUP have provided able assistance to our undergraduate and graduate curriculum and research efforts in the emerging paradigm of the fine grained programmable gate array. Ted Booth, a Xilinx DSP Specialist, has clarified for us the advanced electronic design automation tools provided by Xilinx to assist the student and the professional. Leonard Colavito, Robert Esposito, John Mountney and Andrew Whitworth are my recent graduate students who have contributed to this research at the System Chip Design Center in Electrical and Computer Engineering at Temple University.

This text, though, is dedicated to my wife Kathleen, daughter Marisa and son Matthew without whose love and support my entire efforts pale in comparison.

*Any sufficiently advanced technology is indistinguishable from magic.*

–Arthur C. Clarke

# Table of Contents

## Chapter 3 Programmable Gate Array Hardware



## Chapter 4 Digital Signal Processing, Communications and Control

# Preface

*Embedded Design Using Programmable Gate Arrays* describes the analysis and design of modern embedded processing systems using the field programmable gate array (FPGA). The FPGA has traditionally provided support for embedded design by implementing customized peripherals, controller and datapath constructs and finite state machines (FSM). Although microprocessor-based computer systems have usually been used for the design of larger scale embedded systems, the paradigm of the FPGA now challenges that notion of such a fixed architecture especially with the constraints of *real-time*.

This new paradigm in embedded design utilizes the Verilog hardware description language (HDL) behavioral synthesis of controller and datapath constructs and the FSM for digital signal processing (DSP), communications and control with the FPGA, external interface *hard core* peripherals, custom internal *soft core* peripherals and the *soft core processor*. The transition to embedded design with the parallel processing capabilities and fine grained architecture of the modern FPGA is described by in-part by the translation of C/C++ program segments for real-time processing to a controller and datapath construct or an FSM. However, the emergence of the Xilinx 8-bit PicoBlaze and 32-bit MicroBlaze soft core processors now also challenges the conventional microprocessor with its fixed architecture for embedded design.

*Embedded Design Using Programmable Gate Arrays* features the Xilinx Spartan-3E™ FPGA on the Digilent Basys Board and the Spartan-3E Starter Board evaluation hardware, the Xilinx Integrated Synthesis Environment (ISE) WebPACK electronic design automation (EDA) software tool in the Verilog HDL, the Xilinix CORE Generator for LogiCORE blocks and the auxiliary EDA software for the Xilinx PicoBlaze soft core processor. The complete Xilinx ISE WebPACK projects and Verilog HDL modules described in the Chapters are available (see the Appendix).

*Embedded Design Using Programmable Gate Arrays* is intended as a supplementary text and laboratory manual for undergraduate students in a contemporary course in digital logic and embedded systems. Professionals who have not had an exposure to the fine grained FPGA, the Verilog HDL, an EDA software tool or the new paradigm of the controller and datapath and the FSM will find that this text facilitates an expansive experience with the tenets of DSP, communications and control in embedded design. The References sections at the end of each Chapter contain a list of suitable undergraduate and graduate texts and reference books.

# 1

# Verilog Hardware Description Language

The evolution of the programmable gate array (PGA) from the nascent programmable logic device is facilitated by the concurrent development of a hardware description language (HDL) and electronic design automation (EDA) software tools, such as the Xilinx Integrated Synthesis Environment (ISE) WebPACK™, as described in Chapter 2 Verilog Design Automation. Programmable gate arrays have progressed from logic arrays with a coarse grained architecture of regular macrocells and simple non-volatile local interconnections to the now complex PGA with a fine grained architecture of dissimilar but specialized subsidiary units and advanced but volatile routing interconnections.

An HDL can *model* and *simulate* both the functional behavior and critical timing of digital logic. The Advanced Boolean Expression Language (ABEL) HDL provided combination and sequential logic equations early on (1983) that implemented complex digital logic functions and finite state machines. A contemporary HDL (1995) is Verilog with syntax similar to the C programming language. Verilog HDL concepts and syntax are surveyed here to support the development of an embedded design using programmable gate arrays.

Structural models in the Verilog HDL are presented in this Chapter which portrays a digital logic design as a model similar to a schematic. Several of these simpler structural models can be encapsulated in a high-level architectural description, which provides a less obtrusive structure for implementation. However, a behavioral model is more intuitive and evocative of the process. The finite state machine and the controller and datapath constructs provide this behavioral description in the Verilog HDL design implementation.

The translation of an algorithm in the conversational C language to the Verilog HDL facilitates the implementation of an embedded design using programmable gate arrays. Behavioral synthesis Verilog is both conversational and ultimately executes rapidly at the programmable gate array hardware level. However, the Verilog HDL is not a computer language and common arithmetic functions are problematical.

Finally, the PGA and the microprocessor are functionally compared. Unlike a PGA, operations cannot be performed in parallel by a microprocessor and the throughput rate in real-time processing is then limited by the clock period and the complexity of the task. However, the microprocessor efficiently implements large stored instructions for sequential tasks and can be programmed in the conversational C language. With the availability of the configurable soft core processor within the PGA, embedded designs can advantageously incorporate the sequential processor, essentially a microprocessor, and the inherently parallel and effective controller-datapath construct.

## Programmable Logic Devices

The programmable gate array (PGA) contains basic logic components and programmable interconnections that can be configured to functionally replicate a network of combinational or sequential digital logic. The earliest programmable integrated circuits (IC) (1978) were the programmable array logic (PAL) devices which implemented combinational logic using a *sum-of-products* (AND-OR) configuration. Combination logic and registered output PAL devices were then configured as *macrocells* (1983) which facilitated the design of sequential logic.

The original PAL devices were programmed by electrically destroying the connections in a *fuse map*, which produced a non-volatile logic circuit. The generic array logic (GAL) provided the same logic elements as a PAL device but could be electrically erased and reprogrammed (1987). Complex programmable logic devices (CPLD) combined several macrocells and a programmable interconnection system (1988). The CPLD exhibited a *coarse grained* architecture in which the logic circuit elements were highly regular and implemented as an array of several PAL devices on a single IC.

Although this configuration produces predictable timing delays, the coarse grained architecture of the CPLD results in a restrictive structure with the sum-of-products logic as an input to registered output. As an alternative architecture, the early PGA (1994) provided an array of configurable logic blocks (CLB) dominated by a complex routing scheme. The simplified CLB consists of a multiple input *look-up-table* (LUT), a type D (or data) *flip-flop* as a storage element and a *multiplexer* (MUX) which selects the logic output, as shown in Figure 1.1.

However, the early PGA remained substantially coarse gained with a regular architecture, albeit with a complex interconnection of these CLB elements. The Xilinx Spartan™ field programmable gate array (FPGA) (1998) utilizes a matrix of routing channels surrounding the CLB, a perimeter of programmable input-output blocks (IOB) and a static memory cell for the interconnection (data sheet DS060, *www.xilinx.com*), as shown in the simplified depiction in Figure 1.2. The *boundary scan* hardware is compatible with the IEEE Standard 1149.1 and allows the programming of any number of devices. The oscillator (OSC) is the hardware responsible for the generation and distribution of the synchronizing clock signals. The *start-u*p hardware sets the initial configuration of the flip-flops in the CLBs and the *read back* hardware allows the verification of the programming of the FPGA.

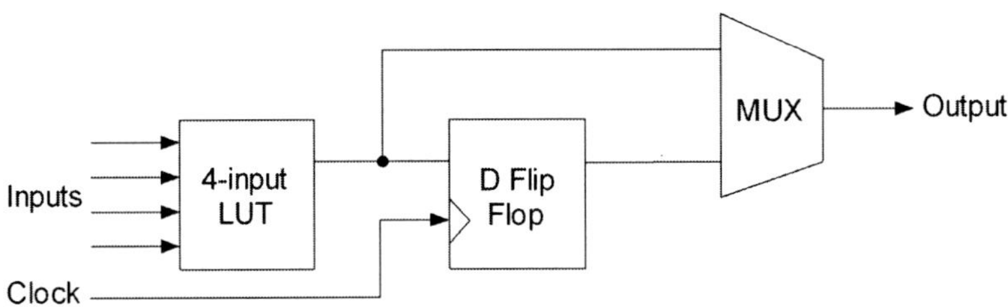


**Figure 1.1** Simplified configurable logic block

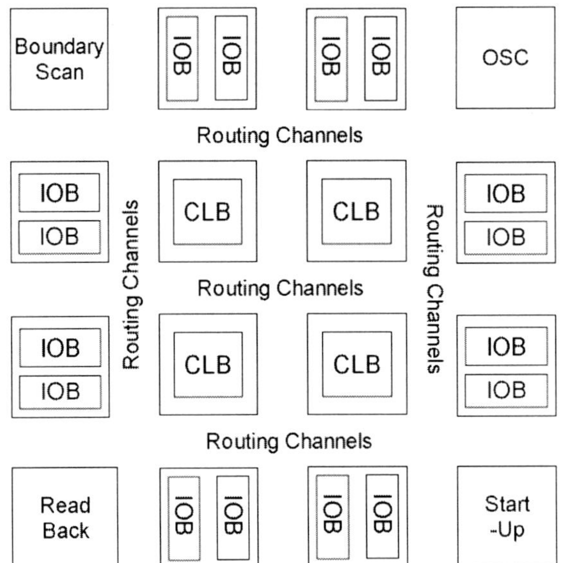


**Figure 1.2** Simplified depiction of the Xilinx Spartan™ FPGA architecture

In the Spartan FPGA the CLB consists of three LUTs which are used as logic function generators, two flip-flops and two groups of multiplexers which select the logic signals. The Spartan CLB also includes the capability to function as a 32 bit distributed random access memory (RAM). The largest Spartan FPGA provides a 28 by 28 CLB matrix with 224 input-output (IO) pins.

The Xilinx Spartan II™ FPGA (2000) introduced dedicated RAM in blocks of 4096 bits and delay-locked loops (DLL) for clock distribution delay compensation (data sheet DS001, *www.xilinx.com*). The DLL can also function as a clock frequency doubler. The largest Spartan II FPGA provides a 28 by 42 CLB matrix with 284 input-output (IO) pins and 56 Kb of block RAM.

The evolutionary *fine grained* architecture of the recent FPGA, exemplified by the Xilinx Spartan™-3E FPGA (2003), now includes these high level logic functions, such as input-output blocks (IOB), digital clock managers (DCM), multipliers, and block and distributed random access memory (RAM) (data sheet DS312 for the Spartan-3E, *www.xilinx.com*), as shown in the simplified depiction of a corner segment in Figure 1.3.

The DCM of the Spartan-3E FPGA supports clock *skew* elimination, phase shift and frequency synthesis. While the DCM incorporates the DLL of the Spartan II FPGA it also includes a digital frequency synthesizer (DFS) and status logic. The DCM is capable of generating a wide range of clock output frequencies by multiplying or dividing the clock input frequency.

In the Spartan-3E FPGA the CLB consists of four *slices* and each slice contains two LUTs and two flip-flops. The two LUTs in each slice can function as either a 16-bit distributed RAM or a 16-bit shift register. The four interconnected slices of the CLB are grouped into a conceptual left and right pair. The left pair of slices supports both logic and memory functions, while the right pair supports only logic functions. The Spartan-3 and the Spartan-3E PGA also have adjacent *dual-ported* 18 kb block RAM and 18-bit multipliers with a 36-bit two's complement product, as shown in Figure 1.3.

The Spartan-3E device is the target FPGA architecture for the embedded design applications here utilizing the evaluation boards, as described in Chapter 3 Field Programmable Gate Array Hardware. These elements of the recent fine-grained FPGA architecture contribute to robust and efficient digital logic but may also increase the difficulty of the design process.

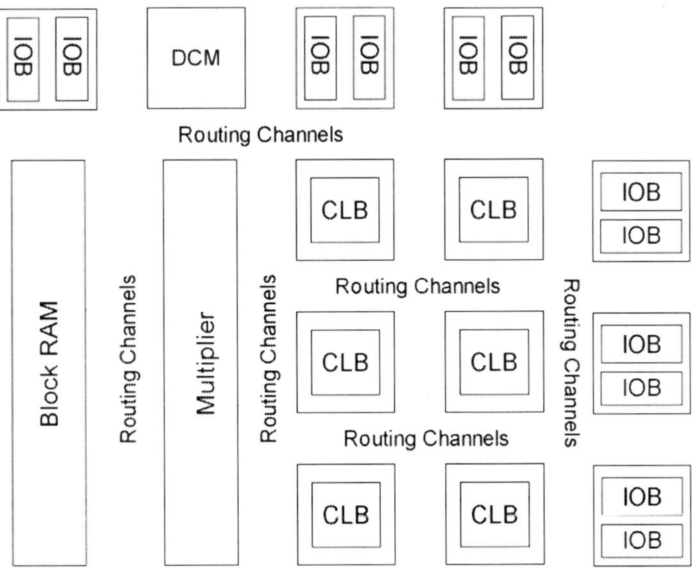


**Figure 1.3** Simplified depiction of a corner segment of the Xilinx Spartan™-3E FPGA architecture

Finally, the interconnection of the either the coarse or fine grained elements of the FPGA is provided by a *switch network*. The pattern of the switch is essentially binary data and is stored in either non-volatile fuses, electrically programmable read-only memory (EPROM), electrically erasable read-only memory (EEPROM), and block-oriented flash memory or volatile static RAM (SRAM). If the volatile SRAM is used internally for the switch configuration, the pattern is *downloaded* to the FPGA on *power-up* from an external device, such as an EEPROM or flash memory device. A standard Joint Test Action Group (JTAG) port for remote programming of the switch configuration by this *bit stream* data is provided on the FPGA, as described in Chapter 2 Verilog Design Automation.

## Hardware Description Languages

The complexity of the CPLD required the early development of a hardware description language (HDL) and an associated set of software electronic design automation (EDA) tools. An HDL can *model* and *simulate* both the functional behavior and critical timing of complex digital logic [Ciletti99]. Although an HDL can portray a digital logic design as an inherent structural model similar to a schematic, it can also describe the intrinsic behavioral function of the logic using constructs and procedures [Botros06]. In either instance, the digital logic signal output of the design generated by the HDL can be simulated with *test vectors* providing the logic signal stimulus input for verification. The resulting HDL model can then be *bound* to the interconnections of the programmable logic device for execution as part of a hardware system, as described in Chapter 2 Verilog Design Automation for the Xilinx ISE WebPACK EDA software tool.

An early HDL is ABEL (Advanced Boolean Expression Language) (1983), which provided combination and sequential logic equations that implemented complex logic functions and finite state machines (FSM). Other early HDLs for PAL and CPLD programmable logic devices include CUPL (1981) and PALASM (1982).

VHDL (Very high speed integrated circuit Hardware Description Language) originally provided documentation for application specific integrated circuits (ASIC). VHDL is a subset of the Ada programming language and, like the Ada language, is case insensitive and strongly typed. Strong typing implies that additional statements are required to convert from one data type to another. VHDL first appeared as IEEE standard 1076 (1987) but was extended (1993) to multi-valued logic and with more consistent syntax.

Verilog is an HDL with syntax similar to the C programming language, but Verilog, unlike the C language, does not have structures, pointers or recursive subroutines. Verilog, like the C language, is also weakly typed. Common to all HDLs but missing in programming languages that execute on sequential processors, Verilog has a different concept of the order of execution. The execution of its modules is not strictly linear, as Verilog has both sequential and concurrent statements [Botros06]. Modules can also execute in parallel and interact through semaphore logic signals and data. Verilog stems from a proprietary HDL (1985) but became IEEE Standard 1364 (1995) and was extended (2001) with an enhanced syntax [Ciletti04]. Verilog 2005 focused only on minor language corrections. A separate EDA development is SystemVerilog (IEEE Standard P1800-2005) which, although based on Verilog 2001, supports design verification.

### ABEL

The structure of the ABEL source file supports a *compiler* that translates its HDL into the appropriate logic element interconnections and routing for the device input and output pins. The output of an HDL compiler is the bit stream of programming instructions for the device. An HDL provides an implicit concurrency of the logic operations that are unlike the instructions of a conversational programming language, such as the C or Java, that execute on sequential processors.

Although an early implementation of an HDL, a survey of the ABEL source file structure is instructive to describe as a precursor of the Verilog HDL that is utilized here for embedded design using the programmable gate array. Both a short description of the ABEL HDL (Xilinx application note XAPP075, *www.xilinx.com*) and a complete language reference [Pellerin94] is available.

An ABEL source file for a combination logic 1-bit adder with carry as a *structural* model, derived from the logic diagram in Figure 1.4, is given in Listing 1.1. The ABEL source file consists of a header, declarations, and logic descriptions. Reserved words or *keywords* indicate the various subsections. The keywords *module* and *end* indicate the extent of the ABEL declaration. Source files can consist of multiple modules with concurrency of the logic operations and communication between the modules. Source file lines are terminated with a semicolon ( ; ) and a comment statement begins with a quotation mark ( " ) and terminates with either another quotation mark or an end-of-line.

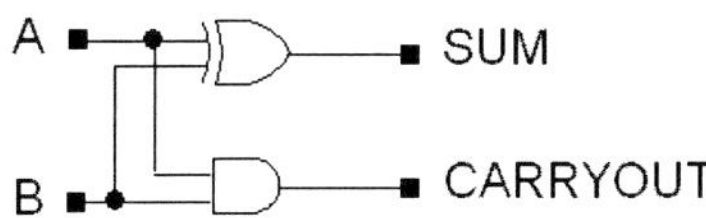


**Figure 1.4** Logic diagram for a combination logic one bit adder with carry

The optional keyword *title* provides a documentation of the module. The optional keyword *device* provides the identification and type of the device to be used (the XC4003E is a Xilinx FPGA). Recent EDA tools specify the type of programmable logic in a project description, rather than an HDL source file, so that the design is more independent of the device, as described in Chapter 2 Verilog Design Automation for the Xilinx ISE WebPACK EDA software tool.

The keyword *pin* instructs the compiler to associate logic signals with external device pins. The keyword *node*, although somewhat superfluous in Listing 1.1, has the same format but is used for an internal logic signal connection. The keywords *istype* and *com* indicate that these signals are combinatorial logic and an alternative description is *reg* for registered logic (a flip-flop output). Recent EDA tools assign the logic signals to pins in a *user constraint file* (UCF), rather than an HDL source file, for flexibility in interfacing the device, again as described in Chapter 2 Verilog Design Automation for the Xilinx ISE WebPACK EDA software tool. Finally, the keyword *equations* indicate the logical operations for the module.

**Listing 1.1** ABEL structural model of a 1-bit adder with carry

```
module onebitadder;
title '1 bit adder';
U1 device 'XC4003E';

" input and output pins
A, B pin 3, 5;
SUM, CARRYOUT pin 15, 18 istype 'com';

" internal node
AandB node istype 'com';

" equations
SUM = (A & !B) # (!A & B);
AandB = A & B;
CARRYOUT = AandB;

end onebitadder;
```

The logic operators (& (AND), ! (NOT), and # (OR)) in the *equations* section for the one bit adder with carry in Listing 1.1 mimic the logic diagram (or schematic) in Figure 1.4 and do not indicate the robustness of ABEL as an HDL. The construct *when-then-else* can be used in the *equations* section to describe a logic function, as given in Listing 1.2. The ABEL *else* clause is optional. The logical equality operator (= =) or inequality operator (! =) specifies the condition for the combinational logic signal assignment (=) here or registered logic signal assignment (: =).

**Listing 1.2** ABEL *when-then-else* logic function

```
when (A == B) then D = !A & !B;
     else when (A != B) then D = A & B;
```

In addition to structural models as equations, ABEL can utilize *behavioral* truth tables and state descriptions for logic design. A partial ABEL module, truth table description for the 1-bit adder with carry combinational logic function is given in Listing 1.3. The keyword *truth_table* and the bracketed term ( [...] ) indicates a *set* of logic signals for the function. The assignment operator (- >) is for a combinational logic output.

**Listing 1.3** ABEL 1-bit adder with carry combinational logic truth table

```
truth_table ([A, B] -> [SUM, CARRYOUT])
     [0, 0] -> [0, 0];
     [0, 1] -> [1, 0];
     [1, 0] -> [1, 0];
     [1, 1] -> [1, 1];
```

The ABEL truth table construct can also be used to describe sequential logic behavior. Listing 1.4 is a three bit counter which generates output logic 1 when the count reaches seven (binary 111).

**Listing 1.4** ABEL three bit counter sequential logic truth table

```
module counter;
title '3 bit counter'
U1 device 'XC4003E';

CLOCK pin 12;
RESET pin 22;

OUTPUT pin 18 istype 'com';
QC, QB, QA pin 14, 15, 16 istype 'reg';

[QC,QB,QA].CLK = CLOCK;
[QC,QB,QA].AR = RESET;

truth_table ([QC, QB, QA] :> [QC, QB, QA] -> OUTPUT)
     [0, 0, 0] :> [0, 0, 1] -> 0;
     [0, 0, 1] :> [0, 1, 0] -> 0;
     [0, 1, 0] :> [0, 1, 1] -> 0;
     [0, 1, 1] :> [1, 0, 0] -> 0;
     [1, 0, 0] :> [1, 0, 1] -> 0;
     [1, 0, 1] :> [1, 1, 0] -> 0;
     [1, 1, 0] :> [1, 1, 1] -> 0;
     [1, 1, 1] :> [0, 0, 0] -> 1;

end counter;
```

The assignment operator (: >) is for a registered logic output. The ABEL *dot extensions* are used to mimic the logic diagram of the counter. The dot extension *.CLK* is the clock input to an edge-triggered flip-flop and *.AR* is the asynchronous reset for the flip-flop. There are also a variety of other dot extensions for specific hardware flip-flop registers [Pellerin94].

Finally, ABEL provides a behavioral finite state machine (FSM) construct for logic design. A partial ABEL module for an arbitrary FSM is given in Listing 1.5. The keywords *state_diagram* and *state* describe an FSM with two input logic signals (Y, and LASTX), one logic output signal (X) and four states (INIT, LOCK, OK and RESET). The construct *if-then-else* is used in the FSM description, rather than *when-then-else* which is used for logic functions. The ABEL *else* clause is optional.

**Listing 1.5** ABEL finite state machine

```
state_diagram datalock

state INIT: if RESET then INIT else LOOK;

state LOCK: if RESET then INIT
      else if (X == LASTX) then OK
      else LOCK;

state OK: if RESET then INIT
      else if Y then OK
      else if (X == LASTX) then OK
      else LOCK;

state RESET: goto INIT
```

The ABEL combinational logic function in Listing 1.1 represents a design process that is allied to the underlying structural model as a hardware logic diagram or schematic. The ABEL HDL there merely translates the structural model to the programming bit stream for the interconnections of the logic elements of the programmable gate array. The ABEL truth table construct in Listing 1.4 approaches a behavioral description of the sequential logic function, but it also utilizes the dot extensions for the specific hardware flip-flop registers. However, the ABEL FSM in Listing 1.5 closely describes the essential behavior of the logic without a requisite structural model. Xilinx ABEL (XABEL) as an HDL is supported by the Xilinx ISE WebPACK EDA software tool.

## Verilog

The Verilog HDL provides all the structural and behavioral models available in the ABEL HDL with additional extensions for a higher level of behavioral abstraction [Ciletti99]. Similarly, the structure of the Verilog source file supports a *compiler* that translates its HDL into the appropriate logic element interconnections and routing for the device input and output pins. Verilog sustains a structured design methodology with a *top-down* or hierarchically representation of a digital logic design as simpler functional units. Alternatively, the verification of the digital logic design can proceed *bottom-up*, in which the lower level of functions are individually simulated and tested.

A Verilog HDL behavioral description consists of procedural statements without reference to available *primitives* which model combinational logic gates. A Verilog source file for the one bit adder with carry as a behavioral description is given in Listing 1.6.

The Verilog source file consists of a module declaration and description. The keywords *module* and *endmodule* indicate the extent of the Verilog HDL declaration. As in the ABEL HDL, source files can consist of multiple modules with concurrency of the logic operations and communication between the modules. Source file lines are terminated with a semicolon ( ; ) and a comment statement begins with a double forward stroke ( // ) and terminate with an end-of-line. Verilog also accepts a comment statement that begins with the C language convention of a forward stroke and asterisk ( /* ) and ends with an asterisk and forward stroke ( */ ).

The revised Verilog HDL standard (Verilog 2001) introduced a less verbose, C language style declaration of the ports of the module which also provided the mode and size (in bits) of the ports [Ciletti04]. The keywords *input* and *output* indicate the direction of the port and the keyword *reg* indicates that the port is an abstraction of a hardware storage element or register. The port size by default is one bit. The logic operators (^ (exclusive or) and & (and)) are not the primitives which directly model combinational logic gates in the Verilog HDL.

The one bit adder with carry in Listing 1.6 is a Verilog behavioral description because its functionality is described by an *event* triggered by changes in the logic levels of the data input signal. The keyword *always@* indicates the concurrent event. The keyword *or* here is part of the event syntax and is not the Verilog primitive for the combinational logic OR gate. The event executes statements contained within the keywords *begin* and *end*.

**Listing 1.6** Verilog behavioral description of a one bit adder with carry

```
module half_add(output reg sum, output reg carry, input a, b);

always@(a or b)            // event
    begin
        sum = a ^ b;       // exclusive OR
        carry = a & b;     // and
    end

endmodule
```

Verilog as an HDL is supported by the Xilinx ISE WebPACK EDA software tool. Rather than a cursory description of structural and behavioral modeling and the implementation of finite state machines, a concise but more systematic discussion of the Verilog HDL is provided in this Chapter. Other more extensive references for Verilog can provide support of these concepts and additional topics [Botros06] [Ciletti99] [Navabi06] [Zeidman99].

## Verilog Syntax and Concepts

Verilog concepts and syntax are surveyed here to support the development of an embedded design using programmable gate arrays. Verilog is a hardware description language (HDL) and, although the syntax is similar to the C language, its concepts are quite different. The description of the Verilog HDL concept and syntax here concentrates on the exacting hardware synthesis in a programmable gate array (PGA) and not the simulation task [Navabi06].

### Number Formats

Verilog is a hardware description language and integer numbers are represented in binary without reference to any fixed organization, such as bytes or words. For convenience in the source depiction, integer numbers can be specified as *<sign><size><base format><number>*. The *<sign>* is optional and is a minus sign ( – ) for a two's complement representation of the number. The *<size>* provides the number of bits and is optional. If the *<size>* is not given, then the number of bits is a default minimum of 32. However, if the number of bits required is greater than that specified then only the least significant bits (LSB) of *<size>* is stored and the most significant bits (MSB) are truncated.

The *<base format>* consist of an apostrophe ( ' ) followed by *b* (binary, base-2), *d* (decimal, base-10), *o* (octal, base-8) or *h* (hexadecimal, base-16) for integer numbers. If the *<base format>* is not used, the base is assumed to be decimal. The <*number*> must contain only digits which are valid for the specified base format. Digits can include *x* (unknown) and *z* (high impedance). Examples of valid and invalid integer number formats are given in Listing 1.7.

Real numbers are only specified in the Verilog HDL for simulation. In hardware synthesis, floating point numbers are usually implemented as signed, single precision 32 bit numbers using the IEEE-754 standard. In this standard there are 23 bits for the mantissa, 8 bits for the exponent and one sign bit. Other real number representations include IEEE-754 double precision floating point with 52 bits for the mantissa, 11 bits for the exponent and one sign bit and a fixed point number (for example, -123.456) with a sign bit and a variable number of bits for the integer and fractional part. The conversion and manipulation of floating point numbers is facilitated by a Xilinx LogiCORE HDL construct, as described in Chapter 2 Verilog Design Automation (data sheet DS335, Floating Point Operator, *www.xilinx.com*).

**Listing 1.7** Examples of valid and invalid integer number formats in Verilog

```
138       // decimal number, 32 bit as 00000000000000000000000010001010
10'd138   // decimal number, 10 bit as 0010001010
6'o74     // octal number, 6 bits as 111100
24'h25F   // hexadecimal number, 24 bit as 000000000000001001011111
8'hxB     // hexadecimal number, 8 bit as xxxx1011
3'b010    // binary number, 3 bits as 010
-6'b101   // 6 bit, two's complement of 000101 or 111011
-10'd15   // 10 bit, two's complement of 0000001111 or 1111110001
5'd124    // decimal number, 5 bits as 11100 since 7 bits are required
12'oF2    // invalid, F is not a octal digit
```

## Signal Data Types

Binary encoded logic signals in the Verilog HDL represent information such as loop indices, input data or a computed value [Cilletti04]. The value of a signal can be either a constant or a variable. Constants in Verilog are declared with the keyword *parameter* and can include arithmetic expressions with other constants, as given in Listing 1.8. Such declared constants are useful to describe the global characteristics of a Verilog module and to facilitate change during development, as in the constructs of the C language.

The keyword *defparam* is used to redefine a parameter within a module [Citelli04]. The redefinition can be specifically applied to the parameters of a specific (M2) nested module (auxbus), as given in Listing 1.8 and described in Chapter 1. However, this parameter redefinition may be evoked unnoticed anywhere within the design hierarchy and thus could cause problems to occur. Other Verilog HDL keywords and constructs that constraint the definition of parameter include the keywords *specparam* and *localparam* and the parameter redefinition by name association (#.*name*(*value*)(*port*);).

**Listing 1.8** Constant *parameter* declaration and *defparam* redefinition with in Verilog

```
parameter BUS_WIDTH = 32;                  // integer
parameter XMAX=640, YMAX = 480;            // integers
parameter START_VALUE = 8'b00001111;       // register
parameter SIZE = XMAX*YMAX;                // arithmetic expression

defparam auxbus.M2.BUS_WIDTH = 16;         // redefinition of BUS-WIDTH in instance M2
                                           // of nested module auxbus
```

Signal variables in Verilog are either of the type *net* or *register*. Net variables provide connectivity between objects in a Verilog module or between modules. Several of the net variable types available are specific for connection of Verilog primitives, which model combinational logic gates, and are not generally used in a Verilog HDL behavioral description [Navabi06]. However, the net variable type *wire* establishes behavioral connectivity with logic values of 0, 1, *x* (unknown) or *z* (high impedance) determined by the module *port* that drives the signal variable. The keyword *wire* declaration is followed by an optional array range and variable name, as given by Listing 1.9.

**Listing 1.9** Net variable *wire* declaration in Verilog

```
wire glbrst;              // scalar net signal
wire mclk, dav;           // scalar net signals
wire [31:0] average;      // 32-bit vector net signal
wire [0:7] adc_value;     // 8-bit vector net signal, reversed MSB
```

Register variables are used in behavioral modeling, are assigned values by procedural statements and store information [Ciletti04]. Register variables that are used in a Verilog HDL behavioral description are declared by the keywords *reg* and *integer*. The keyword *reg* is the abstraction of a hardware storage element and has a default size of one bit and an initial logic value of *x*. The register variables can be declared to utilize signed arithmetic with the keyword *signed*. The keyword *reg* declaration is followed by an optional array range and variable name, as given by Listing 1.10.

**Listing 1.10** Register variable *reg* declaration in Verilog

```
reg clock;                // register signal
reg reset, read_data;     // register signals
reg signed [7:0] sum;     // 7-bit plus sign register signal
reg [15:0] accum;         // 16-bit register signal, reversed MSB
```

The integer type of register variable supports numerical computation in Verilog behavioral synthesis. Integer variables are declared by the keyword *integer*, have a default but fixed size of 32 bits in signed two's complement format and a default initial value of zero [Ciletti04]. Integers are *true abstractions* that must have a numerical value, but the procedures that they comprise are compiled by the Verilog HDL to synthesizable hardware. The keyword *integer* declaration is followed the variable name then by an optional array range, as given by Listing 1.11.

**Listing 1.11** Register variable *integer* declaration in Verilog

```
integer data;             // integer
integer i, j, k;          // multiple integers
integer data[1:1000]      // integer array
```

## Strings and Arrays

Verilog utilizes the register variable with the *reg* declaration to store ASCII character strings as 8-bit values. The string can be initially assigned to the register variable *reg* declaration by enclosing it within quotation marks ( " ), as given by Listing 1.12. If the string assignment uses less than the available number of bits, the unused register variable *reg* declaration bits are filled with zero. The assignment of a string to a register variable with the *reg* declaration is a single address *memory*. Multiple addressable register variables of the same size can be accommodated with the format *<word size><variable name><memory size>*, as given by Listing 1.12.

The revised Verilog HDL standard (Verilog 2001) supports the selection of a word or the contiguous part of a word for net or register variables with the part select operators (<*start bit*> +: <*width*> and <*start bit*> –: <*width*>). The parameter *width* specifies the size of the selection which is obtained by either incrementing ( +: ) or decrementing ( –: ) the index of the bits in the register. In Listing 1.12, the integer register variable i sets the parameter *start bit* as the starting position for the selection [Ciletti04]. A fixed selection of a part of the register utilizes fixed parameters with the separation operator ( : ). Finally, a register variable can be concatenated from either the entire or a portion of two or more smaller register variables. The concatenation is enclosed by the brace symbols ( { } ) and separated by a comma, as given in Listing 1.12.

**Listing 1.12** String and memory register variable *reg* declaration in Verilog

```
parameter STRING_LENGTH = 11;             // parameter declaration
reg [8 * STRING_LENGTH] string_data;      // arithmetic calculation of size
reg [7:0] byte_memory [0:511];            // MSB bit first, 512 byte memory
strdata = "hello world";                  // string assignment to a register
lcddata[7:0] = strdata[i-:8];             // variable selection of a register
lcddata[7:0] = strdata[87:80];            // fixed selection of a register
reg [3:0] data = {adata[1:0], bdata[1:0]}; // concatenation
```

Although Verilog 2001 supports multidimensional array, the Xilinx ISE WebPACK EDA software tool only supports arrays with no more than three dimensions. Listing 1.13 shows a two-dimensional array of register variables with the *reg* declaration and the selection of a fixed word and a fixed part of a word in the array.

**Listing 1.13** Multidimensional arrays and word selection in Verilog 2001

```
reg [7:0] pix_data [0:639] [0:479];          // two dimensional array of bytes
wire [7:0] pixout [120] [330];               // fixed word of pixel (120,330)
wire msb_pix = pix_data [120] [330] [7];     // MSB of pixel (120,330)
```

## Signal Operations

Verilog provides intrinsic signal operations which describe logic symbolically in behavioral synthesis, rather than by Verilog primitives in a structural model which utilize combinational logic gates [Botros06]. The *bitwise* operators combine two signal operands to form a signal result. The symbols utilized in the Verilog HDL standard and the operations are patterned after those in the C language, as given in Table 1.1. The *exclusive not or* bitwise operation is not available in the C language.

**Table 1.1** Bitwise operations in Verilog

| Operator | Operation |
|---|---|
| ~ | Negation (one's complement) |
| & | And |
| \| | Inclusive Or |
| ^ | Exclusive Or |
| ~ ^ | Exclusive Not Or |
| ^ ~ | Exclusive Not Or |

The *reduction* operators produce a scalar with logic values of 0, 1, or *x* (unknown) from a single signal operand, as given in Table 1.2. Each bit of the signal operand participates in the reduction operation to produce the result. For example, if x = 1001, then &x = 0 and |x = 1. The scalar value is *x* (unknown) if the operand contains at least a single bit which is unknown.

The *relational* operators compare two signal operands and produce a scalar with logic values of 0 (false), 1 (true), or *x* (unknown), as given in Table 1.3. The scalar value is *x* (unknown) if either operand contains at least a single bit which is unknown or *z* (high impedance).

The *logical equality* and *logical inequality* operators compare two signal operands *bit-by-bit* and produce a scalar with logic values of 0 (false), 1 (true), or *x* (unknown), as given in Table 1.4. The scalar value is *x* (unknown) if the operand contains at least a single bit which is unknown or *z* (high impedance). If the operands are not the same length, logic 0 is added as the most significant bits of the smaller operand. The *case equality* and *case inequality* operators compare two signal operands *bit-by-bit* utilizing the four logic values (0, 1, *x*, *z*) and produce a scalar with logic values of 0 (false) and 1 (true), as given in Table 1.4.

**Table 1.2** Reduction operations in Verilog

<table>
<tr><td>&</td><td>And</td></tr>
<tr><td>~ &</td><td>Not And</td></tr>
<tr><td>|</td><td>Or</td></tr>
<tr><td>~ |</td><td>Not Or</td></tr>
<tr><td>^</td><td>Exclusive Or</td></tr>
<tr><td>~ ^</td><td>Exclusive Not Or</td></tr>
<tr><td>^ ~</td><td>Exclusive Not Or</td></tr>
</table>

**Table 1.3** Relational operations in Verilog

<table>
<tr><td><</td><td>Less Than</td></tr>
<tr><td>< =</td><td>Less Than or Equal To</td></tr>
<tr><td>></td><td>Greater Than</td></tr>
<tr><td>> =</td><td>Greater Than or Equal To</td></tr>
</table>

**Table 1.4** Equality operations in Verilog

<table>
<tr><td>= = =</td><td>Case Equality</td></tr>
<tr><td>! = =</td><td>Case Inequality</td></tr>
<tr><td>= =</td><td>Logical Equality</td></tr>
<tr><td>! =</td><td>Logical Inequality</td></tr>
</table>

The *logical* operators are similar to the reduction operators but produce a scalar with logic values of 0, 1, or *x* (unknown) from two signal operands, as given in Table 1.5. Each bit of the two signal operands participates in the logical operation to produce the result. For example, if x = 1001 and y = 0110 then x && y = 0 and x || y = 1. The scalar value is *x* (unknown) if either of the operands contains at least a single bit which is unknown. The operation is evaluated from left to right and ends as soon as result is unequivocally true or false [Ciletti04].

**Table 1.5** Logical operations in Verilog

<table>
<tr><td>& &</td><td>Logical And</td></tr>
<tr><td>| |</td><td>Logical Or</td></tr>
<tr><td>!</td><td>Logical Negation</td></tr>
</table>

The logical shift operator shifts the bits in a signal operand to the right or left and fills the vacated bits with a logic value of 0, as given in Table 1.6. For example, if x = 10011100 then x << 2 =

01110000. The revised Verilog HDL standard (Verilog 2001) supports the arithmetic shift operator shifts the bits in a signal to the right or left and fills the vacated bits with the most significant bit (MSB) if a right shift and a logic value of 0 if a left shift, as given in Table 1.7 [Ciletti04]. For example, if x = 10011100 then x >>> 2 = 11100111. The left shift arithmetic operator is functionally the same as the left shift logical operator.

**Table 1.6** Logical shift operations in Verilog

| | |
|---|---|
| >> | Logical Shift Right |
| << | Logical Shift Left |

**Table 1.7** Arithmetic shift operations in Verilog

| | |
|---|---|
| >>> | Arithmetic Shift Right |
| <<< | Arithmetic Shift Left |

## Arithmetic Operations

The common arithmetic operations in Verilog 2001 manipulate the register variable *reg* declaration as signed or unsigned integers of any bit size [Ciletti04]. The keyword *signed* is used to declare that the register variable *reg* declaration is signed, as given in Listing 1.14. The register variable *integer* declaration has a default but fixed size of 32 bits for signed two's complement arithmetic. The common arithmetic operations for the Verilog HDL are listed in Table 1.8.

The division operation ( / ) is somewhat problematical with register variable *reg* declarations because in hardware synthesis the Xilinx ISE WebPACK EDA software tool only supports division by a signed power of two ($\pm 2^n$). Precedence for the arithmetic operations in Verilog, as in the C language, occurs from left to right on a source line with the multiplication, division and modulus operations at a higher precedence level than the addition and subtraction operations.

**Table 1.8** Arithmetic operations in Verilog

| | |
|---|---|
| * | Multiplication |
| / | Division |
| % | Modulus |
| + | Addition |
| – | Subtraction |

The conditional operation in the Verilog HDL can utilize the logic true or false of a Boolean expression to select one of two possible arithmetic expressions, as given in Listing 1.14. The form of the conditional operation is *<Boolean expression>* ? *<result if true>* : *<result if false>*.

**Listing 1.14** Conditional operation in Verilog

```
reg signed [15:0] c;
reg signed [7:0] a;
reg signed [7:0] b;

c = (a > b) ? 1 : 0;            // c will be either 1 or 0
c = (a == b) ? a – b : a + b;   // c will be either a – b or a + b
c = (a – b) > 4 ? a : b;        // c will be either a or b
```

Verilog 2001 provides for assignment width extension for the signed 16-bit register variable *reg* declaration c from the signed 8-bit register variable *reg* declarations a and b in Listing 1.14. If the expression of the right hand side is signed, then the sign bit is used to fill the addition bits of the left hand side of the assignment. If the expression on the right hand side is unsigned, then the additional bits of the left hand side are filled with logic 0.

## Structural Models in Verilog

Structural models in the Verilog HDL portray a digital logic design as a model similar to a schematic [Navabi06]. Several of these simpler structural (or even behavioral) models can be encapsulated in a high-level architectural description, which provides a less obtrusive structure for implementation in an embedded design using a programmable gate array. A simple structural model is a *netlist* or connection of logic gates and is often inferred from an existing design in combinational or sequential digital logic [Zeidman99].

The intrinsic Verilog HDL primitives consists of a logic gate construct that operates on one, two or more 1-bit input signals and provides a one 1-bit output signal, as given in Table 1.9. Verilog primitives can be combined to describe sequential digital logic *flip-flop* operation and other complex functions, such as an arithmetic logic unit (ALU).

**Table 1.9** Intrinsic combinational logic Verilog primitives.

| | |
|---|---|
| and (output, input1, input2) | And |
| nand (output, input1, input2) | Not And |
| or (output, input1, input2) | Or |
| nor (output, input1, input2) | Not Or |
| xor (output, input1, input2) | Exclusive Or |
| xnor (output, input1, input2) | Exclusive Not Or |
| buf (output, input) | Non-inverting Buffer |
| not (output, input) | Inverter |

Additional Verilog HDL primitives are used to model logic operation at the *gate level* of abstraction in hardware synthesis for specific FPGA devices. These additional Verilog primitives are FPGA device and EDA implementation dependent and model *tri-state*, *bi-directional* and *open collector* logic gates [Ciletti99]. The remaining Verilog primitives model the critical path timing of the logic operation at the *switch level* of abstraction, based on the transistor fabrication technique in use.

The Verilog HDL switch level primitives are usually not utilized in an initial structural model of an embedded design using a programmable gate array. The EDA tools provide support for critical path timing determinations. If timing anomalies are detected by the Verilog synthesizer warning messages are produced. Finally, the Verilog gate level primitives can also be inferred in structural FPGA hardware synthesis with the output logic signal *z* (high impedance) that is available for each of the intrinsic combinational logic Verilog primitives, as given in Table 1.9.

### Modules

The Verilog HDL structural (or a behavioral) model consists of declarations beginning with the keyword *module* and ending with the keyword *endmodule*. The declarations specify the signal inputs and outputs of the model at the *port* and the manipulation of the signals using the Verilog primitives. The order of the input and the output ports of the module are irrelevant. However, the order of the output and input signals for the Verilog primitives in Table 1.9 is relevant. The Verilog HDL structural model of the 1-bit adder with carry, similar to that of the one in the ABEL HDL in Listing 1.1, is given in Listing 1.15.

The module name is case sensitive and the names *half_adder* and *Half_adder* are assumed by the Verilog compiler to be different modules. A comparison can be made of this structural model with the Verilog *event-driven* cyclical behavioral description of the one bit added with carry in Listing 1.6. The declarations define whether the module is a structural or a behavioral model or a combination of the two descriptions.

**Listing 1.15** Verilog structural model of a 1-bit adder with carry

```
module half_add(output sum, carry, input a, b);

    xor (sum, a, b);      // exclusive OR
    and (carry, a ,b);    // and

endmodule
```

## Ports

The ports of a Verilog HDL structural (or a behavioral) module describe the interface to other modules or the external environment. The mode of a port can be unidirectional, with the keywords *input* or *output*, or bidirectional, with the keyword *inout*. If the mode of a port is input, then it must appear only as the input of a Verilog primitive (or only on the right hand side of a behavioral assignment statement ( = ) ). If the mode of a port is output, then it must appear only as the output of a Verilog primitive. The port functions as both and input and output if the mode of a port is bidirectional [Botros06]. Finally, the input and output ports of a Verilog HDL module can be modified by the keyword *signed* which declares that signed arithmetic is to be used for the net and register variables and integers of the port.

The ports of a module must be associated in a consistent manner with the declaration of the module. The position and size in bits of the net or register variables are directly mapped in the hierarchical structure of nested modules that form the interconnections. In this *connection by position* option for the ports of a module the formal name in the declaration need not be the same as the actual name evoked in the hierarchical structure, as given in Listing 1.16.

However, this connection by position method with its direct mapping is problematical when the number of the port variables is large and their bit size is diverse. An alternative is the *connection by name* convention in which ports are associated in the nested module port list by the syntax *.formal_name(actualname)*, as given in Listing 1.17. This option connects the *actual_name* to the *formal_name* regardless of the position of the entry in the list [Ciletti04].

## Nested Modules

The *top-down* design of a complex logic architecture implemented in a programmable gate array implies that the system is partitioned into functionally smaller structural (or behavioral) modules. Although these inherently smaller modules are easier to design and test, their utility extends to their *design reuse* and the lessening on the interconnection constraints placed on the place and route operations of the HDL compiler and synthesizer. Nested modules in the Verilog HDL support such a top-down design.

The top-down design methodology using nested modules and Verilog primitives is illustrated by the binary full-adder with carry, as given in Listing 1.16. The Verilog HDL hierarchical model of the design contains two instances of the half_add module that structurally models the 1-bit adder with carry, as given in Listing 1.15, and an *or* gate Verilog primitive [Ciletti04]. Each instance of the half_add module is given a module name (M1 and M2) for identification by the Verilog compiler.

In Listing 1.16 the *wir*e net variable type establish the 1-bit *connection by position* for signals between the half_add modules and the *or* gate Verilog primitive. In the first instance of the half_add module (M1), wire w1 represents the output signal sum, which is connected to the input signal b of the second instance (M2) of the half_add module. This nested module 1-bit full adder with carry can be in turn be nested to form a multiple bit adder.

The ports of the nested Verilog HDL modules in the full_add module in Listing 1.16 are identified as either input or output by their position in the declaration of the module. Although this is adequate for a small number of ports, a large number of ports can generate confusion. Wire w1 is both an output signal to one of the half_add modules and an input signal to the other.

Verilog ports can also use *connection by name* for the unambiguous association of the signals of the half_add modules, which does not require that the port connections be listed in the same relative position. The nested 1-bit full adder with carry module with *connection by name* port mapping is given in Listing 1.17.

**Listing 1.16** Verilog nested structural model of a 1-bit full-adder with carry with port *connection by position*

```
module full_add(output sum_out, carry_out, input a_in, b_in, carry_in);

    wire w1, w2, w3;

    half_add M1 (w1, w2, a_in, b_in);
    half_add M2 (sum_out, w3, carry_in, w1);
    or (carry_out, w2, w3);

endmodule
```

**Listing 1.17** Verilog nested structural model of a 1-bit full-adder with carry with port *connection by name*

```
module full_add(output sum_out, carry_out, input a_in, b_in, carry_in);

    wire w1, w2, w3;

    half_add M1 (.a(a_in), .sum(w1), .b(b_in), .carry(w2));
    half_add M2 (.sum(sum_out), .b(w1), .carry(w3), .a(carry_in));
    or (carry_out, w2, w3);

endmodule
```

## User Defined Primitives

User defined primitives (UDP) in the Verilog HDL facilitate the implementation of complex structural models. The UDP is not a Verilog module and must be incorporated within a module to be instantiated. The UDP is encapsulated by the keywords *primitive* and *endprimitive*, all the input ports must be 1-bit signals and only one 1-bit output port, which must be the first entry, is allowed. Verilog provides the *behavioral* truth table, similar to that in ABEL in Listing 1.3, which is versatile and can describe complex combinational and sequential logic.

The scalar values 0 (false), 1 (true), *x* (unknown) and any transitions between these logic values are indicated in the table. The scalar logic value *z* (high impedance) is not supported by the UDP. The register variable *reg* declaration defines a sequential logic construct, but only the single 1-bit output port can be a *reg* declaration. The keyword *initial* indicates the *power-up* state for the UDP and only the scalar logic values 0 (false), 1 (true) and *x* (unknown) can be used. The default initial state is *x* (unknown).

The Verilog truth table declaration begins with the keyword *table* and ends with the keyword *endtable*. The entries for a combinational logic truth table are specified as *<input logic value>* : *<output logic value>*. A whitespace must separate each input entry in the table. The order of the table entries for the inputs is associated with the declaration of the ports of the Verilog primitive. The question mark ( ? ) entry in the truth table indicates that a *don't care* condition occurs for the scalar logic input value, which can be 0, 1 or *x*. A combinational logic UDP using a truth table for a 2-bit multiplexer is given in Listing 1.18.

The entries for a sequential logic truth table are specified as *<input logic value>* : *<previous logic value>* : *<output logic value>*. The 1-bit output port must be declared a register variable *reg*. Only one input logic signal can have an edge transition for each entry in the truth table. If any input logic signal entry has an edge transition, then all other input logic signal must be have truth table entries to account for their transitions. A sequential logic UDP for a type D (data) flip-flop is given in Listing 1.1

**Listing 1.18** Combinational logic UDP using truth table for a 2-bit multiplexer

```
primitive mux (output y, input a, b, sel);

    table
      // a  b  sel : y
         0  ?   0  : 0;       // select a
         1  ?   0  : 1;       // select a
         ?  0   1  : 0;       // select b
         ?  1   1  : 1;       // select b
    endtable

endprimitive
```

**Listing 1.19** Sequential logic UDP for a type D flip-flop

```
primitive dff (output reg q, input d, clk, rst);

    table
      // d  clk rst:state:q
         ?  ?   0 : ? : 0;      // active low reset
         0  R   1 : ? : 0;      // rising clock edge, data = 0
         1  R   1 : ? : 1;      // rising clock edge, data = 1
         ?  N   1 : ? : -;      // ignore clock negative edge
         *  ?   1 : ? : -;      // ignore all edges on d
         ?  ?   P : ? : -;      // ignore reset positive edge
    endtable

endprimitive
```

In Listing 1.19 the rising edge of the *clock* signal is the desired transition and the negative edge of the clock, the positive and negative edges of the d input and the positive edge of the reset input is ignored. A level sensitive truth table entry, such as that for reset, has precedence over edge sensitive entries. The truth table input entry R indicates a rising transition (logic 0 to 1), N indicates a negative transition (logic 1 to 0, logic 1 to *x* or logic *x* to 0), P indicates a positive transition (logic 0 to 1, logic 0 to *x* or logic *x* to 1), and asterisk ( * ) indicates any possible transition (logic 0 to 1, logic 0 to *x*, logic *x* to 1, logic *x* to 0, logic 1 to 0 or logic 1 to *x*). Here the truth table output entry dash ( – ) indicates no change in state for the register variable.

Other truth table input entries are available but not used in Listing 1.19. The truth table input entry *x* indicates an unknown logic value, B indicates a *don't care* condition if the logic value is 0 or 1 and F indicates a falling transition (logic 1 to 0).

## Behavioral Models in Verilog

Although a digital logic design can be rendered entirely in a structural model in the Verilog HDL, a behavioral model is more intuitive and evocative of the process. Structural models do provide an architectural partition of the design and are often provided as *building blocks* for a more complex behavioral model. The Xilinx LogiCORE building blocks are efficient in the use of FPGA resources and can be integrated with Verilog modules, as described in Chapter 2 Verilog Design Automation. This partitioning facilitates the digital logic design process by allowing different levels of abstraction [Ciletti99] [Navabi06].

### Continuous Assignment

Continuous assignment statements in the Verilog HDL are Boolean equations that describe combination logic using the bitwise signal operations, as given in Table 1.1. These bitwise operations are implicit descriptions of combinational logic and have structural or gate level equivalences and can be easily synthesized. The continuous assignment statement is declared with the keyword *assign* followed by a net variable name and the equal sign ( = ) as the operator. Net variables not only provide connectivity between objects in a Verilog module, using the *wire* construct, but between the output and input objects between Verilog modules.

The Verilog HDL nested structural model of a 1-bit full-adder with carry, as given in Listing 1.16, utilized the wire construct and is difficult to discern without recourse to the schematic. The 1-bit full-adder with carry can also be described with continuous assignment statements for the two output net variables and is more concise, as given in Listing 1.20.

**Listing 1.20** Verilog continuous assignment model of a 1-bit full-adder with carry

```
module full_add(output sum_out, carry_out, input a_in, b_in, carry_in);

    assign sum_out = a_in ^ b_in ^ carry_in;
    assign carry_out = (a_in & b_in) | (b_in & carry_in) | (a_in & carry_in);

endmodule
```

The Verilog continuous assignment statement is not like a program statement in the C language because it is sensitive to the signal variables on the right hand side of the assignment operator and execute concurrently. Whenever a signal variable changes logic state the assignment is reevaluated and the result is updated. The continuous assignment statement can not utilize a register variable *reg* declaration.

**Listing 1.21** Verilog continuous assignment model of a 1-bit full-adder with carry and enable

```
module full_add(output sum, carry_out, input a, b, carry_in, enable);

    assign sum = enable ? a ^ b ^ carry_in : 1'bz;
    assign carry_out = enable ? (a & b) | (b & carry_in) | (a & carry_in) : 1'bz;

endmodule
```

Continuous assignment statements are used with the conditional operator, as given in Listing 1.14, to model *tri-state* digital logic. The Verilog continuous assignment model of a 1-bit full-adder with carry can be augmented with a conditional operator on an *enable* input signal to provide a tri-state output using the *z* (high impedance) logic level, as given in Listing 1.21.

Continuous assignment statements with the conditional operator and feedback can model a logic level sensitive *transparent latch*, as given in Listing 1.22. The output signal of a transparent latch follows the input signal when the latch is enabled. The net variable output signal qout is fedback to itself and will be synthesized in the Verilog HDL as a hardware latch [Zeidman99].

**Listing 1.22** Verilog continuous assignment model of a transparent latch

```
module tlatch(output qout, input data, enable);

    assign qout = enable ? data : qout;

endmodule
```

## Single Pass Behavior

Single pass behavior in the Verilog HDL for simulation utilizes the keyword *initial* which executes the associated statements then expires [Ciletti04]. The single pass behavior is useful for setting the initial value of register variables declared by the keywords *reg* and *integer*, as given in Listing 1.23. However, single pass behavior is not a hardware synthesizable construct and is only used in simulation. For hardware synthesis, register variables can be given a value in a declaration or reset by an external event, such as an input signal derived from a push button [Lee06].

**Listing 1.23** Verilog single pass behavior in simulation using the keyword *initial*

```
reg [4:0] rstate;
reg dav, sclk, ackdata;

initial
    begin
        rstate = 0;
        dav = 0;
        sclk = 0;
        ackdata = 1;
    end
```

## Cyclic Behavior

Cyclic behavior is an abstract model of logic functionality and is not related to gate level primitives in the Verilog HDL. Rather, cyclic behavior executes procedural statements which generate output signals from input signals, as does the C language for variables. However, the difference between the Verilog HDL and any programming language such as C is that the procedural statements can execute either unconditionally and concurrently or with an event control statement [Ciletti04]. The event control statement has a list of input signals that are sensitive to either logic level or edges. The cyclic behavior is declared by the keyword *always* followed by the *at sign* ( @ ) as the event control operator and the comma ( , ) separated sensitivity list of signals in Verilog 2001. The 1-bit full-adder with carry, as given in Listing 1.20, can also be described as a logic level sensitive cyclic behavior, as given in Listing 1.24.

**Listing 1.24** Verilog level sensitive cyclic behavior of a 1-bit full-adder with carry

```
module full_add(output reg sum, output reg carry_out, input a, b, carry_in);

always@(a, b, carry_in)
    begin
        sum = a ^ b ^ carry_in;
        carry_out = (a & b) | (b & carry_in) | (a & carry_in);
    end

endmodule
```

The event control statement is sensitive to changes in the logic level of the net variable input signals a, b, and carry_in. The procedural statements are a block encapsulated by the keywords *begin* and *end* and execute within the block only sequentially here because of the *blocking* assignment operator ( = ). Here the output signal sum is updated only when the logic level changes for either or all of the input signals and before the output signal carry_out is updated. The output signals sum and carry_out are register variable *reg* declarations here which hold the event control statement assignment until they are changed.

Sequential digital logic, such as most implementations of the flip-flop, has edge sensitive cyclic behavior and operates synchronously with a *clock* signal. The Verilog HDL has the qualifiers *posedge* (positive edge) and *negedge* (negative edge) with the event control statement in the sensitivity list to model edge sensitive cyclic behavior. The sequential logic type D flip-flop, as given in listing 1.19 as a UDP, can be also be described as a cyclic behavior, as given in Listing 1.25. The keywords *if* and *else* are Verilog HDL control flow statements.

**Listing 1.25** Verilog cyclic behavioral model for a type D flip-flop

```
module dff (output reg q, input d, clk, rst);

always@(posedge clk)
    begin
        if (rst == 0)
            q = 0;
        else
            q = data;
    end

endmodule
```

The event control statement uses the positive edge of the input signal clk. Here the logic level of the reset input signal rst clears the register variable *reg* declaration of the output signal q with a control flow (*if*) statement. The sequential logic UDP, as given in Listing 1.19, provides more possibilities for the logical functionality of the type D flip than the cyclic behavioral model, as given in Listing 1.25. Thus the UDP type D flip flop model may be more suited for robust embedded designs using programmable gate arrays.

Finally, sequential digital logic designs may require repetitive procedures distributed over multiple clock cycles [Botros06]. This operation can be modeled as *multicyclic* behavior using nested edge sensitive event control statements. A multicyle processor that receives four 4-bit net variable input signals data[3:0]on the positive edge of a clock signal and places them into a 16-bit register variable output signal outdata[15:0] is given in Listing 1.26. The multicyclic behavior is aborted with the reset signal rst by naming the process rcvrloop and using the keyword *disable*. A data available signal dav is used to verify that the complete data transfer has occurred.

**Listing 1.26** Verilog multicyclic behavioral model for a processor

```
module rcvrdata (output reg [15:0] outdata, input [3:0] data,
                              input clk, rst, output reg dav);

always@(posedge clk)
      begin: rcvrloop
            dav = 0;
            if (rst == 0) disable rcvrloop;
                  else outdata[3:0] = data[3:0];
            @(posedge clk)
                  if (rst == 0) disable rcvrloop;
                        else outdata[7:4] = data[3:0];
            @(posedge clk)
                  if (rst == 0) disable rcvrloop;
                        else outdata[11:8] = data[3:0];
            @(posedge clk)
                  if (rst == 0) disable rcvrloop;
                        else
                              begin
                                    outdata[15:12] = data[3:0];
                                    dav = 1;
                              end
      end

endmodule
```

## Blocking and Non-Blocking Assignments

Verilog HDL statements that use the procedural assignment operator ( = ) execute in the order listed which is a *blocking assignment* [Botros06]. The net or register variable blocking assignment occurs immediately but before the next statement executes. Although this can be interpreted as a sequential execution as in the C language, Verilog also has a concurrent procedural assignment operator ( <= ) which is a *non-blocking assignment*.

**Listing 1.27** Verilog structural model of a four bit shift register with blocking assignments

```
module shiftreg (output reg [3:0] Q, input data, clk);

always@(posedge clk)
      begin
            Q[0] = Q[1];
            Q[1] = Q[2];
            Q[2] = Q[3];
            Q[3] = data;
      end

endmodule
```

However, the Xilinx ISE WebPACK EDA software tool provides an error on compilation if a net or register variable or different portions (*bit slices*) of a register variable are assigned through both blocking and non-blocking statements.

The group of net or register variable non-blocking assignments occur in parallel and the order listed implies no precedence of any kind. Positive or negative edge sensitive cyclic behavior, as in sequential logic, is usually described by non-blocking assignments, while combinational logic is described with blocking assignments. This practice prevents a logic *race condition* from occurring [Ciletti04]. However, there are specific exceptions to this practice. Listing 1.27 is a sequential logic four bit serial shift register using blocking assignments to assure the correct transfer of data for this structural model.

## Control Flow

Verilog HDL statements for the control of execution flow are similar to those found in the C language [Ciletti99]. The keyword *case* has a counterpart in the C language *switch* statement and searches for the first occurrence of an exact match between the *case expression*, which is a net or register variable, and the *case item*, which can be expressed as logic values of 0 (false), 1 (true), *x* (unknown) or *z* (high impedance). The *case* construct is terminated with the keyword *endcase*.

Listing 1.28 is a four channel multiplexer using the *case* statement with an expression as the net variable select to output one of four signal inputs if an exact bit-wise match occurs. Since only four *case items* (00, 01, 10, and 11) of the sixteen possible (00, 01, 10, 11, 0x, 0z, 1x, 1z, x0, z0, x1, z1, xx, xz, zx, and zz) are provided in Listing 1.27, the keyword *default* determines what the output of the multiplexer should be is no match occurs.

**Listing 1.28** Verilog behavioral model of a four channel multiplexer using the *case* statement

```
module mux4ch (output reg data, input [1:0] select, input a, b, c, d);

always@(a or b or c or d or select)
        case (select)
            0: data = a;
            1: data = b;
            2: data = c;
            3: data = d;
            default data = 1'bz;
        endcase

endmodule
```

The Xilinx ISE WebPACK EDA software tool may provide unpredictable results with unsized integers in the *case expression*. In Listing 1.28 the net variable select is sized to two bits, as in Listing 1.7.

The exact match requirement of the *case* statement can be eased by the Verilog HDL keywords *casex* and *casez*. The *casex* statement ignores values in the case expression or the case item which are *x* (unknown) or *z* (high impedance) and effectively considers these values are *don't cares*. The *casez* statement ignores values in the case expression or the case item which are *z* (high impedance). The *casez* also utilizes the *question mark* ( ? ) as an explicit don't care. Listing 1.29 is an address decoder which provides a RAM output enable register variable ram_oe for a 12-bit range of valid addresses for the 16-bit net variable input address signal addr, event triggered by the positive edge of the address latch enable net variable input signal ale.

**Listing 1.29** Verilog behavioral model of an address decoder using the *casez* statement

```
module addrdec (output reg ram_oe, input [15:0] addr, input ale);

always@(posedge ale)
        casez (addr)
            16'b0100????????????: ram_oe = 1;
            default: ram_oe = 0;
        endcase

endmodule
```

The Verilog HDL keywords *if...else* and *if...else...if* for flow control have a direct counterpart in the C language. The use of these conditional statements, which alter the sequence of activity, is illustrated for cyclic and multicyclic behavior in Listing 1.245 and Listing 1.26.

The Verilog HDL keyword *for* supports repetitive flow control also has a direct counterpart in the C language. The initial statement of the *for loop* construct executes once to set a register variable declared by the keywords *reg* or *integer*. If the *end of loop* expression is true, the statement or block of statements is executed and afterwards the *loop update* expression is executed. The common C language construct k++ for post-incrementation of a variable is not a valid syntax in the Verilog HDL and the construct k = k +1 is used instead in the *for loop* in Listing 1.30.

The *for loop* construct executes repeatedly until the *end of loop* expression is no longer true [Botros06]. Listing 1.30 is an odd parity generator using the *for loop* construct with the integer variable k which counts the number of logic 1s in an 8-bit net variable input signal (data). The odd parity output register variable parity is logic 1 if data contains an even number of logic 1s and logic 0 if otherwise.

**Listing 1.30** Verilog behavioral model of an odd parity generator using the *for loop* construct

```
module oddparity_for (output reg parity, input [7:0] data);

integer k;

always@(data)
    begin
        parity = 1;
        for (k = 0; k <= 7; k = k+1)
        begin
            if (data[k] == 1)
                parity = ~parity;
        end
    end

endmodule
```

The Verilog HDL keyword *repeat* supports repetitive flow control for a specific number of times determined by the *repeat expression* and has no direct counterpart in the C language. Listing 1.31 is the odd parity generator of Listing 1.30 using the *repeat loop* construct. The *repeat expression* must be a constant in the Xilinx ISE WebPACK EDA software tool. Note that the integer variable k is still declared and incremented as an index separately here.

**Listing 1.31** Verilog behavioral model of an odd parity generator using the *repeat loop* construct

```
module oddparity_repeat (output reg parity, input [7:0] data);

integer k;

always@(data)
	begin
		parity = 1;
		k = 0;
		repeat (8)
			begin
				if (data[k] == 1)
					parity = ~parity;
				k = k + 1;
			end
	end

endmodule
```

The Verilog HDL keyword *while* supports repetitive flow control as long as the *while expression* is true. Listing 1.32 is a generator that outputs a gated clock register variable gclk derived from the net variable signals clk and clkgate.

However, such valid Verilog HDL constructs as the *while loop* do not have an explicit termination, are not necessary synthesizable in the hardware of an FPGA and should be avoided. The repetitive *for loop* and *repeat loop* constructs with an explicit termination can also be terminated early by using the keyword *disable*. An external or internal signal, such as the net variable reset signal rst in Listing 1.26, can be used to abort a repetitive flow control loop by naming the process and using the keyword *disable* [Ciletti04].

**Listing 1.32** Verilog behavioral model of a gated clock generator using the *while loop* construct

```
module gated_clock (output reg gclk, input clk, clkgate);

always@(clk or clkgate)
	begin
		while (clkgate)
			gclk = clk;
	end

endmodule
```

## Functions and Tasks

The Verilog HDL can organize and improve the rendering of a structural or behavioral description with a function or a task [Botros06]. Verilog functions and tasks are intended to support blocks used in multiple instances. A Verilog function is declared within a module but exhibits only combinational behavioral. A function returns a value, must have at least one input argument and cannot declare an output or bidirectional input and output argument.

The keywords *function* and *endfunction* encapsulate the function, which contains a declaration of the implied output, followed by the inputs associated by the order in which they are declared and optional local variables. Verilog functions cannot invoke a Verilog task nor recursively call other functions in the Xilinx ISE WebPACK EDA software tool. Listing 1.33 is a Verilog module with a function which returns the greater of two 8-bit signed net variables as an argument.

**Listing 1.33** Verilog module with a function which returns the greater of two 8-bit signed variables

```
module greater (input signed [7:0] a, b, output reg signed [7:0] c);

always @(a, b)
    begin
        c = great(a, b);
    end

function [7:0] great (input signed [7:0] x, y);
    begin
        if (x >= y)
            great = x;
        else
            great = y;
    end
endfunction

endmodule
```

A Verilog task is declared within a module and can have parameters passed to it and one or more results returned. The arguments of the task are passed by value and associated by the order in which they are declared. The keywords *task* and *endtask* encapsulate the task, which contains a declaration of the implied output, input and optional local variables. Verilog tasks cannot recursively call other functions in the Xilinx ISE WebPACK EDA software tool. Listing 1.34 is a Verilog module of a 1-bit adder, as in Listing 1.20, but here with a task which implements two 1-bit half adders.

**Listing 1.34** Verilog module of a 1-bit full adder with a task implementing two 1-bit half adders

```
module full_add(output reg sum, output reg carry_out, input a, b, carry_in);

reg psum, p1carry, p2carry;

always@(a, b, carry_in)
    begin
        half_adder(psum, p1carry, b, carry_in);
        half_adder(sum, p2carry, psum, a);
        carry_out=p1carry | p2carry;
    end

task half_adder(output half_sum, half_carry, input x, y);
    begin
        half_sum=x ^ y;
        half_carry=x & y;
    end

endtask

endmodule
```

## Finite State Machines

The finite state machine (FSM) is used in embedded design in Verilog using the field programmable gate array (FPGA) to represent sequential behavior. In the traditional FSM the logic output depends not only on a combination of the current logic input but on the *sequence* of past logic inputs. This past sequence or present *state* completely characterizes the FSM without regard to the past logic inputs [Mano07]. The number of states is assumed to be *finite* and can be described by the contents of a multiple bit state register [Navabi06].

The traditional FSMs are described by the Moore State Machine and the Mealy State Machine, as shown by the block diagrams in Figure 1.5 and Figure 1.6. The *next state* (NS) of either of the traditional FSMs are formed from the logic inputs and the state register stored value of the *present state* (PS). Every *state transition* must be logically explicit or the FSM will exhibit unpredictable behavior. [Zeidman99]. The logic outputs of the Moore Machine depend only upon the state register which is synchronized to the input clock. However, the outputs of the Mealy Machine depend on the state register and the logic inputs which may not be synchronized to the input clock.

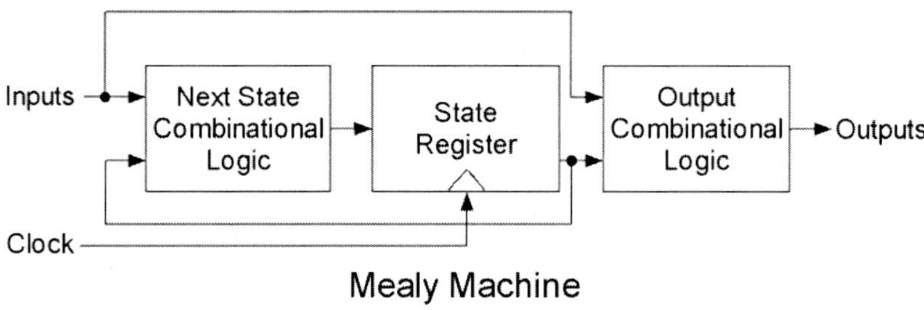


**Figure 1.5** Block diagram of the traditional Mealy FSM

Listing 1.35 is a Xilinx ISE Language Template for a 4-state Mealy FSM with *binary state encoding* using a 2-bit state register. Binary state encoding utilizes the minimum number of bits require for the number of states. However, binary state encoding requires more combination logic decoding than the *one-hot state encoding*, as given in Listing 1.36, which uses a bit for each state [Mano07].

**Listing 1.35** Verilog Language Template of the Mealy FSM with binary state encoding

```
parameter <state1> = 2'b00;
parameter <state2> = 2'b01;
parameter <state3> = 2'b10;
parameter <state4> = 2'b11;

reg [1:0] state = <state1>;

always@(posedge <clock>)
      begin
            if (<reset>
                  state <= <state1>;
            else
                  case (state)
                        <state1> : begin
                                          if (<condition>)
                                                state <= <next_state>;
```

```
                else if (<condition>)
                    state <= <next_state>;
                else
                    state <= <next_state>;
            end
        <state2> : begin
                if (<condition>)
                    state <= <next_state>;
                else if (<condition>)
                    state <= <next_state>;
                else
                    state <= <next_state>;
            end
        <state3> : begin
                if (<condition>)
                    state <= <next_state>;
                else if (<condition>)
                    state <= <next_state>;
                else
                    state <= <next_state>;
            end
        <state4> : begin
                if (<condition>)
                    state <= <next_state>;
                else if (<condition>)
                    state <= <next_state>;
                else
                    state <= <next_state>;
            end
    endcase
    assign <output1> = <logic_equation_based_on_states_and_inputs>;
    assign <output2> = <logic_equation_based_on_states_and_inputs>;
end
```

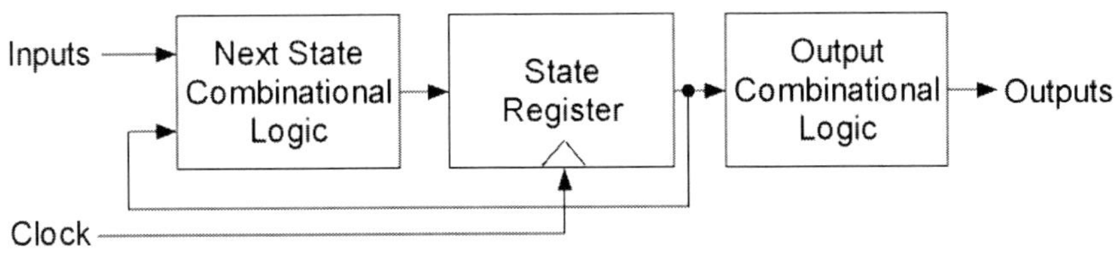

**Figure 1.6** Block diagram of the traditional Moore FSM

Listing 1.36 is a Xilinx ISE Language Template for a 4-state Moore FSM with one-hot state encoding using a 4-bit state register. The Xilinx ISE WebPACK EDA software tool also provides the StateCAD graphical interface to draw the *state diagram* of an FSM, as shown in Figure 1.7. StateCAD has a tutorial and a State Machine Wizard which facilitates the implementation and produces a Verilog HDL source file. Although useful for state machines with a small number of states, the graphical interface becomes difficult to discern for complicated processes and the Verilog HDL behavioral description, as given in Listing 1.35 and Listing 1.36, is then used.

**Listing 1.36** Verilog Language Template of the Moore FSM with one-hot state encoding

```
parameter <state1> = 4'b0001;
parameter <state2> = 4'b0010;
parameter <state3> = 4'b0100;
parameter <state4> = 4'b1000;

reg [3:0] state = <state1>;

always@(posedge <clock>)
    begin
        if (<reset>)
            begin
                state <= <state1>;
                    <outputs> <= <initial_values>;
            end
        else
            case (state)
                <state1> : begin
                            if (<condition>)
                                state <= <next_state>;
                            else if (<condition>)
                                state <= <next_state>;
                            else
                                state <= <next_state>;
                            <outputs> <= <values>;
                        end
                <state2> : begin
                            if (<condition>)
                                state <= <next_state>;
                            else if (<condition>)
                                state <= <next_state>;

                            else
                                state <= <next_state>;
                            <outputs> <= <values>;
                        end
                <state3> : begin
                            if (<condition>)
                                state <= <next_state>;
                            else if (<condition>)
                                state <= <next_state>;
                            else
                                state <= <next_state>;
                            <outputs> <= <values>;
                        end
                <state4> : begin
                            if (<condition>)
                                state <= <next_state>;
                            else if (<condition>)
                                state <= <next_state>;
```

```
                                else
                                     state <= <next_state>;
                                <outputs> <= <values>;
                           end
           endcase
     end
```

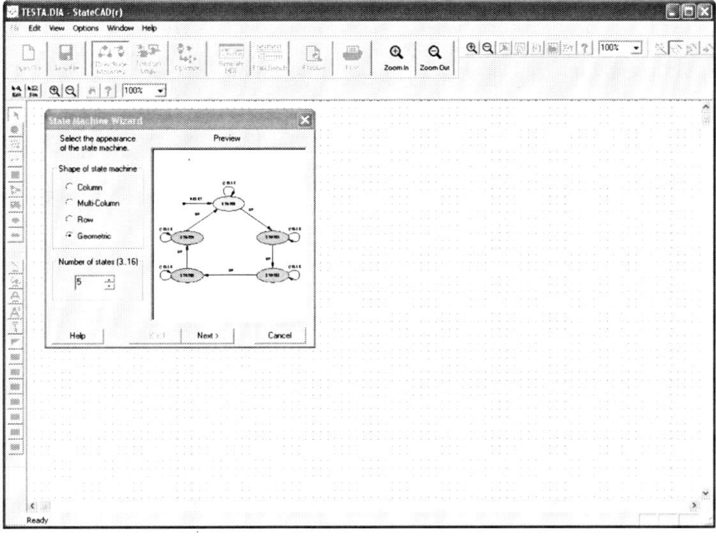


**Figure 1.7** Xilinx StateCAD graphical interface for finite state machine design

An arbitrary FSM as a Moore Machine with five states, where each state transition is logically explicit, is shown in Figure 1.8. The asynchronous RESET signal is used to set the present state to State 1. The synchronous CLOCK signal, which is not explicitly shown, initiates the determination of a state transition. The input logic signals are A, B, C and D and the output logic signals are X, Y and Z. State transitions occurs with a synchronous clock and the assertion of the combinational logic of the signals as AND ( & ), OR ( | ) and NOT ( ! ) . The equivalent Verilog logic operations can process input logic signals even with multiple bits and use the symbols &&, | | and !, as listed in Table 1.5.

The positive edge clock event driven Verilog HDL module for behavioral synthesis of the arbitrary Moore FSM of Figure 1.8 is given in Listing 1.39. The non-blocking assignment operator ( <= ) is used here to concurrently designate the output logic signals and the next state. The state register state requires three bits to provide binary state encoding for the five states. The statement default is required to insure that the unused state register values (0, 6, and 7) are determined. The Xilinx ISE WebPACK EDA software tool configures an optimum decoding of the state register in Verilog HDL and one-hot state encoding is not explicitly required.

**Listing 1.39** Verilog HDL of the arbitrary Moore FSM

```
reg [2:0] state;

  always@(posedge CLOCK)
     begin
          if (RESET)
               begin
                    state <= 1;
                    X <= 0;
```

```
                Y <= 0;
                Z <= 0;
            end
        else
            case (state)
                1:  begin
                        if (A || B)
                            state <= 2;
                        X <= 0;
                        Y <= 0;
                        Z <= 0;
                    end
                2:  begin
                        if (!A && C)
                            state <= 1;
                        else if (A && D)
                            state <= 3;
                        X <= 1;
                        Y <= 1;
                        Z <= 0;
                    end
                3:  begin
                        if (!A || !C)
                            state <= 5;
                        else if (A && C)
                            state <= 4;

                        X <= 0;
                        Y <= 0;
                        Z <= 1;
                    end
                4:  begin
                        if (D)
                            state <= 5;
                        X <= 0;
                        Y <= 1;
                        Z <= 0;
                    end
                5:  begin
                        if (!B && D)
                            state <= 3;
                        else if (B)
                            state <= 1;
                        X <= 1;
                        Y <= 0;
                        Z <= 0;
                    end
                default: state <= 1;
            endcase
    end
```

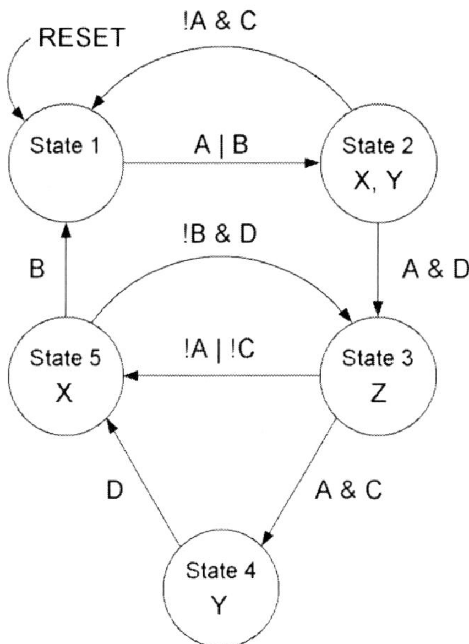


**Figure 1.8** An arbitrary state diagram of a Moore FSM

## Controller-Datapath Construct

Embedded design in Verilog using programmable gate arrays can utilize controller and datapath modules to facilitate the implementation of complex task. The controller module accepts external control signals and status signals from the datapath module and uses one or more finite state machine (FSM) to coordinate the process. The controller module sets the datapath module control input signals that route input data, perform any processing on the data and output the data to a functional unit [Navabi06]. The datapath module stores and manipulates data in registers using combinational logic and can use one or more FSM to output data but not autonomously. The controller can also accept external control signals from and return status signals to a sequential processor to augment the performance of the embedded system, as described in Chapter 5 Embedded Soft Core Processors.

The controller and datapath construct partition the design into modules that can be separately verified in simulation, as described in Chapter 2 Verilog Design Automation. Rather than one module that encapsulates the entire process, the controller and datapath modules then each have a reduced number of interconnections which facilitates the Verilog behavioral synthesis into programmable gate array hardware. Datapath modules also support the concept of *design reuse*, as described in Chapter 3 Programmable Gate Array Hardware.

The controller module can be easily modified to accommodate a new task, which can then even include additional datapath modules. The configuration of a typical controller and datapath construct is shown in Figure 1.9. The synchronous clock input schedules the state transitions of the FSMs the controller and datapath. Registers can be initialized by a global reset signal, a local reset signal or by a declaration in the behavioral synthesis of the controller and datapath [Lee06]. The reset signal is not shown in Figure 1.9.

The controller has *control input* logic signals that initiate the process and *status output* logic signals that signify the completion of the process. The datapath has only *data* as an input and output and no external process *control* logic signals other than those derived from the controller [Zeidman99]. The datapath outputs *status* logic signals to the controller to coordinate the process. A clock signal is used to provisionally evoke a state transition in the finite state machines (FSM) if utilized in the controller and datapath. However, the controller control signals and the datapath status signals are required to have a state transition.

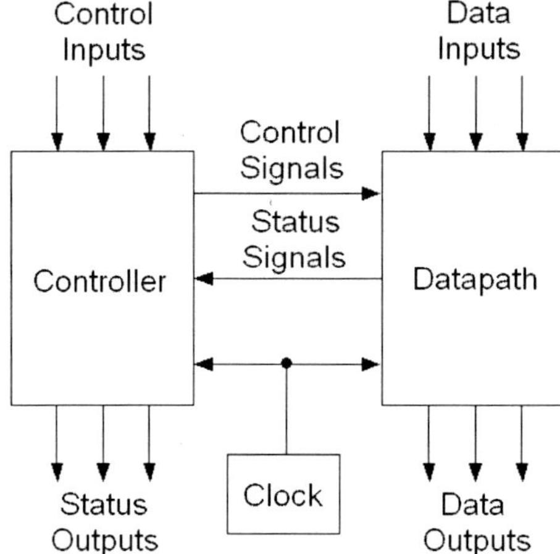


**Figure 1.9** Configuration of a typical controller and datapath construct

## C to Verilog Translation

Embedded design has often relied upon the vast experience of utilizing the conversational language C to affect a process. If real-time processing is required, as in digital signal processing, digital communication or digital control, assembly language subroutines are used to augment the C language routines [Brown94]. However, although assembly language routines execute rapidly and are virtually at the microprocessor hardware level, they are not as discernable.

The translation of C language algorithms to the Verilog HDL can facilitate embedded design because behavioral synthesis Verilog is both conversational and ultimately executes rapidly at the programmable gate array hardware level [Smith00]. Selected algorithms available in the conversational C language can be readily translated to the and finite state machines (FSM) and then to the controller and datapath construct in the Verilog HDL, as shown in Figure 1.9. An example is the C algorithm for the greatest common denominator [Vahid02], as given in Listing 1.40.

**Listing 1.40** Greatest common denominator C language algorithm

```
int gcd (int xin, int yin)
{
int x, y;

    x = xin;
    y = yin;
    while(x != y)
        {
        if (x < y)
            y = y – x;
        else
            x = x – y;
        }
    return x;
}
```

The statements in the algorithm are first classified as assignment, loop or branch [Vahid02]. Assignment statements are mapped to an initial state which is immediate mapped to the next state in the FSM of the controller, as shown in Figure 1.10. Loop statements, derived from the C constructs *for* or *while*, are conditional statements which are mapped to either the loop statements and then a return to the conditional statement or to the next state if the condition is satisfied, as also shown in Figure 1.10. Finally, branch statements, derived from the C constructs *if-then-else* or *case*, are mapped separately to statements and then the next statement in FSM, as shown in Figure 1.11.

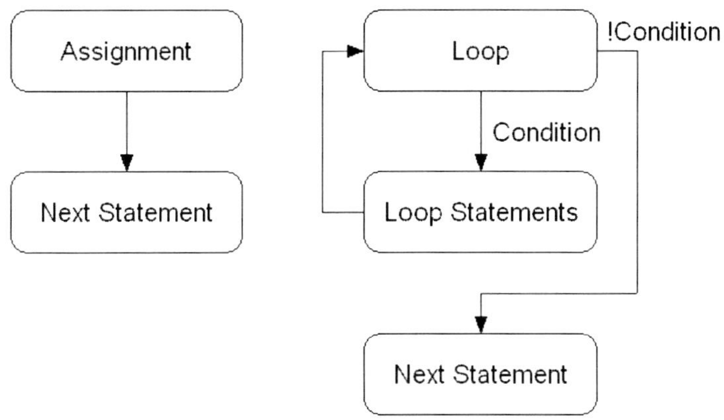


**Figure 1.10** C language assignment and loop statement construct

The resulting five state FSM for the greatest common denominator algorithm is shown in Figure 1.12. Although this simple algorithm certainly does not require this level of complexity, for illustration the FSM can now be converted to Verilog HDL controller and datapath modules. The controller module incorporates the FSM and outputs commands to and inputs status signals from the datapath module, as shown in Figure 1.9. The datapath module is responsible for the input data (x and y) and output data (x), testing the loop statement (x ! = y) and the branch statement (x < y) and the data subtractions (y – x and x – y).

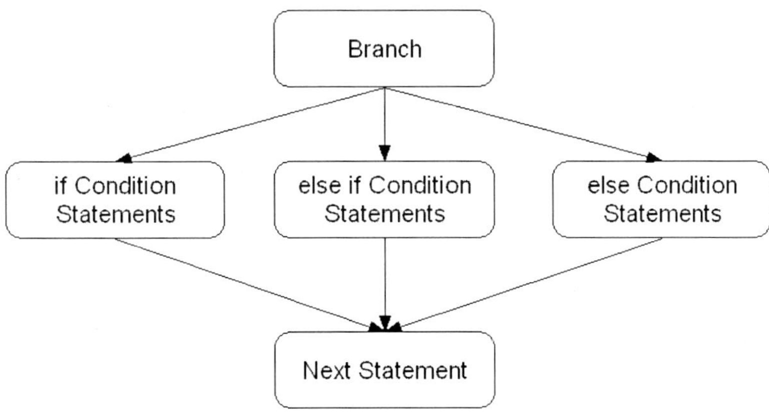


**Figure 1.11** C language branch statement construct

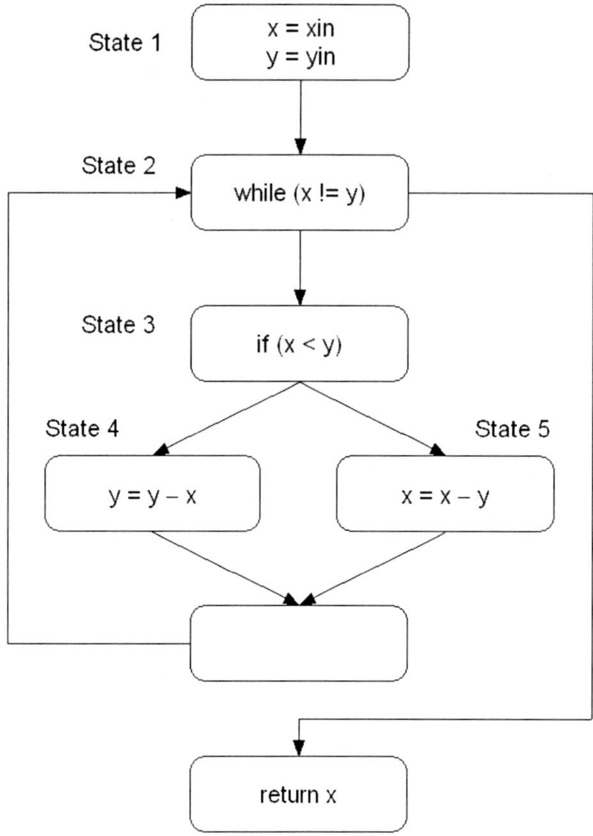


**Figure 1.12** FSM for the greatest common denominator algorithm

The Verilog HDL controller and datapath modules for the greater common denominator algorithm are given in Listing 1.41 and Listing 1.42. The controller module requires an external command signal that indicates that the process can start from another module. For the C language algorithm the subroutine merely is *called* from another routine but for the controller module an external signal gcddata is inputted for coordination. The controller module outputs the signal datagcd to the datapath module and the module that provided the external signal gcddata.

**Listing 1.41** Greatest common denominator C language algorithm Verilog controller module

```
module gcdcontroller (input clock, gcddata, lddata, xneqy, xlty, input xysub, yxsub,
                      output gcdinit, datagcd, datald, subxy, subyx)

reg [2:0] gcdstate;

always@(posedge clock)
    begin
        if (gcddata == 0)
            begin
                gcdstate = 1;
                datagcd = 0;
                gcdinit = 1;
```

```
                datald = 0;
            end
        else
            case(gcdstate)
                1:  begin
                        gcdinit = 0;
                        datald = 1;
                        if (lddata)
                            begin
                                datald = 0;
                                gcdstate =  2;
                            end
                    end
                2:  begin
                        if (xneqy)
                            gcdstate = 3;
                        else
                            gcdstate = 6;
                    end
                3:  begin
                        if (xlty)
                            gcdstate = 4;
                        else
                            gcdstate = 5;
                    end
                4:  begin
                        subyx=1;
                            if (yxsub)
                                begin
                                    subyx = 0;
                                    gcdstate = 2;
                                end
                     end
                5:  begin
                        subxy = 1;
                        if (xysub)
                            begin
                                subxy = 0;
                                gcdstate = 2;
                            end
                    end
                6:  begin
                            datagcd = 1;
                            gstate = 6;
                        end
                    default: gstate = 6;
                endcase
    end

endmodule
```

**Listing 1.42** Greatest common denominator C language algorithm Verilog datapath module

```
module gcddatapath (input clock, gcdinit, datagcd, datald, input subxy, subyx, output lddata,
        output xneqy, xlty, xysub, yxsub,  inout [15:0] xdata, input signed ydata)

reg [15:0] x;
reg [15:0] y;

always@(posedge clock)
    begin
        if (gcdinit)
            begin
                lddata = 0;
                xney = 0;
                xlty = 0;
                xysub = 0;
                yxsub = 0;
            end

        if (datald)
            begin
                x = xdata;
                y = ydata;
                if (x != y)
                    xney = 1;
                if (x < y)
                    xlty = 1;
                lddata = 1;
            end

        if (subxy)
            begin
                x = x – y;
                if (x == y)
                    xney=0;
                if (x < y)
                    xlty = 1;
                xysub = 1;
            end

        if (subyx)
            begin
                y = y – x;
                if (x == y)
                    xney = 0;
                if (x < y)
                    xlty = 1;
                yxsub = 1;
            end
```

```
        if (datagcd)
            xdata = x;
    end

endmodule
```

The command and status signals of the controller and datapath modules are shown in Figure 1.13. The controller signal lddata commands the datapath module to load the data xin and yin into the register variables x and y.

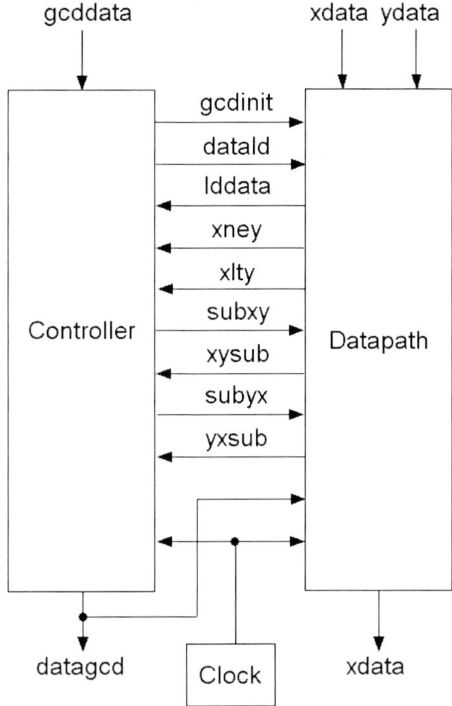


**Figure 1.13** Controller and datapath modules for the greatest common denominator algorithm

The datapath module sends the semaphore signal datald to the controller module to indicate that the data load has occurred. The other controller and datapath signals, subxy and xysub and subyx and yxsub, are for the arithmetic subtraction of the datapath register variables. The datapath modules also provides the xney and xlty signals to the controller module for the loop and branch statements.

Although this example of the greatest common denominator algorithm can be implemented as a controller and datapath construct in the Verilog HDL, an assembly or C language routine executing on a hard or soft core processor could be more discernable in practice because of the vast experience of utilizing these software paradigms. However, the controller utilizes only a small number of event clock cycles, can execute a portion of the algorithm in parallel in each event state and does not require the multiple clock cycles of the instruction load, decode, and execute phase of a single processor instruction. Obviously, the design tradeoff and selection of the controller and datapath construct verses a hard or soft core processor is task specific in embedded design.

The ubiquitous nature of the conversational C programming language and the emerging paradigm of the programmable gate array in embedded design have provided the impetus to seek a hardware description language within the C syntax. SystemC is a system description language available as a set of library routines which makes it possible to simulate concurrent processes (*www.systemc.org*). SystemC has a hardware synthesis class subset and is now IEEE Standard 1066-2005.

The Impulse C design automation environment facilitates the execution of parallel algorithms that target programmable gate array and embedded, hard or soft core microprocessors for high performance computing (*www.impulsec.com*). The Impulse C compiler generates an HDL for synthesizable hardware that implements streams, signals and memories. Impulse C extends standard ANSI C to support a communicating process parallel programming model (Pellerin05). Impulse C supports the Xilinx 32-bit embedded soft core MicroBlaze™ processor on the Spartan-3E FPGA, as described in Chapter 2 Verilog Design Automation, and the hard core PowerPC™ processor on the Xilinx Vertex II FPGA (Xilinx application note XAPP901, *www.xilinx.com*).

## Arithmetic Functions

The intrinsic arithmetic capabilities of the programmable gate array are similar to the conventional microprocessor because both are register based architectures [Smith00]. Signed and unsigned binary addition and subtraction are directly supported. Unlike the conventional microprocessor, the register size of the FPGA can be set to accommodate the task at hand. Microprocessors with an 8-bit register architecture, such as the Xilinx PicoBlaze soft core processor for the Xilinx Spartan-3E FPGA, require multiple byte arithmetic when processing analog-to-digital converter (ADC) and digital-to-analog converter (DAC) data with greater than 8 bits of resolution. However, the FPGA can provide registers for integer addition and subtraction suitable for the size in bits of the binary data and the arithmetic manipulation of the data.

Integer multiplication and division are often supported by conventional microprocessors. The 16-bit Intel 8086 microprocessor provides signed and unsigned integer multiplication, operational codes (*op codes*) IMUL and MUL, and division, op codes IDIV and DIV, through a *microcode* operation which requires multiple clock cycles [Brey05]. The Xilinx Spartan-3 and Spartan-3E FPGA support rapid integer multiplication with the available of 18-bit hardware multipliers. Integer division, except by powers-of-two ($2^n$), is also supported by the use of the Xilinx CORE Generator and the LogiCORE blocks, as described in Chapter 2 Verilog Design Automation. The Multiplier LogiCORE block provides a parallel integer multiplier.

Floating-point arithmetic manipulations are usually not directly supported by conventional microprocessors. The Intel 8087 floating point coprocessor was an early adjunct to the 16-bit Intel 8086 microprocessor [Brey05]. The Xilinx Spartan-3E FPGA provide floating point operations with the Xilinx CORE generator LogiCORE block, as described in Chapter 2 Verilog Design Automation. These numbers are presented in both IEEE-754 Floating-Point Standard 32-bit and 64-bit format and non-standard size format. A floating-point number is represented using a sign bit, a $w_e$-bit exponent and a $w_f$-bit fraction, as shown in Figure 1.14. The value v of the floating-point number is determined by Equation 1.1.

$$v = (-1)^S\, 2^E\, b_0 b_1 b_2 \ldots b_{w_f - 1} \qquad (1.1)$$

In Equation 2.1 S is the sign bit, E is the binary value of the $w_e$-bit exponent, and $b_i = 2^{-i}$. The sign bit causes the floating-point value to be negative if S = 1. The value of the biased unsigned exponent field e is determined by Equation 1.2.

$$e = \sum_{i=0}^{w_e - 1} e_i\, 2^i \qquad (1.2)$$

In Equation 1.2 $e_i$ represents the bits of the exponent, as shown in Figure 1.14. The exponent E in Equation 2.1 is obtained by removing the bias as determined by Equation 1.3.

$$E = e - (2^{w_e - 1} - 1) \qquad (1.3)$$

The most significant bit $b_0$ is a *normalized* constant equal to 1 and is not represented in the floating-point number, as shown in Figure 1.14. The IEEE-754 Standard specifies a 32-bit format with a 24-bit fraction, an 8-bit exponent and a sign bit but only 23 bits are required for the fraction since $b_0$ is not represented. The IEEE-754 Standard also specifies a 64-bit format with a 53-bit fraction, an 11-bit exponent and a sign bit but again only 52 bits are required for the fraction.

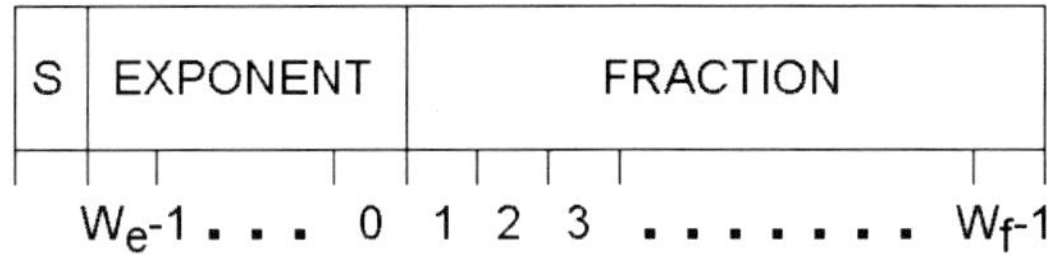


**Figure 1.14** Floating-point number representation using a sign bit, a $w_e$-bit exponent and a $w_f$-bit fraction

A fixed-point number is represented by a two's complement number weighted by a fixed power of two and an implied decimal point ( ■ ), as shown in Figure 1.15. The w-bit fixed-point number has a $w_f$-bit fraction, a $(w - w_f - 1)$-bit integer and a sign bit. The value u of the fixed-point number is determined by Equation 1.4 where $b_i = 2^{i - w_f}$. A signed integer number occurs when the fractional bit width is zero.

$$u = (-1)^S \, 2^{w-w_f-1} + b_{w-2}b_{w-3}\ldots b_{w_f+1}b_{w_f}\, b_{w_f-1}b_{w_f-2}\ldots b_1 b_0 \qquad (1.4)$$

S | INTEGER ■ | FRACTION

W-1 . . . . . Wf Wf-1 . . . . . . . . . . 0

**Figure 1.15** Fixed-point number representation using a sign bit, a $(w - w_f - 1)$-bit integer and a $w_f$-bit fraction

The Xilinx CORE Generator LogiCORE Floating-Point block provides the floating-point operations of add, subtract, divide, square root and comparison and fixed-point to floating-point and floating-point to fixed-point conversion. The Coordinate Rotation Digital Computer (CORDIC) LogiCORE block provides vector rotation and translation, sine, cosine and tangent, arctangent, hyperbolic sine and cosine and square root calculations. The Multiplier and Pipelined Divider LogiCORE blocks support fixed-point arithmetic manipulations.

These LogiCORE blocks for arithmetic functions and other LogiCORE blocks are described in Chapter 2 Verilog Design Automation. Chapter 4 Digital Signal Processing, Communications and Control present Xilinx ISE WebPACK projects which survey the application of the Verilog HDL and programmable gate arrays in embedded design and use the FIR Compiler, Sine-Cosine Look-Up Table and the Direct Digital Synthesis Compiler LogiCORE blocks.

## Programmable Gate Array and Microprocessor Comparison

The performance of a programmable gate array finite state machine (FSM) or controller and datapath construct for a task is quite noticeable when compared that of a conventional microprocessor. Microprocessors utilize operational codes (*op codes*) or instructions and data to sequentially perform the tasks required for an embedded design. The microprocessor is event driven by a master clock but requires several clock cycles to retrieve an op code instruction from random access or read-only memory and decode it, retrieve data if required, process the instruction and store the result. Usually operations cannot be performed in parallel by a conventional microprocessor and the throughput rate in real-time processing is then limited by the clock period and the complexity of the task.

Parallel operation of the Verilog HDL modules executing on a field programmable gate array (FPGA) facilitate real-time processing of diverse tasks. An example of an embedded design in Verilog for the Spartan-3E Starter Board, which utilizes a Xilinx Spartan-3E FPGA, is an application which sets the gain of the integral programmable amplifier, initiates a conversion of an analog input voltage to a 14-bit number and displays the results on the liquid crystal display (LCD). Five Verilog HDL modules are essentially operating simultaneously here, as given in Listing 1.43 The complete application is described in Chapter 3 Field Programmable Gate Array Hardware and given in Listing 3.31.

The programmable gain amplifier and analog-to-digital converter (ADC) Verilog modules s3eprogamp.v and s3eadc.v are operating in parallel with the LCD Verilog modules adclcd.v and lcd.v while being supervised by the Verilog module genampadc.v. In a microprocessor each of these tasks must be processed in sequence which often requires complex interrupt service routines [Vahid02].

**Listing 1.43** Verilog HDL modules operating in parallel in an embedded design application

```
s3eadc M0 (CCLK, adcdav, davadc, adc0data, adc1data, adcsck, adcspod, conad);
s3eprogamp M1 (CCLK, ampdav, davamp, ampcmd0, ampcmd1, ampsck, ampspid, csamp, sdamp);
adclcd M2 (CCLK, BTN0, resetlcd, clearlcd, homelcd, datalcd, addrlcd, initlcd, lcdreset, lcdclear,
        lcdhome, lcddata, lcdaddr, lcddatin, digitmux, data);
lcd M3 (CCLK, resetlcd, clearlcd, homelcd, datalcd, addrlcd, lcdreset, lcdclear, lcdhome, lcddata,
    lcdaddr, rslcd, rwlcd, elcd, lcdd, lcddatin, initlcd);
genampadc M4 (CCLK, SW0, SW1, SW2, SW3, ampdav, davamp, ampcmd0, ampcmd1, adcdav,
        davadc, adc0data, adc1data, digitmux, data);
```

The variable register size and bitwise logic operations of the FPGA also provide a performance increase compared to the fixed register size and register based logic operations of a conventional microprocessor. As an example of the performance comparison, the sequential transfer of 12 bits of data by a 16-bit microprocessor (Intel 8086) from the most significant bits (MSB) of a 16-bit general purpose register (bx) to a 12-bit digital-to-analog converter (DAC) input-output (I/O) peripheral in assembly language mnemonics (ASM) is given in Listing 1.44. This is a single task process that would be called as a subroutine.

The DAC here is a 12-bit serial peripheral at I/O address 400h (hexadecimal) that inputs data (SDATA) at bit 0 of an 8-bit general purpose register (al) on the negative edge logic transition of the clock (SCLK) at bit 1 of the same register. The 8-bit I/O port is written to by the ASM command out dx,al and the ASM command mov is the register transfer instruction [Brey05]. The register bit manipulation ASM commands and and or are similar to the intrinsic combinational logic Verilog HDL primitives, as listed in Table 1.9, but not as versatile as the bitwise logic operations, as listed in Table 1.1.

**Listing 1.44** Sequential processor data transfer using Intel 8086 assembly language mnemonics

```
;SDATA from bx to DAC
;
sdatoproc near
          mov  ax,400H        ;IOW 400H
          mov  dx,ax
          mov  cl,12          ;count for 12 SDATA bits
sdatlp:   rol  bx,1
          mov  al,bl
          and  al,1           ;SDATA bit 0, SCLK=0 bit 1
          out  dx,al
          or   al,2           ;SDATA bit 0, SCLK=1 bit 1
          out  dx,al
          and  al,0FDH
          out  dx,al          ;SDATA bit 0, SCLK=0 bit 1
          dec  cl             ;decrement count
          jnz  sdatlp         ;jump if count is not zero
          ret
sdatoendp
```

The finite state machine (FSM) in the Verilog HDL intrinsically provides a sequence of operations that is suited to convert sequential machine assembly language code that executes on a conventional microprocessor. If the DAC peripheral device sclk and sdata signals were interfaced to two I/O pins of an FPGA the resulting clock event driven Verilog HDL module for behavioral synthesis is rendered as given in Listing 1.45.

The I/O location of the logic signals for behavioral synthesis in an FPGA is set by a User Constraints File (UCF), as described in Chapter 2 Verilog Design Automation. The data transfer is affected by the positive edge of the signal clock and the semaphores dacdav and davdac are the controller and datapath signals. The integer variable i in the Verilog HDL behavioral model is not only the data transfer count but accesses the proper bit of the 12-bit input data. The data transfer is determined by this count and the FSM which uses the dacstate state register. When the data transfer is completed the datapath sets the staus signal dacdac for the controller to logic 1.

**Listing 1.45** Verilog HDL module for the data transfer

```
module sdat0(input clock, dacdav, input [11:0] data, output reg sdata, output reg sclk,
              output reg davdac);

reg [1:0] dacstate;
integer i;

always@(posedge clock)
      begin
            if (dacdav == 0)
                  begin
                        i = 12;
                        dacstate = 0;
                        sclk = 1;
                  end
```

```
        else
            begin
                davdac = 0;
            case (dacstate)
                0:  begin
                        i = i – 1;
                        if (i == 0)
                            dasstate = 3;
                        else
                            sdata = data[i – 1];
                            dacstate = 1;
                    end
                1:  begin
                        sclk = 0;
                        dacstate = 2;
                    end
                2:  begin
                        sclk = 1;
                        dacstate = 0;
                    end
                3:  begin
                        davdac = 1;
                        dacstate = 3;
                    end
                endcase
            end
    end

endmodule
```

A comparison of the data transfer in Intel 8086 assembly language in Listing 1.43 and the Verilog HDL module in Listing 1.44 shows that the conversational nature of Verilog produces an easily discernable process. While a microprocessor could also employ a conversational language such a C to affect the same data transfer, the number of op codes and resulting execution time and may be even greater.

The command and status signals dacdav and davdac and the controller and datapath modules for the data transfer are shown in Figure 1.16. The logic signals dacdav and davdac provide parallelism and synchronization of the data transfer that is not available in a microprocessor executing a single task.

A microprocessor would need to utilize a *multi-tasking operating system* to have this level of process coordination that is inherent in the Verilog HDL. Finally, the data to be transfer is inputted in parallel to the required number of bits and routing to the Verilog HDL module in the fine grained elements of the FPGA is provided by a *switch network* rather than a fixed path within a microprocessor. Such fixed paths readily led to bus contention and data transfer *stalls* [Smith00].

However, the both the microprocessor and the soft core processor is a traditional architecture that is familiar to practitioners of embedded design. The limited internal program store for such a processor, utilizing the Spartan-3E FPGA block random access memory (RAM) as described in Chapter 5 Embedded Soft Core Processors, could be augmented by external non-volatile memory.

The soft core processor can utilize the datapath construct efficiently to *hand-off* such tasks as servicing both external hard core and internal soft core peripherals by effectively becoming the controller, as shown in Figure 1.16. Ultimately, several soft core processors could be instantiated in an embedded design on a single FPGA for either independent or cooperative operation.

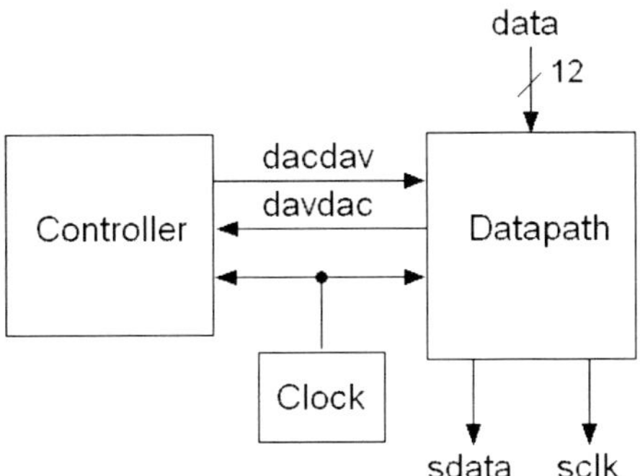


**Figure 1.16** Controller and datapath modules for data transfer

The freely downloadable Xilinx ISE WebPACK EDA software tool does not includes the Xilinx Embedded Development Kit (EDK) for the Xilinx 32-bit MicroBlaze soft core processor which can be programmed using the conversational C language. Auxiliary electronic design automation (EDA) software for the Xilinx 8-bit PicoBlaze soft core processor is available, however programming is accomplished in assembly language. Embedded design using the Xilinx PicoBlaze soft core processor and its EDA is described in Chapter 5 Embedded Soft Core Processors.

## Summary

In this Chapter the evolution of the Xilinx field programmable gate array (FPGA) and the concurrent development of a hardware description language (HDL) are described. Verilog HDL concepts and syntax are surveyed to support the development of an embedded design using programmable gate arrays. The finite state machine (FSM) and the controller and datapath constructs are presented as a behavioral description. The translation of an algorithm in the conversational C language to the Verilog HDL facilitates the embedded design implementation. Behavioral synthesis Verilog is shown to be both conversational and ultimately executes rapidly at the FPGA hardware level. Finally, the conventional microprocessor and the FPGA are functionally compared. Unlike an FPGA, operations cannot be performed in parallel by a microprocessor and the throughput rate in real-time processing is then limited by the clock period and the complexity of the task.

Chapter 2 Verilog Design Automation presents the operation of the Xilinx Integrated Synthesis Environment (ISE) WebPACK electronic design automation (EDA) tool for embedded design. Chapter 3 Programmable Gate Array Hardware describes two available Xilinx Spartan-3E FPGA hardware evaluation boards in operation by complete Xilinx ISE WebPACK projects. Chapter 4 Digital Signal Processing, Communications and Control presents complete Xilinx ISE WebPACK projects in digital filtering, digital modulation in communication systems, data communication and digital process control. Chapter 5 Embedded Soft Core Processors describes the auxiliary EDA software for the Xilinx PicoBlaze 8-bit soft core processor and its comparison to the Verilog controller and datapath constructs and Xilinx LogiCORE blocks in the Xilinx ISE WebPACK. These four Chapters survey the application of the Verilog HDL, the soft core processor and programmable gate aays in embedded design.

## References

[Botros06] Botros, Nazeih M., *HDL Programming Fundamentals*. Thomson Delmar, 2006.

[Brey05] Brey, Barry M., *The Intel Microprocessors*. Prentice Hall, 2005.

[Brown94] Brown, John F., *Embedded Systems Programming in C and Assembly*. Van Nostrand Reinhold, 1994.

[Ciletti99] Cilletti, Michael D., *Modeling, Synthesis and Rapid Prototyping with the Verilog HDL*. Prentice Hall, 1999.

[Ciletti04] Cilletti, Michael D., *Starter's Guide to Verilog 2001*. Prentice Hall, 2004.

[Lee06] Lee, Sunggu., *Advanced Digital Logic Design*. Thomson, 2006.

[Mano07] Mano, M. Morris and Michael D. Cilletti, *Digital Design*. Prentice Hall, 2007.

[Navabi06] Navabi, Zainalabedin, *Verilog Digital System Design*. McGraw-Hill, 1999.

[Pellerin94] Pellerin, David and Michael Holley, *Digital Design using ABEL*. Prentice Hall, 1994.

[Pellerin05] Pellerin, David and Scott Thibault, *Practical FPGA Programming in C*. Prentice Hall, 2005.

[Smith00] Smith, David R. and Paul D. Franzon, *Verilog Styles for Synthesis of Digital Systems*. Prentice Hall, 2000.

[Vahid02] Vahid, Frank and Tony Civargis, *Embedded System Design*, Wiley 2002.

[Zeidman99] Bob Zeidman, *Verilog Designer's Library*. Prentice Hall, 1999.

# 2

# Verilog Design Automation

Embedded design in the Verilog hardware description language (HDL) using a field programmable gate array (FPGA) is facilitated by software electronic design automation (EDA) tools which provide both simulation and hardware synthesis. The complexity of either the coarse or fine grained architecture of the FPGA, as described in Chapter 1 Hardware Description Language Verilog Hardware Description Language, requires EDA tools which can *route* the constructs of the Verilog HDL as an efficient, synthesizable interconnection of diverse logical elements.

The Xilinx (*www.xilinx.com*) Integrated Synthesis Environment (ISE) is one such suite of EDA tools for the synthesis of embedded system applications in the FPGA from the Verilog HDL. This Chapter introduces the Xilinx ISE WebPACK™ EDA in a quick-start manner, with an emphasis on the project environment and the processes required to produce an embedded design for the Xilinx Spartan™-3E FPGA evaluation boards. The Verilog source and project file structure for the Xilinx ISE WebPACK and FPGA hardware synthesis and configuration is described for a *stop watch* example as a practical application. The FPGA evaluation boards and the *stop watch* project are further described in Chapter 3 Field Programmable Gate Array Hardware.

This Chapter demonstrates the use of several additional components of the Xilinx ISE WebPACK EDA. The Xilinx CORE Generator creates Verilog HDL modules as LogiCORE blocks that can be instantiated into a project to facilitate embedded design. A binary to binary coded decimal (BCD) converter Xilinx ISE WebPACK project features the Divider Generator LogiCORE block. The Xilinx Floorplanner is an advanced software EDA tool that can in some instances improve the performance of an automatically placed and routed FPGA design. The Floorplanner is particularly useful for placing *critical path* logic to optimize timing performance in structural models and datapath modules. The Xilinx ISE Simulator provides behavioral and timing verification without hardware synthesis and implementation to the FPGA.

The Xilinx ISE Verilog Language Templates simply the process of implementing an embedded system design in an FPGA by providing a language implementation and syntax reference. The Xilinx Architecture Wizard is an advanced software electronic design automation (EDA) tool that augments other Verilog HDL module creation tools such as the Xilinx CORE Generator. Finally, the Xilinx Timing Analyzer Xilinx Timing Analyzer provides a description of the timing constraint in error and design implementation suggestions to alleviate the error.

The Verilog source modules and project files are provided in the *Chapter 2* folder as subfolders identified by the name of the appropriate module. The complete contents and the file download procedure are described in the Appendix.

## Xilinx ISE WebPACK

The registered and licensed EDA provided by Xilinx is the Integrated Synthesis Environment (ISE) Foundation™ which supports the complete Xilinx complex programmable logic devices (CPLD) and field programmable gate arrays (FPGA). Xilinx also provides the registered but unlicensed ISE WebPACK™, which features the same functionality as ISE Foundation but with limited complex programmable logic device (CPLD) and FPGA support. However, the Xilinx Spartan™-3E FPGA devices are supported by ISE WebPACK. The evaluation boards, as further described in Chapter 3 Programmable Gate Array Hardware, both utilize the Xilinx Spartan-3E FPGA.

The Xilinx ISE WebPACK EDA is presented here in a *quick start* manner to begin the investigation of embedded system design in Verilog targeting the Xilinx Spartan-3E FPGA and the evaluation boards. The complete description of the Xilinx ISE WebPACK is also available as either a Quick Start Tutorial or an In-Depth Tutorial (*www.xilinx.com*). The Xilinx ISE WebPACK is freely available for download with software registration (*www.xilinx.com*). The initial download for ISEWebPACK is relatively small but it administers the download of the entire application. Disk storage space can be saved if only support for the Spartan-3E FPGAs is evoked.

The Xilinx ISE WebPACK supports variety of design entry, simulation, synthesis, implementation, programming and verification tools including the HDL Editor, the Xilinx Synthesis Technology (XST), the CORE Generator™, the iMPACT device programming environment, the FloorPlanner™, the ISE Simulator and a version of the Xilinx Embedded Development Kit (EDK) for the simple 8-bit PicoBlaze soft core processor. However, not supported as part of the freely downloadable ISE WebPACK are the Xilinx System Generator for DSP design entry tool, which is integrated with MATLAB/Simulink (*www.mathworks.com*), the complete Xilinx EDK, the architectural PlanAhead™ implementation tool and the ChipScope™ Pro verification tool.

The complete Xilinx EDK provides support for the advanced 32-bit MicroBlaze soft core processor, which pursues an alternative paradigm for embedded design using the conversational C language. The Xilinx PlanAhead EDA facilitates a hierarchical design methodology which minimizes routing congestion and interconnecting complexity for improved performance. The Xilinx ChipScope™ Pro EDA inserts logic analyzer, bus analyzer and virtual input-output (I/O) cores so that any internal signal or node can be captured or stimulated at system speed. The captured logic signals are then outputted through the programming interface without using normal I/O pins.

After installation in the suggested directory *C:\EDPGA* and initial execution of the application the Xilinx ISE WebPACK *Project Navigator* design window is displayed, as shown in Figure 2.1. The bottom of the Project Navigator window shows a folder selection for console display of all process messages (Console), a display of only the fatal error messages (Error), a display of only warning messages (Warning), the Tool Command Language (Tcl) display which can be used to *script* the ISE WebPACK EDA and the result of the *Find in Files* display which is evoked from the *Edit...Find in Files* menu command, as shown in Figure 2.2.

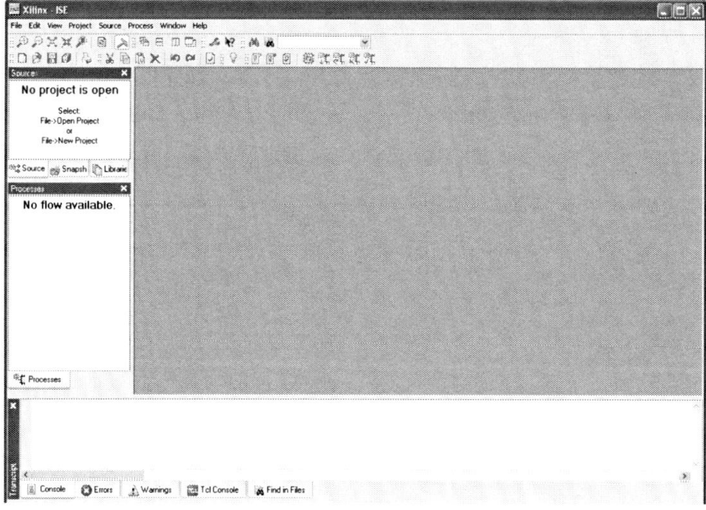


**Figure 2.1** Xilinx ISE WebPACK Project Navigator design window

The Xilinx ISE WebPACK provides an on-line Help facility. Function key F1 provides Help for any highlighted tool or function and the *Help* menu command of the Project Navigator design window provides manuals and tutorials.

## Project Creation

A new project can be created in the Xilinx ISE WebPACK Project Navigator from the menu command *File...New Project* which opens the New Project Wizard – Create New Project window, as shown in Figure 2.3. The name of the Xilinx ISE project is *elapsedtime* (spaces are not allowed) and the location is the *C:\EDPGA\Chapter 2\elapsedtime* folder.

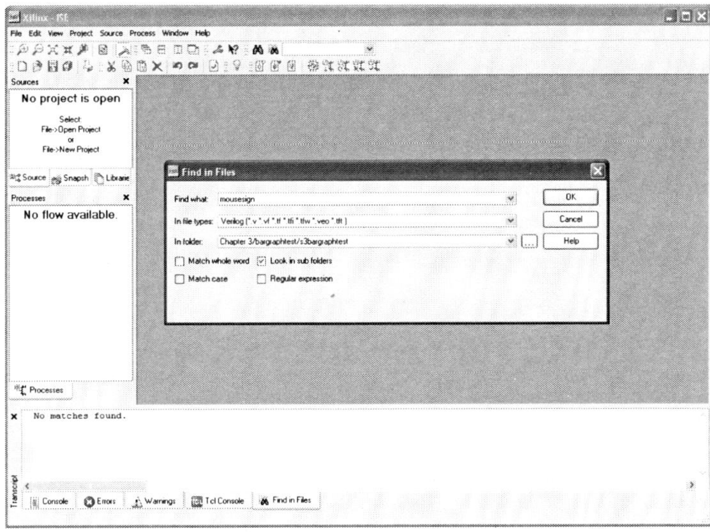


**Figure 2.2** Find in Files menu command and the result display window

The Spartan-3E Starter Board (*www.digilentinc.com*) elapsed time or *stop watch* Verilog project is further described Chapter 3 Programmable Gate Array Hardware and given in Listing 3.16 but is used here to elucidate the requisite steps in the EDA process. The *Chapter 2\elapsedtime* folder is created automatically. The *Top-Level Source Type* should be HDL.

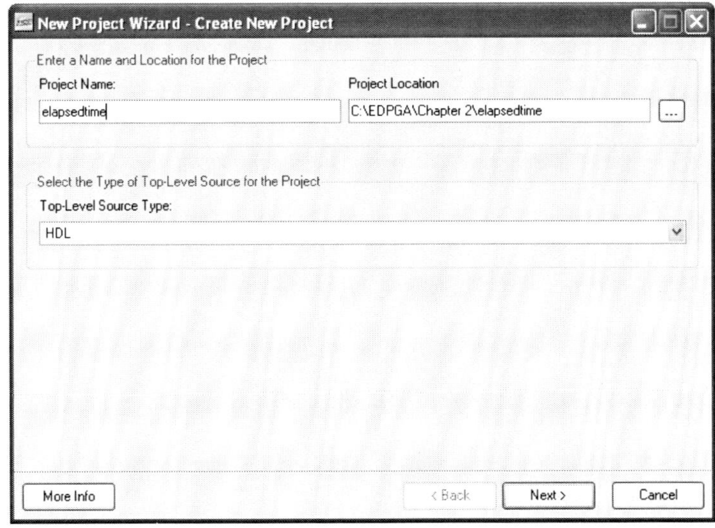


**Figure 2.3** New Project Wizard – Create New Project window

Clicking *Next* opens the New Project Wizard – Device Properties window, as shown in Figure 2.4. The FPGA device properties must match those of the device on the specific evaluator board, as described in Chapter 3 Programmable Gate Array Hardware. Here the Spartan-3E Starter Board is used and the *Family* is Spartan3E, the *Device* is XCS500E (a Spartan-3E XC3S500E FPGA), the *Package* is FG320 (a fine pitch ball grid array package with 320 pins) and the *Speed* is -4 (standard speed grade). The *Top Level Source Type* is Verilog, the *Synthesis Tool* is XST and the *Simulator* is the ISE Simulator. Checking *Enable Enhanced Design Summary* provides a complete describe of the design resources in use for the project.

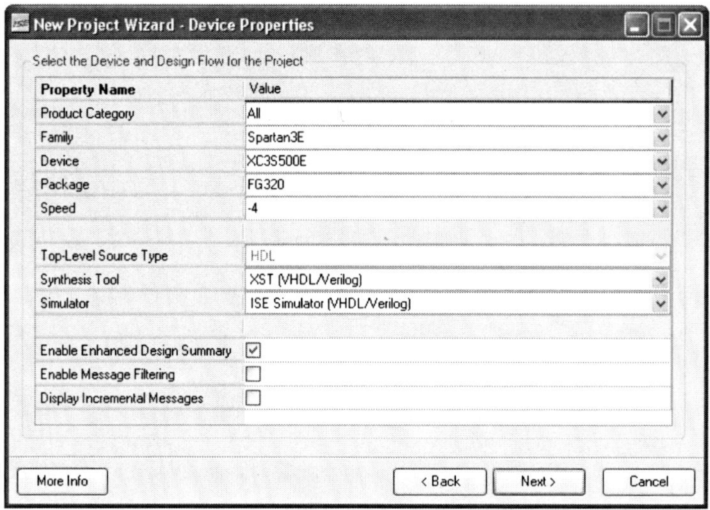


**Figure 2.4** New Project Wizard – Device Properties window

The Digilent Basys Board (*www.digilentinc.*com), described in Chapter 3 Programmable Gate Array Hardware, utilizes the Spartan-3E XCS100E device in the VQ100 package (a very thing quad flat pack package with 100 pins) and speed grade of -4 (a Spartan-3E XC3S100E-VQ100 FPGA).

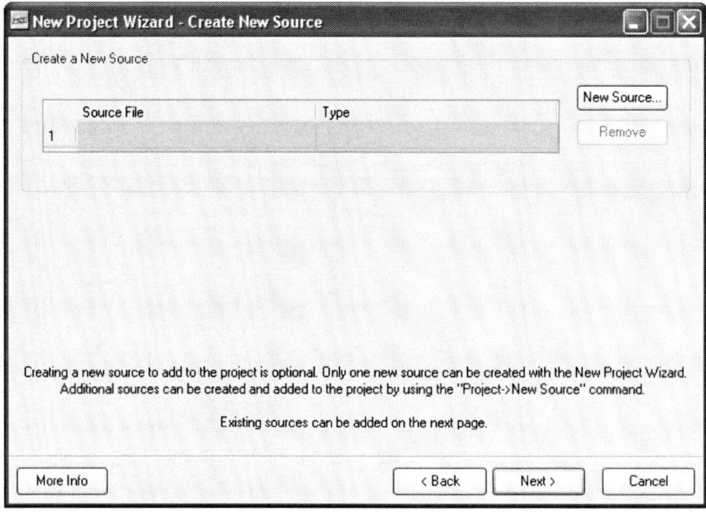


**Figure 2.5** New Project Wizard – Create New Source window

Clicking *Next* opens the New Project Wizard – Create New Source window, as shown in Figure 2.5. Since only one new source file can be created and is optional, this window can be bypassed. If the Create New Source window is not bypassed, then the New Source Wizard – Define Module window opens and the Verilog port names and direction must be specified at this time, which may be problematical at an early stage of a project.

Clicking *Next* opens the New Project Wizard – Add Existing Sources window, as shown in Figure 2.6. Adding existing source files at this point is optional and sources can be added later, so this window can also be bypassed.

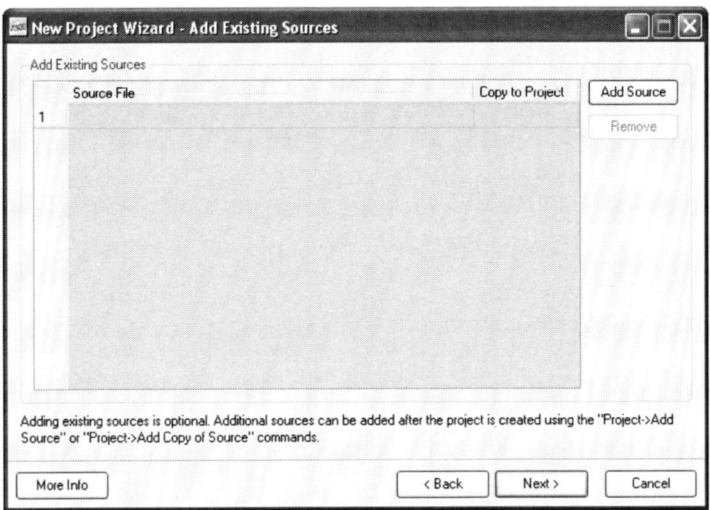


**Figure 2.6** New Project Wizard – Add Existing Sources window

Clicking *Next* opens the New Project Wizard – Project Summary window, as shown in Figure 2.7. A summary of the specifications can be examined in this window before the project files are generated.

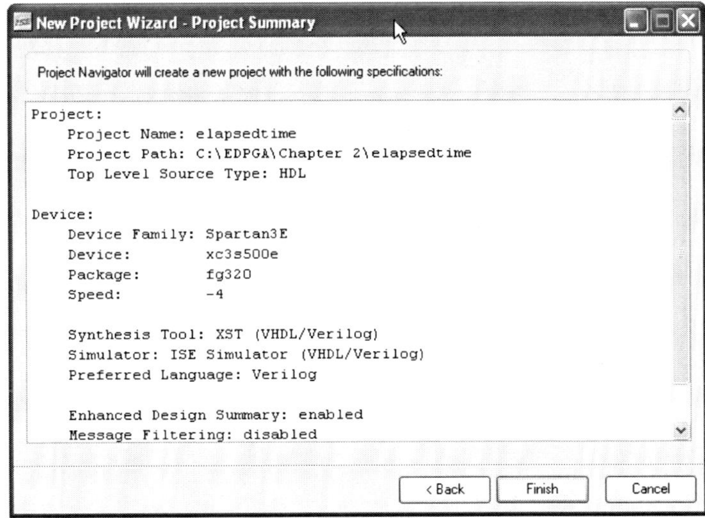


**Figure 2.7** New Project Wizard – Project Summary window

Clicking *Finish* generates the project files and opens the Xilinx ISE WebPACK Project Navigator design window with the rudiments of the elapsedtime.ise project installed, as shown in Figure 2.8.

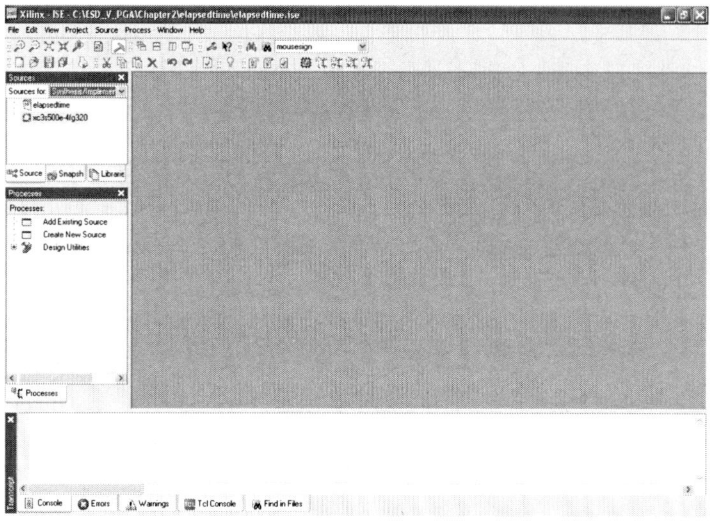


**Figure 2.8** Project Navigator design window for the ISE project elapsedtime.ise

Verilog source files can be created for the project by clicking on *File...New* which opens the select Text File window, as shown in Figure 2.9.

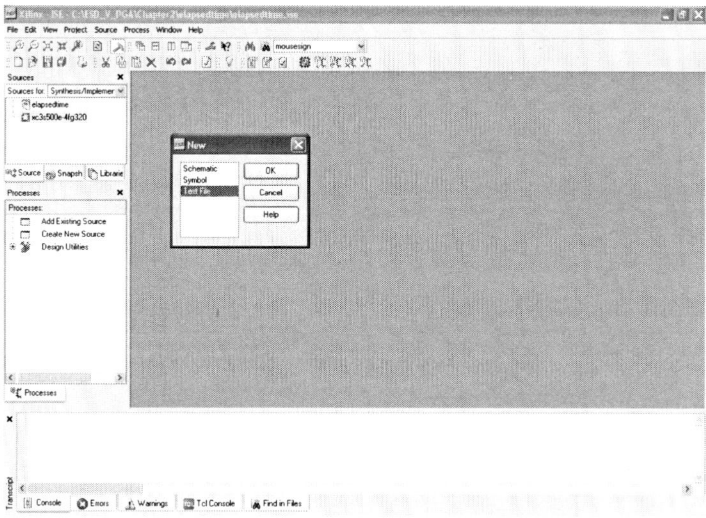


**Figure 2.9** Project Navigator design window select Text File window

The editor of the Xilinx ISE WebPACK Project Navigator design window then opens for the Verilog source file, as shown in Figure 2.10. The Verilog source text files must be added to the project after editing and saving them, as shown in Figure 2.11 and Figure 2.12. The Verilog source file association is for both *synthesis/implementation* and *simulation*. The project has a designated *top module* and modules are implemented either by editing a new source or, through design reuse, adding an existing Verilog module. The top module can be set by highlighting then *right mouse button clicking* on the name of the module in the Sources window or from the Source menu of the taskbar of the Project Navigator.

Here the source modules for the elapsed time project are in folders and the *Project...Add Source* menu is used to add the top module s3eelapsedtime.v and the UCF elapsedtimes3esb.ucf from the *Chapter 2\elapsedtime* folder and the clock.v and the datapath lcd.v modules from the *Chapter 3 \peripherals* folder. The controller etlcd.v module is in the top module s3eelapsedtime.v.

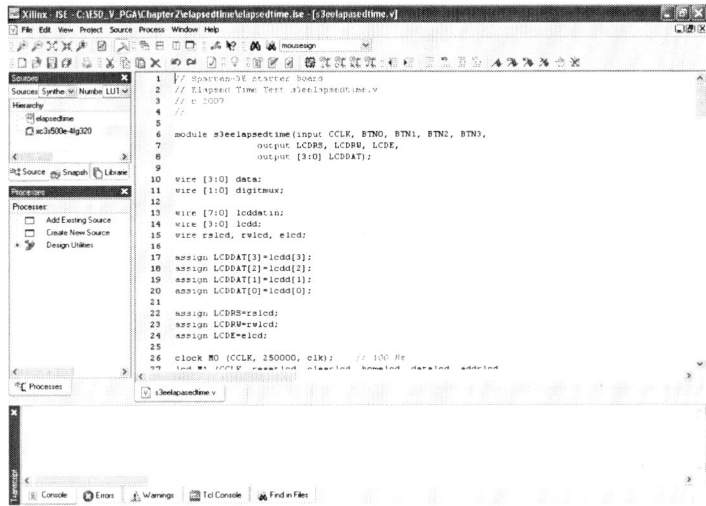


**Figure 2.10** Project Navigator design window source editor

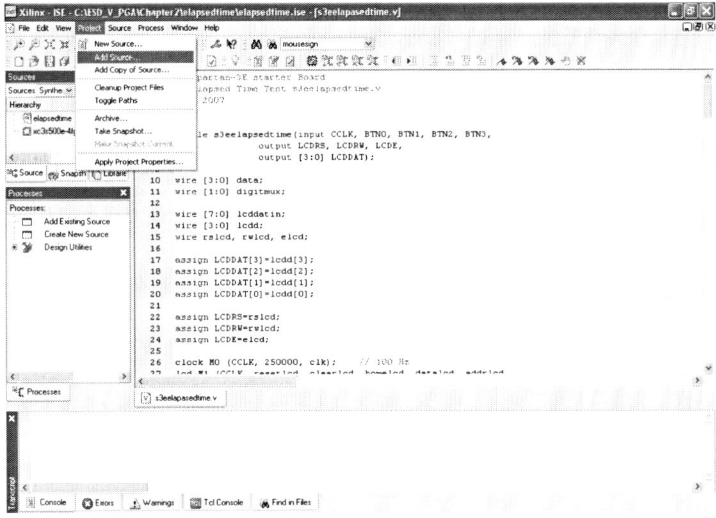


**Figure 2.11** Project Navigator - Add Source... menu

The User Constraints File (UCF) must also be edited and added to the project to assure that the input and output signals are routed to the proper pins, as described in Chapter 3 Programmable Gate Array Hardware. Constraints can specify not only the naming, direction and pin placement of signals but can include implementation and timing considerations [Ciletti04]. The ISE WebPACK provides a Xilinx Constraint Editor with advanced design features for the UCF, as shown in Figure 2.13.

However, the *template* UCF for the Spartan-3E FPGA evaluation boards, as described in Chapter 3 Programmable Gate Array Hardware in Listing 3.1 for the Basys Board and in Listing 3.2 for the Spartan-3E Starter Board, can also be opened as a file in the Project Navigator design window source editor, as shown in Figure 2.14. This simple direct editing of the ISE project UCF can be used here because load, timing and implementation constraints are already included in the template UCF.

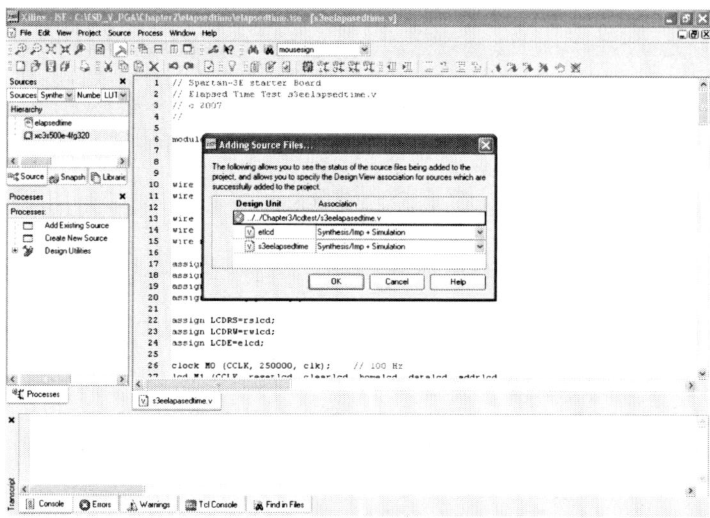


**Figure 2.12** Project Navigator - Adding Source Files… window

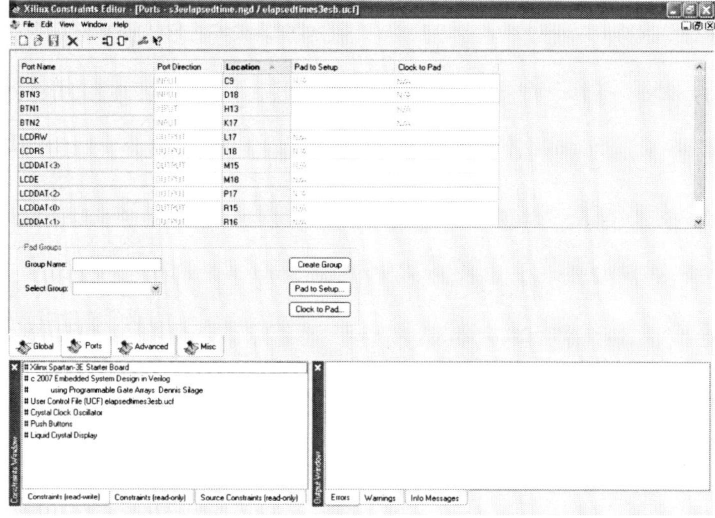


**Figure 2.13** Xlinx Constraint Editor for the ISE project UCF

The signals in the UCF used in the ISE project are uncommented (by removing the initial #) and the default signal names can be used in the Verilog module. The UCF is then saved with an appropriate new file name in the directory of the ISE project. The ISE project UCF must be added to the project, as shown in Figure 2.11 and Figure 2.12. The template UCF for the Spartan-3E Starter Board is in the *Chapter 3\ucf* folder as the s3esb.ucf file. The template UCF for the Basys Board, the other Spartan-3E FPGA evaluation board, is also in the same folder as the basys.ucf file. After expanding Synthesis – XST in the Process window of the Project Navigator, the syntax of the Verilog modules next can be checked by highlighting and *left mouse button clicking* the Check Syntax process, as shown in Figure 2.15.

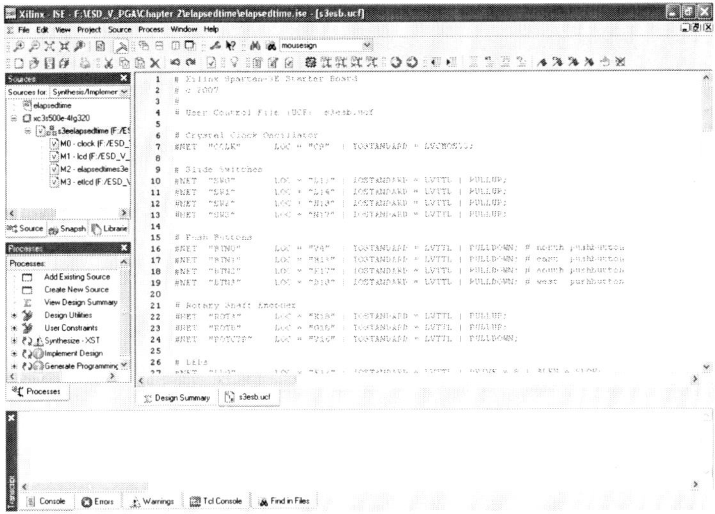

**Figure 2.14** The template UCF in the Project Navigator design window source editor

**Figure 2.15** Project Navigator - Synthesize - XST…Check Syntax design process

## Project Implementation

Although any of the intermediate steps in the FPGA hardware synthesis process can be taken next, including design simulation, the project can readily be implemented as a configuration or *bit file* here targeting the specific Spartan-3E Starter Board. The process for the Basys Board is similar but not described here. The project UCF specifies the pin locations for the external signals that interface the Spartan-3E FPGA and its evaluation board peripherals, as described in Chapter 3 Programmable Gate Array Hardware. After expanding Generate Programming File in the Process window of the Project Navigator, the project can be synthesized and the FPGA can be programmed by highlighting and *left mouse button clicking* the Configure Device (iMPACT) process, as shown in Figure 2.16.

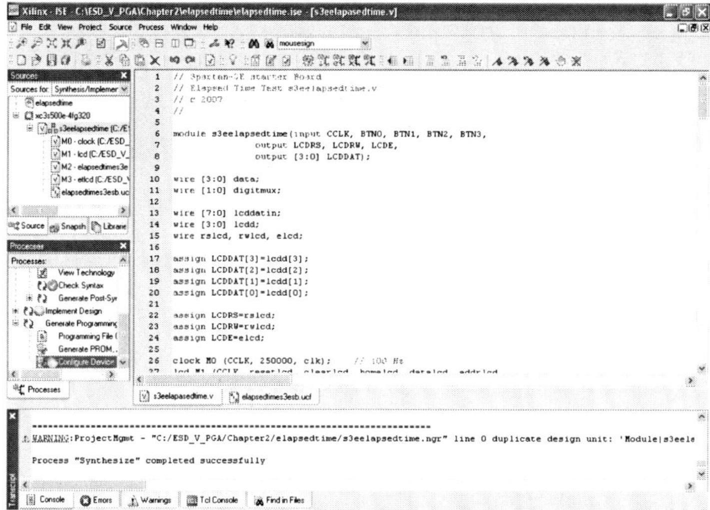


**Figure 2.16** Project Navigator – Generate Programming File…Configure Device (iMPACT)

The Verilog structural or behavioral synthesis of a project is a time-consuming process with numerous steps and intermediate file and console outputs. The progress and any warnings or errors of the hardware synthesis process can be followed in the Project Navigator console display, as shown in Figure 2.1. If the FPGA hardware synthesis process is successful the Xilinx iMPACT configuration and programming EDA tool is started, as shown in Figure 2.17.

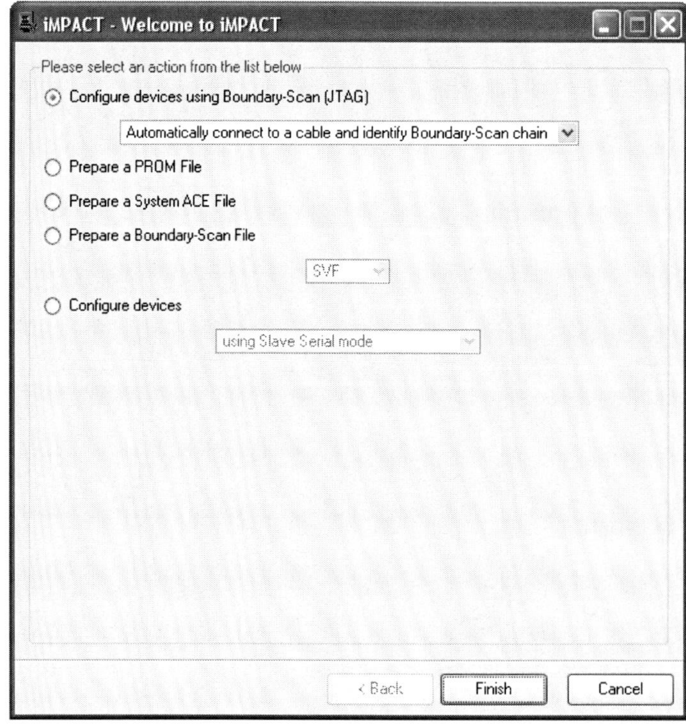


**Figure 2.17** Xilinx iMPACT configuration and programming tool

Xilinx iMPACT utilizes a standard a Joint Test Action Group (JTAG) or *boundary scan* mode (IEEE 1149.1) to communicate with a *chain* of compatible devices. The Spartan-3E Starter Board used here has three such devices, as described in Chapter 3 Programmable Gate Array Hardware. A USB cable from the host to the USB port of the Spartan-3E Starter Board provides the JTAG mode. A green light emitting diode (LED) near the USB port indicates when the cable is connected properly and the JTAG mode is activated by iMPACT.

iMPACT identifies the JTAG chain of a Xilinx XC3S500E Spartan-3E FPGA, a Xilinx XCF04S 4 Mb programmable read only memory (PROM) and a Xilinx XC2C64A CoolRunner-II complex programmable logic device (CPLD) for the Spartan-3E Starter Board. Only the FPGA will be targeted for a volatile configuration file here, although a non-volatile bit file can be programmed into the PROM and then downloaded into the FPGA on *power up*. Highlighting and right clicking on the Xilinx XC3S500E Spartan-3E FPGA opens the iMPACT Assign New Configuration File window for the bit configuration file s3eelapsedtime.bit, as shown in Figure 2.18.

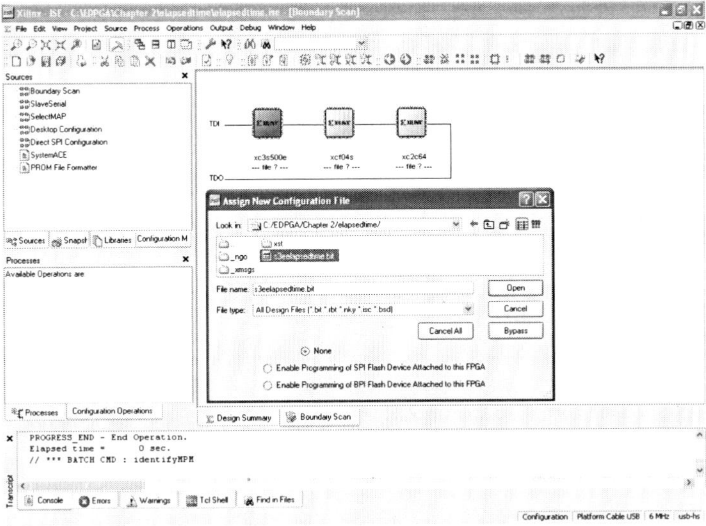

**Figure 2.18** iMPACT Assign New Configuration File window

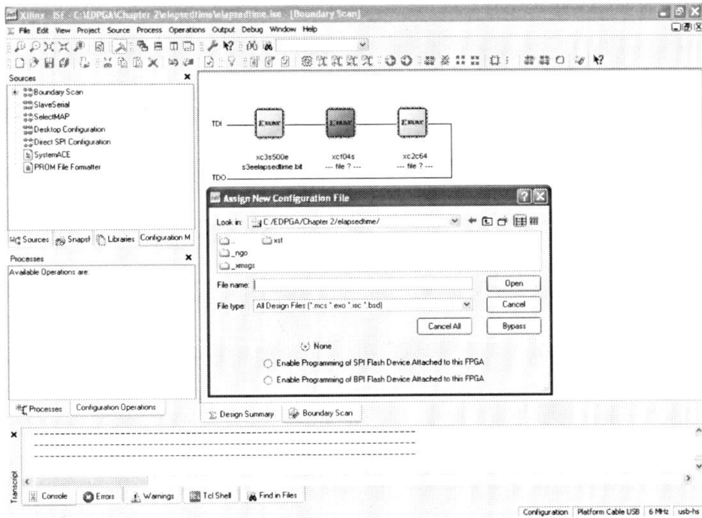

**Figure 2.19** iMPACT iterative assignment of configuration files to the PROM and CPLD

After the assignment of the configuration file to the FPGA, iMPACT attempts to assign bit files to the PROM and CPLD of the Spartan-3E Starter Board, as shown in Figure 2.19. This iterative assignment is not necessary here and can be bypassed. Highlighting and right clicking on the FPGA opens the operations dialog box, as shown in Figure 2.20. The Program process downloads the assigned configuration file to the targeted device via a USB cable to the USB port of the Basys Board or the Spartan-3E Starter Board which is configured as the JTAG port.

iMPACT next opens the Programming Properties window which has extensive options to support a variety of devices from other manufacturers, as shown in Figure 2.21. However, default parameters for the Xilinx FPGA, PROM and CPLD devices for the Spartan-3E Starter Board are available in iMPACT and the options can be ignored.

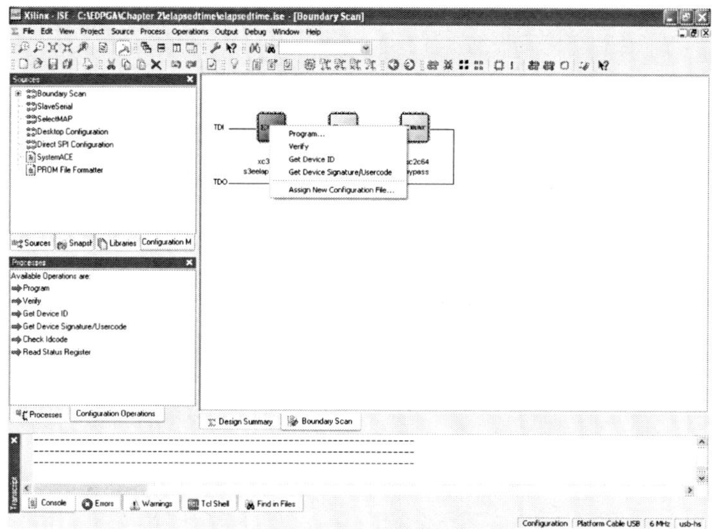


**Figure 2.20** iMPACT Operations dialog box to program the FPGA

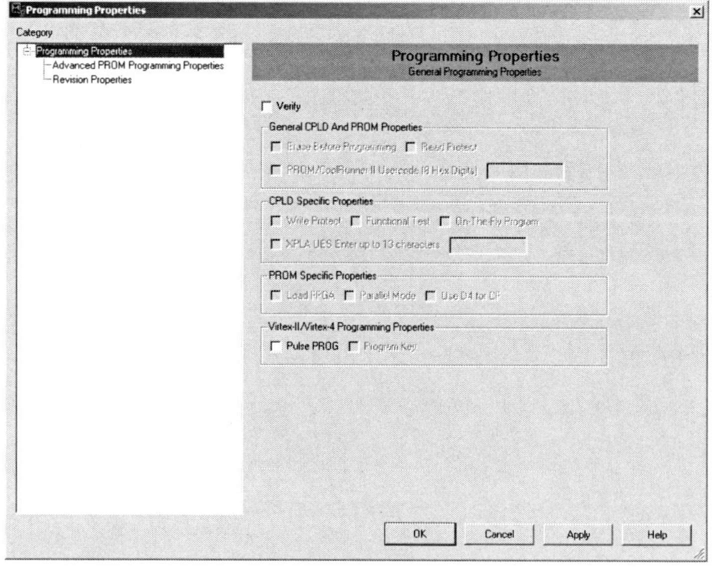


**Figure 2.21** iMPACT Programming Properties window

The iMPACT Progress Dialog box displays the JTAG transfer of the configuration file to the FPGA, as shown in Figure 2.22. Finally, iMPACT displays the Programming Succeeded message if there are no errors in the JTAG transfer of the configuration file, as shown in Figure 2.23.

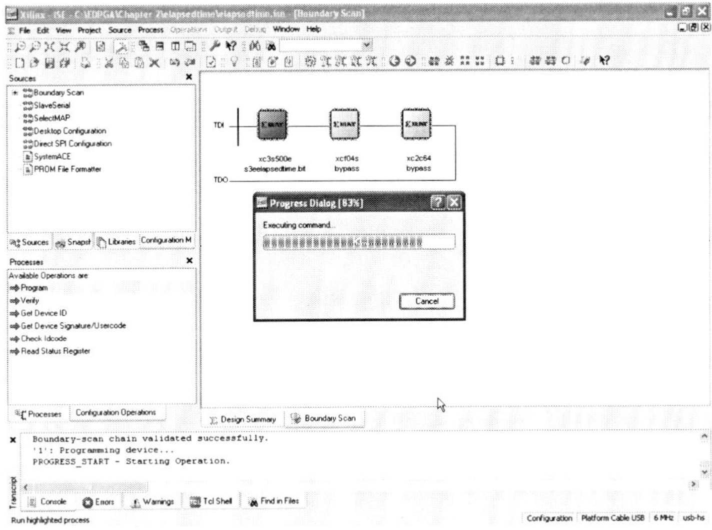


**Figure 2.22** iMPACT Progress Dialog box for programming the FPGA

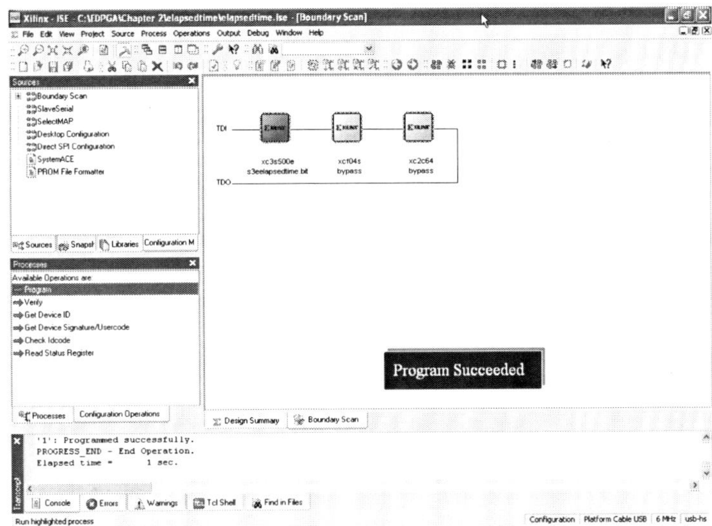


**Figure 2.23** iMPACT Programming Succeeded message for transfer of the bit configuration file to the FPGA

The Spartan-3E Starter Board is now configured with the ISE project and begins executing. If the Xilinx ISE WebPACK project source files or the UCF is modified it is only necessary to highlight and *left mouse button click* the Generate Programming File process to repeat the FPGA hardware synthesis process, as shown in Figure 2.16. The iMPACT programming tasks can also evoked from the Processes menu and the Boundary Scan tab menu, as shown in Figure 2.20.

## Design Summary

The Xilinx ISE WebPACK Project Navigator window provides a Design Summary tab for the synthesized ISE project, as shown in Figure 2.24. The Project Status window gives the name of the project, the Verilog hierarchical *top module* file name, the synthesis FPGA target device, the ISE version number, links to any errors or warnings and the date and time of the most recent project synthesis. The Device Utilization Summary window gives the logic utilization and distribution for the project synthesis.

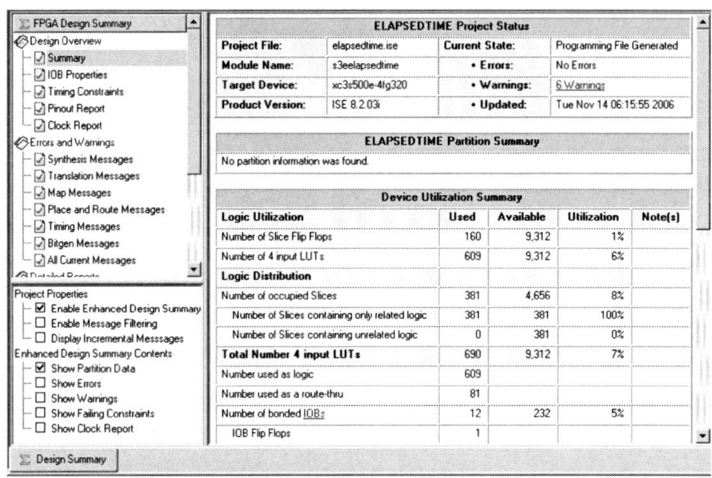


**Figure 2.24** Project Navigator – Design Summary initial window

The logic utilization is given as the number used and the percentage of the total available *slice flip flops* and 4-input *look up tables* (LUT) of the specific Spartan-3E FPGA, as described in Chapter 1 Hardware Description Language Verilog Hardware Description Language. The logic distribution is given as the number used and the percentage of the total available *slices* and reported as those containing related and unrelated logic. The number of 4-input LUTs are reported as those used for logic, *route-through* and input-output blocks (IOB).

FPGA Design Summary
Design Overview: Summary, IOB Properties, Timing Constraints, Pinout Report, Clock Report
Errors and Warnings: Synthesis Messages, Translation Messages, Map Messages, Place and Route Messages, Timing Messages, Bitgen Messages, All Current Messages
Project Properties: Enable Enhanced Design Summary, Enable Message Filtering, Display Incremental Messsages
Enhanced Design Summary Contents: Show Partition Data, Show Errors, Show Warnings, Show Failing Constraints, Show Clock Report

| Total Number 4 input LUTs | 690 | 9,312 | 7% | |
|---|---|---|---|---|
| Number used as logic | 609 | | | |
| Number used as a route-thru | 81 | | | |
| Number of bonded IOBs | 12 | 232 | 5% | |
| IOB Flip Flops | 1 | | | |
| Number of GCLKs | 1 | 24 | 4% | |
| Total equivalent gate count for design | 5,623 | | | |
| Additional JTAG gate count for IOBs | 576 | | | |

Performance Summary

| Final Timing Score: | 0 | Pinout Data: | Pinout Report |
|---|---|---|---|
| Routing Results: | All Signals Completely Routed | Clock Data: | Clock Report |
| Timing Constraints: | All Constraints Met | | |

Detailed Reports

| Report Name | Status | Generated | Errors | Warnings | Infos |
|---|---|---|---|---|---|
| Synthesis Report | Current | Sat Nov 11 10:34:24 2006 | 0 | 5 Warnings | 5 Infos |
| Translation Report | Current | Sat Nov 11 10:34:56 2006 | 0 | 0 | 0 |
| Map Report | Current | Sat Nov 11 10:35:22 2006 | 0 | 0 | 3 Infos |
| Place and Route Report | Current | Sat Nov 11 10:37:03 2006 | 0 | 1 Warning | 2 Infos |
| Static Timing Report | Current | Sat Nov 11 10:37:25 2006 | 0 | 0 | 2 Infos |
| Bitgen Report | Current | Sat Nov 11 10:38:02 2006 | 0 | 0 | 0 |

Design Summary

**Figure 2.25** Project Navigator – Design Summary scrolled window

The scrollable Design Summary window provides the total equivalent gate count for the synthesized ISE project, as shown in Figure 2.25, which is a useful metric for comparison. The Performance Summary and the Detailed Reports provide links to more ISE project information.

The All Current Messages window provides a listing of the warnings and errors encounters in the ISE project synthesis and other salient information, as shown in Figure 2.26. The messages are grouped by the EDA tool that generated them. Warning and informational messages can be useful to the design process. Warning message describe several signals as not changing or not used in the ISE project and could be eliminated. The Xilinx XST HDL Advisor informational messages describe automatic steps taken in the ISE project synthesis.

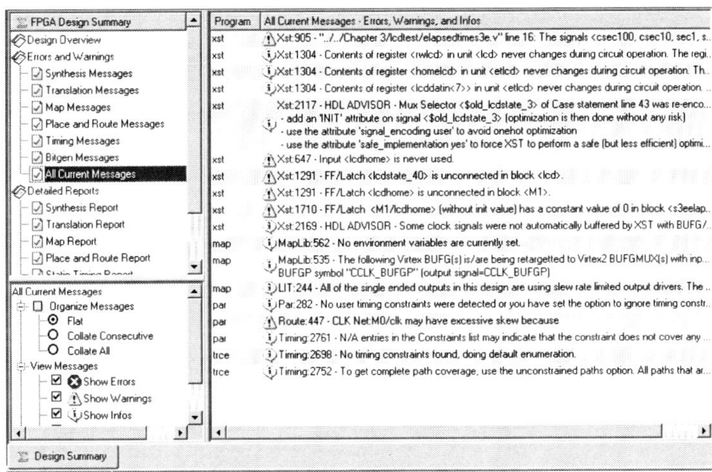


**Figure 2.26** Project Navigator – Design Summary – All Current Messages window

# Xilinx CORE Generator

The Xilinx CORE Generator™ System provides an optimized set of *building blocks* for common and specialized functions that facilitate embedded system design in Verilog using programmable gate arrays. The CORE Generator System provides access to the Xilinx LogiCORE™ intellectual property (IP) products and the AllianceCORE products from *third-party* IP developers. Several of the LogiCORE blocks for the Spartan-3E FPGA of the Basys Board and the Spartan-3E Starter Board are freely available and described in a Xilinx application note (XAPP474, *www.xilinx.com*). The implementation and inclusion of one such LogiCORE block into an ISE project is described here. Each LogiCORE product is provided with a descriptive data sheet, Verilog instantiation and a module description.

## LogiCORE Creation

The name of the Xilinx ISE WebPACK project is *binBCDdivider* and the location is the *Chapter 2\binBCDdivider* folder. This project reads the setting of the Spartan-3E Starter Board slide switches as the four least significant bits (LSB) and the pushbuttons as the four most significant bits (MSB) of an 8-bit binary number. The 8-bit binary number is converted by a LogiCORE *pipelined divider* to a three digit (000 to 225) binary coded decimal (BCD) number, which is then outputted to the liquid crystal display (LCD). The 8-bit binary number is also outputted to the light emitting diodes (LED). Further details of this project implementation as Verilog modules in support of the slide switches (SW), pushbuttons (BTN), LED and the LCD of the Spartan-3E Starter Board are described in Chapter 3 Programmable Gate Array Hardware.

The CORE Generator is installed as a part of the Xilinx ISE WebPACK software electronic design automation (EDA) tools and is accessible from the *Project...New Source* menu command, which opens New Source Wizard – Select Source Type window, as shown in Figure 2.27. Select *IP (Coregen & Architecture Wizard)* and enter the file name (*binbcddivider*) and project location for the LogiCORE Verilog instantiation and a module description. The file name cannot contain uppercase letters or start with a digit. The location for the LogiCORE module should be the same directory as that of the ISE project and the *Add to project* box should be checked.

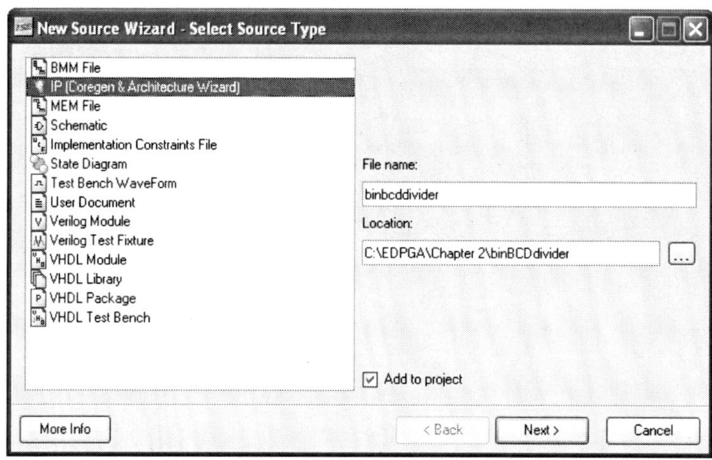


**Figure 2.27** New Source Window – Select Source Type window

Clicking *Next* opens the New Source Wizard – Select IP window, as shown in Figure 2.28. Opening the *Math Function...Dividers* directory folder shows the two LogiCORE products that are available. Select the Divider Generator 1.0 LogiCORE product for the project here.

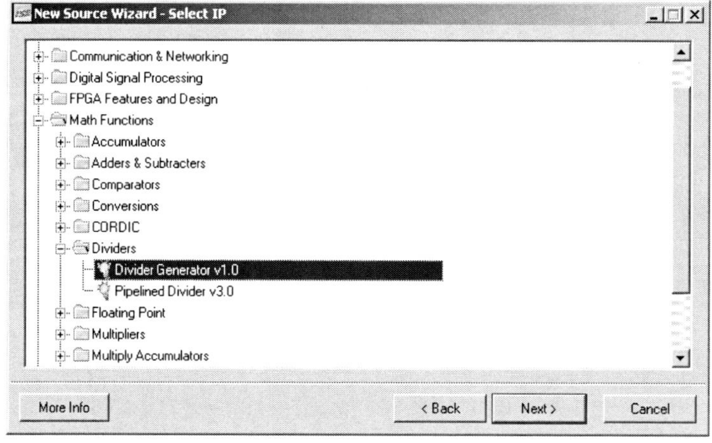


**Figure 2.28** New Source Wizard – Select IP window

Clicking *Next* produces the New Source Wizard – Summary window, as shown in Figure 2.29. The summary describes the LogiCORE product selected, the file name and location and is customized for the Spartan-3E FPGA specified in the ISE project, as shown in the New Project Wizard – Device Properties window in Figure 2.4.

Clicking *Finish* opens the first of the Divider Generator v1.0 LogiCORE design windows, as shown in Figure 2.30. The Divider Generator LogiCORE block is a fixed-point divider based on radix-2 non-restoring division or a floating-point divider by repeated multiplications and is described in the data sheet DS530 (*www.xilinx.com*). The data sheet can be viewed (as a *pdf* file) by clicking the *View Data Sheet* button. The Divider is implemented as a fixed-point divider with a *clock enable* (CE) input signal, while the *asynchronous clear* (ACLR) and the *synchronous clear* (SCLR) input signals are not *checked* and are not used here. The *ready for data* (RFD) output signal indicates the clock cycle in which input data is sampled. However, the RFD signal is only applicable if the number of cycles for the pipelined division is not 1.

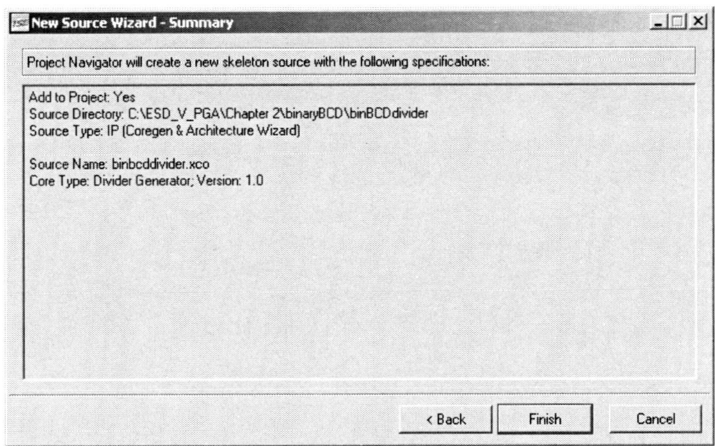


**Figure 2.29** New Source Wizard – Summary window

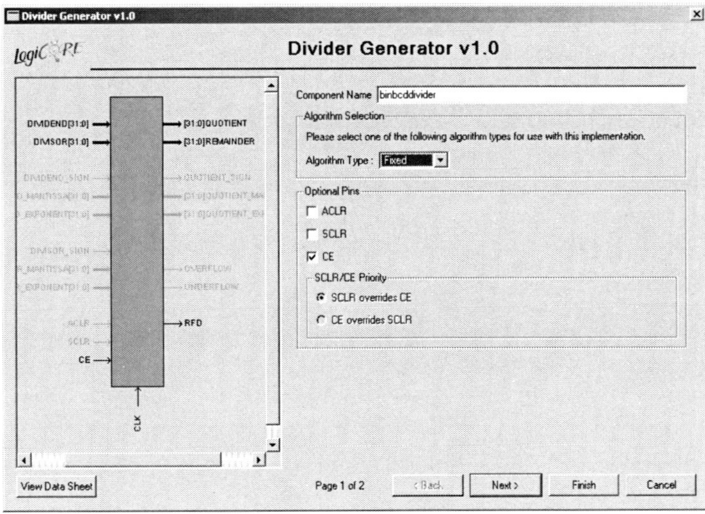


**Figure 2.30** First design window of the LogiCORE Divider Generator v1.0

Clicking *Next* opens the second of the Divider Generator v1.0 LogiCORE product windows, as shown in Figure 2.31. The dividend and divisor bus widths are both selected to be 8-bit, the number of clocks per division is 1, the operands are unsigned, and the remainder option is set.

The generation of the selected LogiCORE block in the Xilinx CORE Generator provides a *Readme File* with a description of the components. Here the file name is binbcddivider_readme.txt in the ISE project directory. The Verilog components include the LogiCORE parameter file binbcddivider.xco, the instantiation template file binbcddivider.veo and the (so-called) wrapper file binbcddivider.v for functional simulation. The template file for each generated core must be instantiated into the parent design. To instantiate a LogiCORE block into an Xilinx ISE project in Verilog, *copy and paste* the instantiation template file (.veo) into the appropriate area of the top module of the project. The module instantiation template file binbcddivider.veo, as shown in Listing 2.1, is in the project directory and opened in the Project Navigator design window by clicking on *File...Open.*

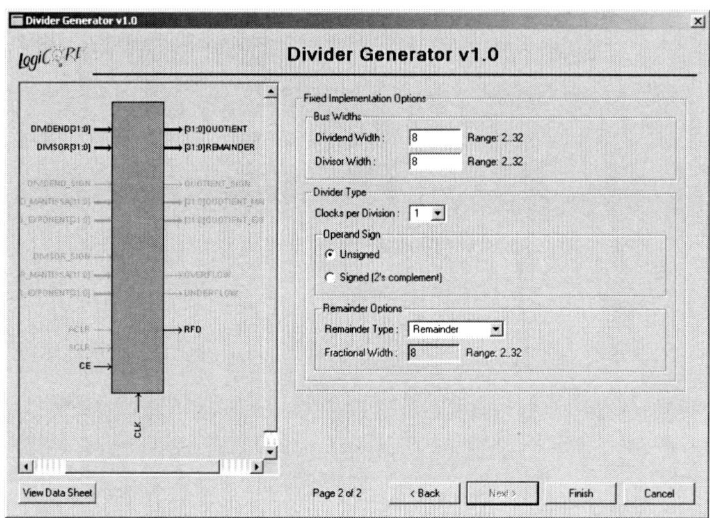


**Figure 2.31** Second design window of the LogiCORE Divider Generator v1.0

**Listing 2.1** LogiCORE instantiation template file binbcdcddivider.veo

```
// The following must be inserted into your Verilog file for this
// core to be instantiated. Change the instance name and port connections
// (in parentheses) to your own signal names.

//----------- Begin Cut here for INSTANTIATION Template ---// INST_TAG
binbcddivider YourInstanceName (
	.clk(clk),
	.ce(ce),
	.dividend(dividend),
	.divisor(divisor),
	.quotient(quotient),
	.remainder(remainder),
	.rfd(rfd));
// INST_TAG_END ------ End INSTANTIATION Template ---------
```

*YourInstanceName*, a dummy name from the instantiation template binbcdcddivider.veo in Listing 2.1, must be changed and the port connections may need to change to reflect the actual connections in the ISE project. The *connection by name* is used in the instantiation template for the port connections, as described in Chapter 1 Verilog Hardware Description Language. The Xilinx CORE generator parameter file binbcddivider.xco is automatically added to the ISE project, as shown in Figure 2.32.

The LogiCORE binary to BCD pipelined divider module binbcddivider.xco is verified by the Verilog top module binBCD.v in Listing 2.2 which is in the *Chapter 2\binBCDdivider* folder. The instantiated LogiCORE module is copied from Listing 2.1 and inserted and modified, as shown in Listing 2.2. The module name for the pipelined divider is M1 in the top module and the default port connection names are used. The complete Xilinx ISE WebPACK project uses a User Constraints File (UCF) which uncomments the signals CCLK, BTN0, BTN1, BTN2, BTN3, SW0, SW1, SW2, SW3, LD0, LD1, LD2, LD3, LD4, LD5, LD6, LD7, LCDDAT<0>, LCDDAT<1>, LCDDAT<2>, LCDDAT<3>, LCDRS, LCDRW, and LCDE in the Spartan-3E Starter Board UCF s3esb.ucf in Listing 3.2. The three Verilog modules operate in parallel and some independently in the top module, as described in Chapter 3 Programmable Gate Array Hardware.

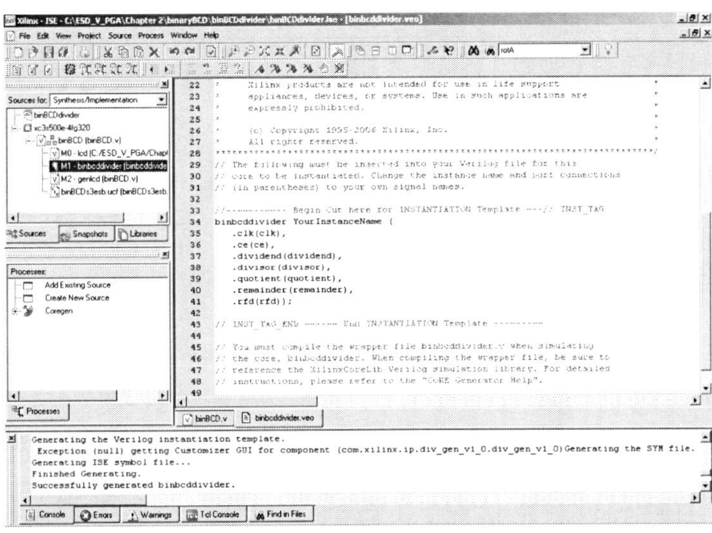

**Figure 2.32** Project Navigator window for the ISE project binbcddivider.ise

**Listing 2.2** LogiCORE pipelined divider verification top module binBCD.v

```
// Spartan-3E Starter Board
// LogiCORE Binary to BCD binBCD.v
// c 2008 Embedded Design using Programmable Gate Arrays  Dennis Silage

module binBCD (input CCLK, BTN0, BTN1, BTN2, BTN3, SW0, SW1, SW2, SW3,
               output LD0, LD1, LD2, LD3, LD4, LD5, LD6, LD7, LCDRS, LCDRW, LCDE,
               output [3:0] LCDDAT);

wire [7:0] lcddatin;
wire [3:0] lcdd;
wire rslcd, rwlcd, elcd;

wire [7:0] bindata, databin, divisor, dividend;
wire [7:0] quotient, remainder;

assign LCDDAT[3] = lcdd[3];
assign LCDDAT[2] = lcdd[2];
assign LCDDAT[1] = lcdd[1];
assign LCDDAT[0] = lcdd[0];
```

```
assign LCDRS = rslcd;
assign LCDRW = rwlcd;
assign LCDE = elcd;

assign LD7 = bindata[7];
assign LD6 = bindata[6];
assign LD5 = bindata[5];
assign LD4 = bindata[4];
assign LD3 = bindata[3];
assign LD2 = bindata[2];
assign LD1 = bindata[1];
assign LD0 = bindata[0];

lcd M0 (CCLK, resetlcd, clearlcd, homelcd, datalcd, addrlcd, lcdreset, lcdclear, lcdhome, lcddata,
        lcdaddr, rslcd, rwlcd, elcd, lcdd, lcddatin, initlcd);
binbcddivider M1 (.clk(CCLK), .ce(ce), .dividend(dividend), .divisor(divisor), .quotient(quotient),
        .remainder(remainder), .rfd(rfd));
genlcd M2 (CCLK, BTN0, BTN1, BTN2, BTN3, SW0, SW1, SW2, SW3, bindata, ce, rfd, dividend,
        divisor, quotient, remainder, resetlcd, clearlcd, homelcd, datalcd, addrlcd, initlcd,
        lcdreset, lcdclear, lcdhome, lcddata, lcdaddr, lcddatin);

endmodule
```

The structure and function of the LCD controller module genlcd.v and the LCD datapath module lcd.v are described for other specific embedded design applications in Chapter 3 Programmable Gate Array Hardware. The Xilinx ISE WebPACK project here is intended to introduce the Xilinx CORE Generator System. The concept of a controller and datapath Verilog behavioral synthesis modules in embedded design is discussed in Chapter 1 Verilog Hardware Description Language.

The Design Utilization Summary for the top module binBCD.v for the LogiCORE pipelined divider shows the use of 279 slice flip-flops (2%), 631 4-input LUTs (6%), 414 occupied slices (8%) and a total of 7397 equivalent gates in the XC3S500E Spartan-3E FPGA synthesis.

## Implementation Comparison

The conversion of a binary number to a BCD representation is prevalent in embedded design applications [Botros06]. The Xilinx XST Verilog HDL compiler only supports division ( / ) by signed powers-of-two ($\pm 2^n$) so a direct conversion by dividing by powers-of-ten ($10^n$) is not possible [Ciletti04]. The LogiCORE pipeline divider, as given in Listing 2.2, is a general solution that operates on any divisor and can produce its result in one clock cycle. Although the FPGA resource allocation of slice flip-flops and 4-input LUTs is large, it can be reduced if 2, 4 or 8 clock cycles are specified for the pipelined divider.

However, embedded design in the Verilog HDL often requires a reasonable search for alternative structures that either may require less FPGA resources or present other desirable properties such as the data throughput rate [Ciletti99]. A modular design for an ISE project implies that other algorithms can be convenient inserted and synthesized.

The first alternative is the behavioral model in the Verilog HDL for binary to BCD conversion by iterative subtraction. The name of the Xilinx ISE project is *binBCDiterative* and the location is the *Chapter 2\binBCDiterative* folder. The iterative subtraction algorithm is verified by the top module binBCDiter.v, which is similar to the top module for the LogiCORE pipelined divider binBCD.v, as given in Listing 2.2.

The behavioral model iterative subtraction Verilog module for binary to BCD conversion iterative.v is the datapath to the controller module genlcditer.v and is given in Listing 2.3. The controller-datapath signals are dav and dataav, as described in Chapter 1 Hardware Description Language Verilog Hardware Description Language. The binary input data bindata is sequentially subtracted by powers-of-ten while incrementing the 4-bit BCD most significant and middle digit counters msdigit and middigit. The least significant digit lsdigit is the remainder. The 8-bit input signal bindata must be copied to an internal 8-bit register value for processing.

**Listing 2.3** Iterative subtraction module for binary to BCD conversion iterative.v

```
// Spartan-3E Starter Board
// Binary to BCD Iterative Conversion iterative.v
// c 2008 Embedded Design using Programmable Gate Arrays  Dennis Silage

module iterativediv (input dav, input [7:0] bindata, output reg [3:0] msdigit, output reg [3:0] middigit,
                     output reg [3:0] lsdigit, output reg dataav);

reg [7:0] value;
integer i;

always@(posedge dav)
    begin
        dataav = 0;
        value = bindata;
        msdigit = 0;                    // most significant digit
        for (i = 1; i <= 9; i = i + 1)
            begin
                if (value >= 100)
                    begin
                        msdigit = msdigit + 1;
                        value = value – 100;
                    end
            end
        middigit = 0;                   // middle digit
        for (i = 1; i <= 9; i = i + 1)
            begin
                if (value >= 10)
                    begin
                        middigit = middigit + 1;
                        value = value – 10;
                    end
            end
        lsdigit = value;                // least significant digit
        dataav = 1;
    end

endmodule
```

The Design Utilization Summary for the top module binBCDiter.v for the iterative subtraction algorithm for binary to BCD conversion shows the use of 115 slice flip-flops (1%), 574 4-input LUTs (6%), 338 occupied slices (7%) and a total of 4739 equivalent gates in the XC3S500E Spartan-3E

FPGA synthesis. This FPGA resource utilization is less than half that of the top module binBCD.v for the LogiCORE pipeline divider that accommodates any divisor.

The iterative subtraction algorithm produces its results for all three BCD digits in one clock cycle, although it uses a fixed scheme. The controller-datapath signals dav and datav are assigned to one of the 4-bit output ports (6-bit header) of the Spartan-3E Starter Board as *test points*, as described in Chapter 3 Programmable Gate Array Hardware. The latency of the controller-datapath signals can be monitored by a logic analyzer or oscilloscope to assess its performance.

The second alternative for binary to BCD conversion in the Verilog HDL is the shift-and-add-3 behavioral model [Wakerly00]. The name of the Xilinx ISE project is *binBCDshiftadd3* and the location is the *Chapter 2\binaryBCDshiftadd3* folder. The shift-and-add-3 binary to BCD conversion is verified by the top module binBCDshiftadd3.v, which is similar to the top module for the LogiCORE pipelined divider binBCD.v, as given in Listing 2.2.

The shift-and-add-3 Verilog module for binary to BCD conversion shiftadd3.v is itself a top module for the behavioral module shiftadd.v, which illustrates the concept of a hierarchical structure of nested modules, as described in Chapter 1 Verilog Hardware Description Language. The shiftadd.v datapath module is event driven on change in the input signal indata.

The genlcdshift.v module is the controller but here is modified not to send or receive explicit controller-datapath signals. The binary input data bindata is processed by a *cascade* of seven shift-and-add-3 modules to produce the most significant msdigit. middle middigit and least significant lsdigit digits. Here the 8-bit input signal bindata does not need to be copied to an internal register for processing and continuous assignment statements route the net variable signals.

**Listing 2.4** Shift-and-add-3 module for binary to BCD conversion shiftadd3.v

```
// Spartan-3E Starter Board
// Binary to BCD Shift and Add 3 Conversion shiftadd3.v
// c 2008 Embedded Design using Programmable Gate Arrays  Dennis Silage

module shiftadd3 (input [7:0] bindata, output [3:0] msdigit, output [3:0] middigit, output [3:0] lsdigit);

wire [3:0] result1, result2, result3, result4, result5;
wire [3:0] result6, result7;
wire [3:0] value1, value2, value3, value4, value5;
wire [3:0] value6, value7;

assign value1 = {1'b0, bindata[7:5]};
assign value2 = {result1[2:0], bindata[4]};
assign value3 = {result2[2:0], bindata[3]};
assign value4 = {result3[2:0], bindata[2]};
assign value5 = {result4[2:0], bindata[1]};
assign value6 = {1'b0,result1[3], result2[3], result3[3]};
assign value7 = {result6[2:0], result4[3]};

shiftadd M1 (value1, result1);
shiftadd M2 (value2, result2);
shiftadd M3 (value3, result3);
shiftadd M4 (value4, result4);
shiftadd M5 (value5, result5);
shiftadd M6 (value6, result6);
shiftadd M7 (value7, result7);
```

```
assign msdigit[1:0] = {result6[3], result7[3]};
assign msdigit[3:2] = 0;
assign middigit = {result7[2:0], result5[3]};
assign lsdigit = {result5[2:0], bindata[0]};

endmodule

module shiftadd(input [3:0] indata, output reg [3:0] outdata);

always@(indata)
    case(indata)
        0:    outdata = 0;
        1:    outdata = 1;
        2:    outdata = 2;
        3:    outdata = 3;
        4:    outdata = 4;
        5:    outdata = 8;
        6:    outdata = 9;
        7:    outdata = 10;
        8:    outdata = 11;
        9:    outdata = 12;
        default: outdata = 0;
    endcase

endmodule
```

The Design Utilization Summary for the top module binBCDshiftadd3.v for the shift-and-add-3 algorithm for binary to BCD conversion shows the use of 89 slice flip-flops (1%), 564 4-input LUTs (6%), 310 occupied slices (6%) and a total of 4490 equivalent gates in the XC3S500E Spartan-3E FPGA synthesis. This FPGA resource utilization is approximately the same as that of the top module binBCDiter.v for the iterative subtraction algorithm and less than half that of the top module binBCD.v for the LogiCORE pipeline divider for binary to BCD conversion. The shift-and-add-3 algorithm produces its results for all three BCD digits in one clock cycle, although it uses a fixed scheme.

The third alternative is the structural model for the serial conversion of binary to BCD in a complex programmable logic device (CPLD), as described in a Xilinx application note (XAPP029, *www.xilinx.com*) but not implemented here. The binary to BCD conversion is performed in a modified shift register that sequentially shifts the binary data serially into the register and successively doubles its contents. However, if during the conversion a BCD digit or 5 or greater is encountered, the shift register contents must be converted to the proper BCD value of its doubled contents and a logic 1 shifted into the next higher digit register [Navabi06].

Finally, the fourth alternative binary to BCD conversion is the *table look-up*. The 74185 medium scale integration (MSI) integrated circuit (IC) implemented a 6-bit binary to BCD conversion by using a 256-bit read-only memory (ROM). Although the 74185 integrated circuit (IC) device is now obsolete it illustrates another possible implementation for binary to BCD conversion.

The block RAM of the Spartan-3E FPGA can be used for the conversion of 8-bit binary to three BCD digits (10 bits are required for the BCD digits) and only $256 \times 10 = 2560$ bits would be required. The table look-up concept can also be used for other applications in embedded design such as the linearization of an analog transducer signal output after quantization.

## Xilinx Floorplanner

The Xilinx ISE Floorplanner™ is an advanced software electronic design automation (EDA) tool that can in some instances improve the performance of an automatically placed and routed FPGA design. The Floorplanner is particularly useful for placing *critical path* logic to optimize timing performance in structural models and datapath modules, as described in Chapter 1 Hardware Description Language Hardware Description Language. Although the User Constraints File (UCF) determines the placement of logic to output pins, the Floorplanner can add additional constraints on the routing, timing, grouping, initialization, synthesis and mapping. For example, the Floorplanner can be used to select the placement of logic in the control logic blocks (CLB), the location of CLBs on the FPGA, and the maximum delay between storage elements.

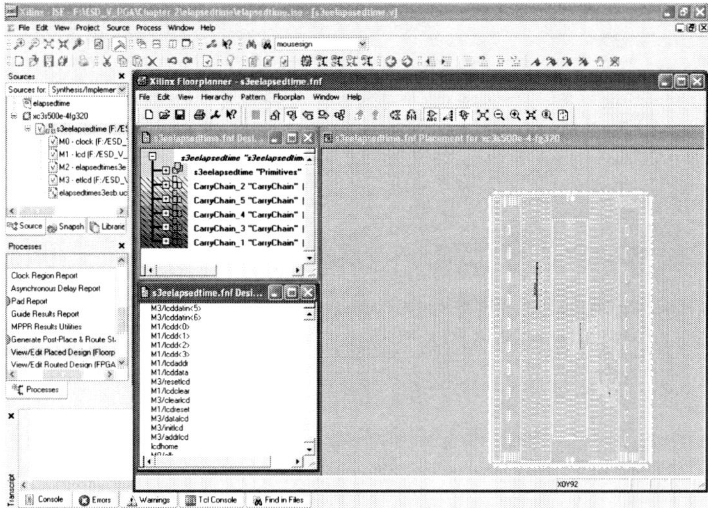


**Figure 2.33** Floorplanner window for the ISE project elapsedtime.ise

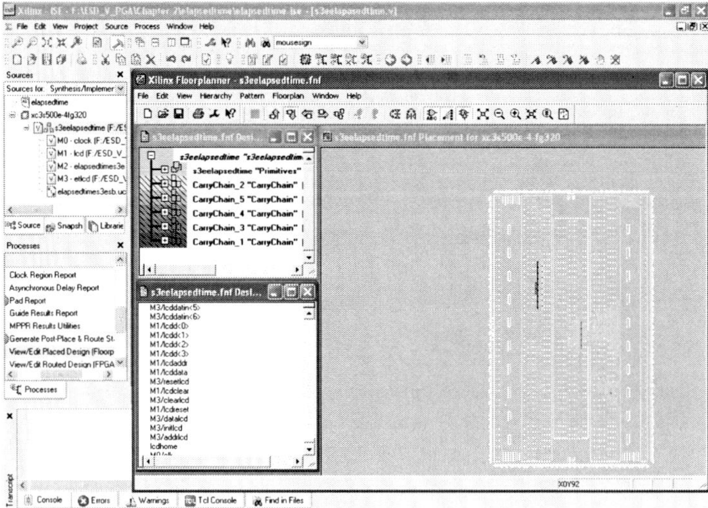


**Figure 2.34** Floorplanner design window for the routing and slice and flip-flop utilization for lcddata in module M1 of the ISE project elapsedtime.ise

The Floorplanner is installed as a part of the Xilinx ISE WebPACK software EDA tools and is accessible by expanding the *Processes...Implement Design...Place & Route* and left clicking on *View/Edit Placed Design (Floorplanner)* tabular listing command, which here opens Xilinx Floorplanner window for the ISE project elapsedtime.ise, as shown in Figure 2.33.

However, to utilize the Floorplanner to its best extent requires a detailed knowledge of the FPGA architecture and resources and the underlying structural model being implemented. The Floorplanner uses a manual and iterative approach to achieve realizable goals for routing, timing and synthesis. These factors become increasingly critical for real-time processes or an embedded design that utilizes a high degree of FPGA resources.

Figure 2.34 shows the routing of a selected portion of the ISE project elapsedtime.ise from the *Chapter 2\elapsedtime* folder. The Floorplanner also displays the utilization of the FPGA slice and flip-flops for the specific register variable lcddata in the M1 module of the hierarchical nested modules in the ISE project.

Figure 2.35 shows the package and pin placement for the ISE project which is determined by the UCF. If the embedded system design has not fixed the input-output (IO) pins then the Floorplanner can be used to reroute the signals to meet the critical path specifications. However, the Spartan-3 and Spartan-3E evaluation boards used here have fixed IO pins. This intricate process of *floorplanning* can be obviated in many projects in embedded system design and is not used here. Additional details of the Floorplanner are available in either the Xilinx ISE Quick Start Tutorial or in the Xilinx ISE In-Depth Tutorial (*www.xilinx.com*).

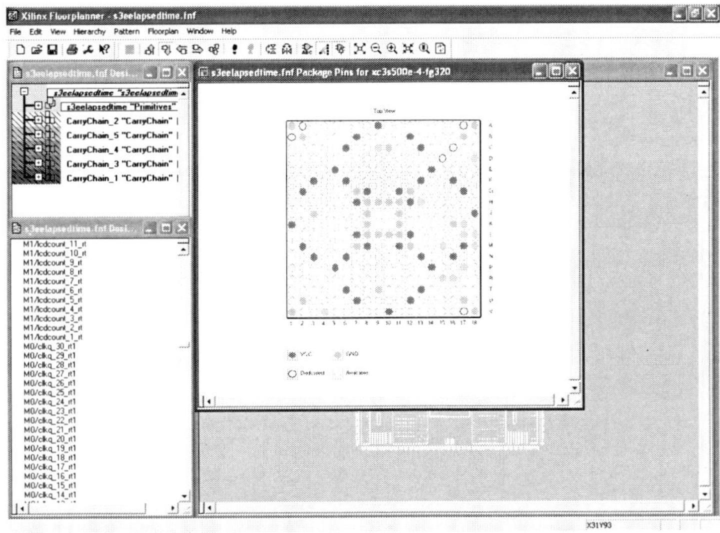


**Figure 2.35** Floorplanner design window for IO pin placement

## Xilinx Simulator

The Xilinx ISE Simulator can provide behavioral and timing verification without hardware synthesis and implementation to the FPGA [Botros06]. The elapsed time module for the Spartan-3E Starter Board elapsedtimes3e.v in the Xilinx ISE WebPACK project elapsedtime.ise in the *Chapter 2 \elaspsedtime* folder is used for the simulation, as given in Listing 3.17. Open the ISE Project Navigator and select the elapsedtimes3e.v Verilog HDL file in the *Sources* window, as shown in Figure 2.36.

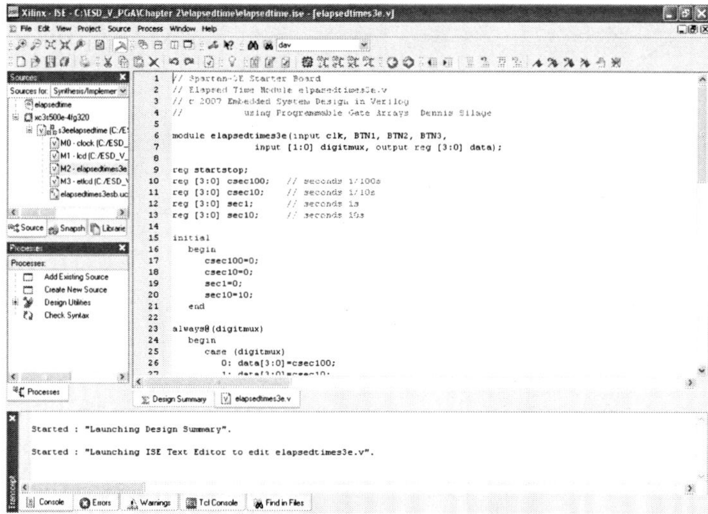


**Figure 2.36** Project Navigator selection of the elapsedtimes3e.v module to simulate

Create a new source by selecting *Project...New Source* and the New Source Wizard – Select Source Type window opens, as shown in Figure 2.37. Select *Test Bench Waveform* as the source type and input a file name (testbench is used here). Clicking *Next* and the New Source Wizard – Associate Source window opens which indicates which source file is associated with the testbench waveforms, as shown in Figure 2.38.

Clicking *Next* New Source Wizard – Summary window opens, as shown in Figure 2.39. Clicking *Finish* and the ISE Project Navigator creates the source file testbench.tbw with the Verilog HDL source file association indicated.

The Initial Timing and Clock Wizard – Initialize Timing window sets the timing and clock parameters of the associated elapsedtimes3e.v module for the simulation, as shown in Figure 2.40. Although the original clock period for the signal clk in the elapsed time application was 10 milliseconds (msec) or a frequency of 100 Hz, 200 nanoseconds (ns) with a 50% *duty cycle* or a *Clock High Time* and a *Clock Low Time* of 100 ns is used for the simulation.

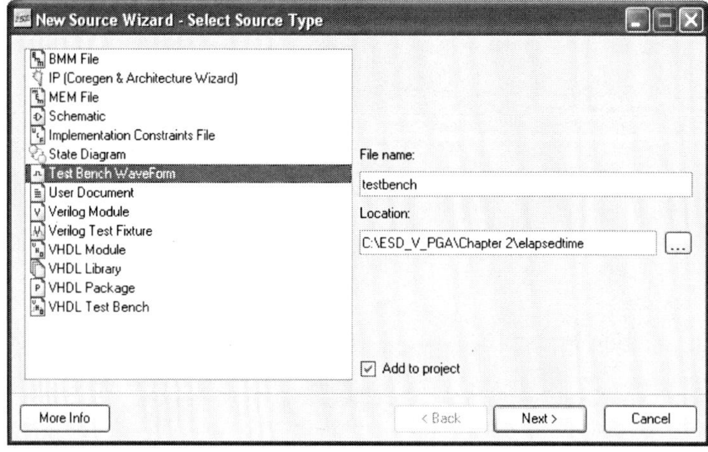


**Figure 2.37** New Source Wizard – Select Source Type window

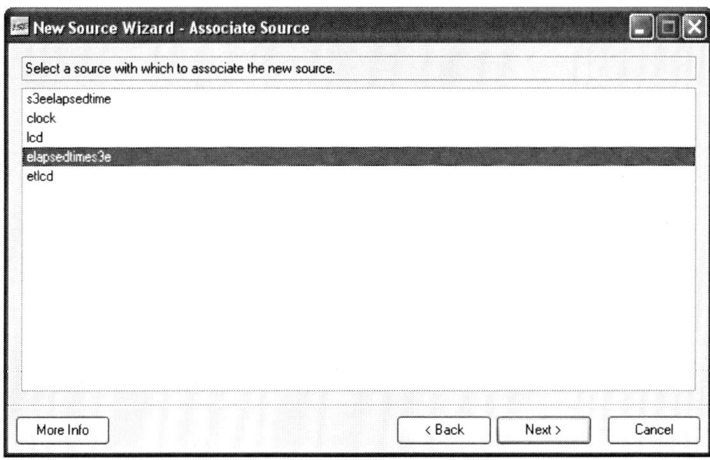


**Figure 2.38** New Source Wizard – Associate Source window

The default *Input Setup Time* and *Output Valid Delay* is 10 ns. Check the *GSR (FPGA)* box (Global Set/Reset) and a default *High for Initial* of 100 ns is automatically added to the nominal *Offset*. The *Single Clock* box is checked for the signal clk and the *Initial Length of Test Bench* is set to 10 000 nanoseconds (ns) but can be changed during the simulation.

After saving the ISE project, select the testbench.tbw in the *Sources* for *Behavioral Simulation* window of the ISE Project Navigator, as shown in Figure 2.41. The blue shaded areas of the test bench waveform editor represent the *input setup time* specified, as shown in Figure 2.41. For this simulation of the elapsed time application the input transitions occur on the positive edge of the clock signal clk.

The remaining input stimuli are the push button signals BTN1 (stop), BTN2 (start) and BTN3 (reset) and the 2-bit digit multiplexer signal digitmux. The output signal data for the simulation of the elapsedtimes3e.v module is the multiplexed binary coded decimal (BCD) tens, units, tenths or hundreds digit of the elapsed time count.

New Source Wizard - Summary

Project Navigator will create a new skeleton source with the following specifications:

Add to Project: Yes
Source Directory: C:\ESD_V_PGA\Chapter 2\elapsedtime
Source Type: Test Bench WaveForm
Source Name: testbench.tbw

Association: elapsedtimes3e

< Back | Finish | Cancel

**Figure 2.39** New Source Wizard – Summary window

**Figure 2.40** Initial Timing and Clock Wizard – Initialize Timing window

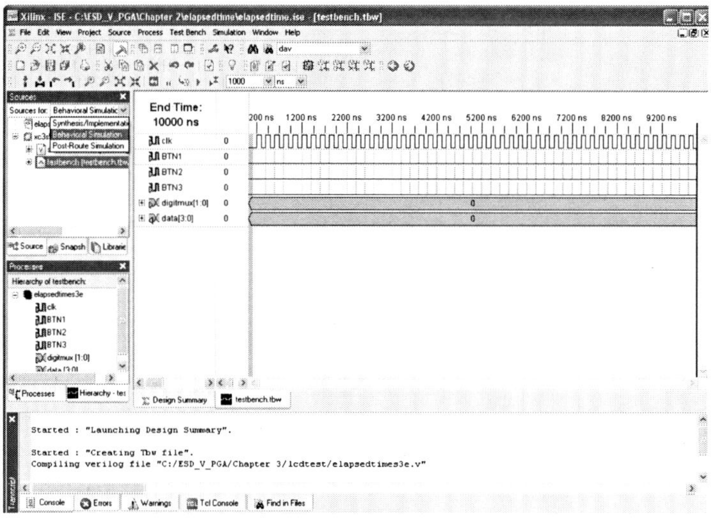


**Figure 2.41** ISE Simulator test bench window

Other than the clock signal clk, the 1-bit push button input stimuli are set manually by clicking on the blue cell which toggles the signal at that point. The 2-bit multiplexer signal digitmux can be set by expanding this bus signal and then clicking on the individual signals of the bus, as shown in Figure 2.42.

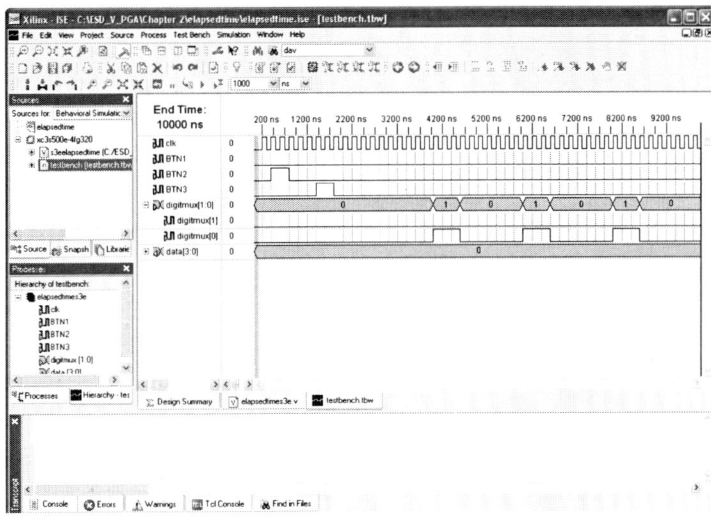


**Figure 2.42** ISE Simulator test bench waveform editor window

The 2-bit multiplexer signal digitmux can also be set by clicking on the bus signal which opens the Pattern Wizard, as shown in Figure 2.43.

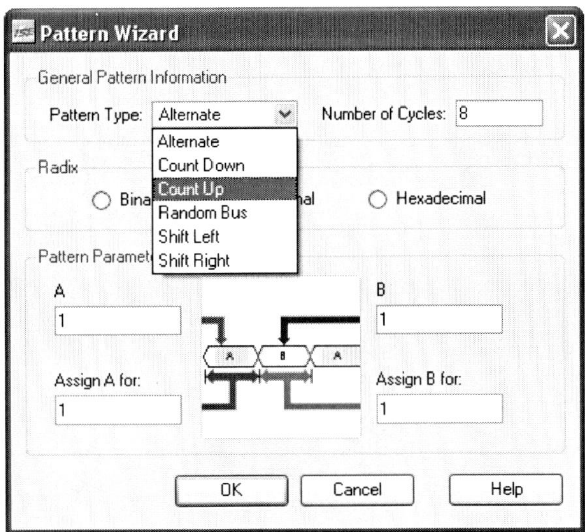


**Figure 2.43** Pattern Wizard window for a bus signal

After saving the ISE project, select *Simulate Behavioral Model* in the *Processes…Xilinx ISE Simulator* window of the ISE Project Navigator, as shown in Figure 2.44. The simulation executes to the specified initial length of the test bench (10 000 ns) as shown in Figure 2.40. The resulting signals are displayed in the *Simulation* window of the ISE Simulator within the ISE Project Navigator.

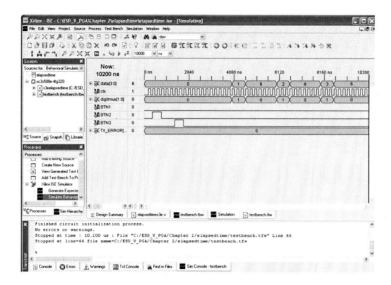

Figure 2.44 Simulation of the elapsedtimes3e.v module

The *Simulation* waveform viewer can zoom-in or zoom-out and apply single or double markers for signal measurement using the ISE Simulator taskbar, as shown in Figure 2.45.

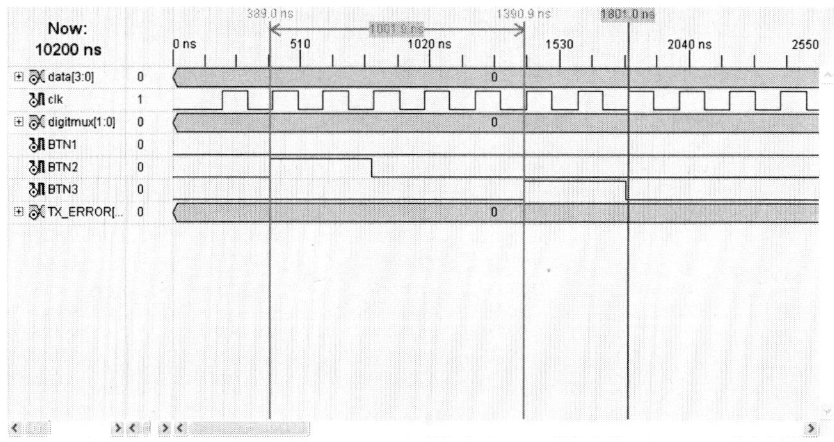


**Figure 2.45** Presynthesis simulation of the elapsedtimes3e.v module

The ISE Simulator utilizes a Verilog HDL stimulus file is automatically generated by the test bench waveform editor window, as shown in Figure 2.42. The stimulus file testbench.tfw is inserted into the Xilinx ISE WebPACK project to verify the performance of the elapsedtimes3e.v module, as given in Listing 2.5. The generation of the simulation stimulus file is transparent and uses Verilog complier directives (such as `timescale), time step instructions (#OFFSET), file input-output tasks ($fopen), console display tasks ($display) and simulation control ($stop) which are not used for FPGA hardware synthesis [Navabi06].

**Listing 2.5** ISE Simulator Verilog HDL testbench.tfw stimulus file

```
// Copyright (c) 1995-2003 Xilinx, Inc.
// All Right Reserved.
`timescale 1ns/1ps
```

```
module testbench;

reg clk = 1'b0;
reg BTN1 = 1'b0;
reg BTN2 = 1'b0;
reg BTN3 = 1'b0;
reg [1:0] digitmux = 2'b00;
wire [3:0] data;
parameter PERIOD = 200;
parameter real DUTY_CYCLE = 0.5;
parameter OFFSET = 100;

initial   // Clock process for clk
      begin
            #OFFSET;
            forever
                  begin
                        clk = 1'b0;
                        #(PERIOD-(PERIOD*DUTY_CYCLE)) clk = 1'b1;
                        #(PERIOD*DUTY_CYCLE);
                   end
       end

elapsedtimes3e UUT (
      .clk(clk),
      .BTN1(BTN1),
      .BTN2(BTN2),
      .BTN3(BTN3),
      .digitmux(digitmux),
      .data(data));

integer TX_FILE = 0;
integer TX_ERROR = 0;

initial
      begin       // Open the results file...
            TX_FILE = $fopen("results.txt");
            #10200     // Final time:  10200 ns
            if (TX_ERROR == 0)
                  begin
                        $display ("No errors or warnings.");
                        $fdisplay (TX_FILE, "No errors or warnings.");
                  end
            else
                  begin
                        $display ("%d errors found in simulation.", TX_ERROR);
                        $fdisplay (TX_FILE, "%d errors found in simulation.",
                                          TX_ERROR);
                  end
            $fclose(TX_FILE);
            $stop;
      end
```

```
initial begin
	// ------------ Current Time: 390ns
	#390;
	BTN2 = 1'b1;
	// ------------------------------------
	// ------------ Current Time: 790ns
	#400;
	BTN2 = 1'b0;
	// ------------------------------------
	// ------------ Current Time: 1390ns
	#600;
	BTN3 = 1'b1;
	// ------------------------------------
	// ------------ Current Time: 1790ns
	#400;
	BTN3 = 1'b0;
	// ------------------------------------
	// ------------ Current Time: 3990ns
	#2200;
	digitmux = 2'b01;
	// ------------------------------------
	// ------------ Current Time: 4590ns
	#600;
	digitmux = 2'b00;
	// ------------------------------------
	// ------------ Current Time: 5990ns
	#1400;
	digitmux = 2'b01;
	// ------------------------------------
	// ------------ Current Time: 6590ns
	#600;
	digitmux = 2'b00;
	// ------------------------------------
	// ------------ Current Time: 7990ns
	#1400;
	digitmux = 2'b01;
	// ------------------------------------
	// ------------ Current Time: 8590ns
	#600;
	digitmux = 2'b00;
	// ------------------------------------
	end

task CHECK_data;

input [3:0] NEXT_data;

	#0 begin
		if (NEXT_data !== data)
			begin
				$display("Error at time=%dns data=%b, expected=%b", $time,
								data, NEXT_data);
```

```
            $fdisplay(TX_FILE, "Error at time=%dns data=%b,
                                expected=%b", $time, data, NEXT_data);
            $fflush(TX_FILE);
            TX_ERROR = TX_ERROR + 1;
        end
    end

endtask

endmodule
```

Behavioral simulation of a Verilog HDL embedded system design can be done on each module of an ISE project for verification of performance [Ciletti04]. Note that only the input and output signals are displayed and that internal net and register variables are not accessible, as described in Chapter 1 Verilog Hardware Description Language.

However, simulation requires test data entered in the waveform editor window, as shown in Figure 2.42. This step is somewhat time-consuming and, unless the test data is comprehensive, may not verify all possible conditions of operation [Navabi06].

The simulation of the elapsedtimes3e.v module here performs the same regardless of the clock frequency, although the ISE Simulator does include the input setup time and the output valid delay as a parameter, as shown in Figure 2.40. A characteristic of behavioral simulation (also known as *presynthesis simulation*) is that the clock, gate and routing propagation delays are not included [Navabi06].

Behavioral simulation demonstrates that the Verilog HDL module functions as expected although the actual performance can only be verified after the embedded system design is synthesized to the FPGA. A timing simulation (also know as *postsynthesis simulation*) utilizes signal propagation information that is available after the *Implement Design...Place & Route* process from the *Sources...Synthesis/Implementation* of the ISE Project Navigator is executed. After the Xilinx ISE XST implementation of the design select *Simulate Post-Place & Route Model* in the *Processes...Xilinx ISE Simulator* window from the *Sources...Post-Route Simulation* of the ISE Project Navigator, as shown in Figure 2.46.

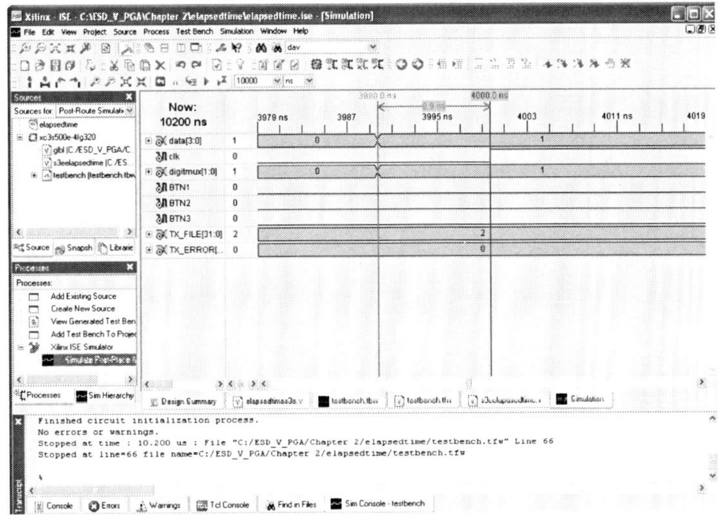


**Figure 2.46** Postsynthesis simulation of the elapsedtimes3e.v module

## Xilinx Verilog Language Templates

A Verilog language or *inference* template menu is evoked from the *Edit...Language Templates* taskbar of the Xilinx ISE Project Navigator, as shown in Figure 2.47.

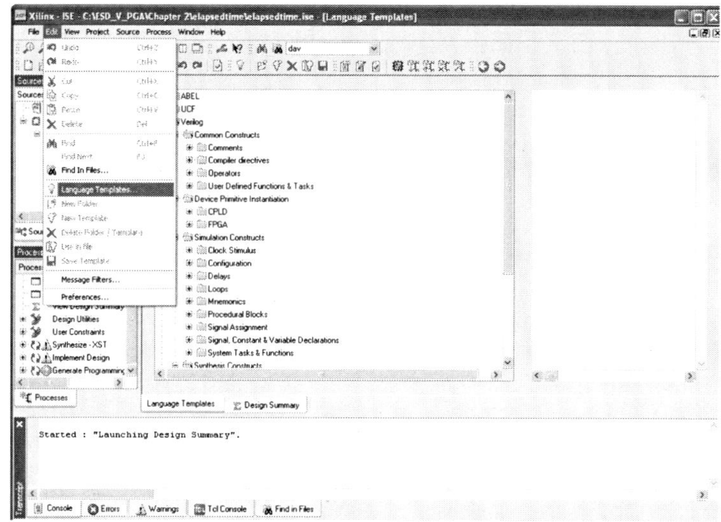

**Figure 2.47** Language Template menu in the ISE Project Navigator

The Xilinx ISE WebPACK provides such Language Templates in the Verilog HDL to simply the process of implementing an embedded design in an FPGA. The *Verilog...Common Constructs* language template menu provides terse reference material and Verilog HDL module segments for comment statements, compiler directives, arithmetic, logical and bitwise operators, and tasks and functions. The *Verilog...Device Primitive Instantiation...FPGA* provide arithmetic functions, clock components, configuration components, input-output (IO) components, RAM and ROM components, register, latches, look-up table (LUT) shift registers and slice and control logic block (CLB) primitives.

The *Verilog...Simulation Constructs* provide clock stimulus, configuration, delays, loops, mnemonics, procedural blocks, signal assignment, signal, constant and variable declarations and system task and functions. The *Verilog...Synthesis Constructs* provide *always* statement constructs and port, signal, constant and coding examples. The coding example synthesis constructs include accumulators, bidirectional IO, gates, counters, encoders and decoders, flip flops, logic shift, multiplexers, block and distributed RAM, ROM and LUT, shift registers, finite state machines (FSM) and tristate buffers.

An example of a 2-bit bidirectional IO module Verilog HDL language template with output enable is given in Listing 2.6. These language templates provide reasonable constructs and serve as a reference for incorporation into a Xilinx ISE project in embedded system design in Verilog using the FPGA.

**Listing 2.6** Xilinx Language Template for a 2-bit bidirectional IO with output enable

```
inout [1:0] <top_level_port>;
reg [1:0] <input_reg>, <output_reg>, <output_enable_reg>;

assign <top_level_port>[0] = <output_enable_reg>[0] ? <output_reg>[0] : 1'bz;
assign <top_level_port>[1] = <output_enable_reg>[1] ? <output_reg>[1] : 1'bz;
```

```
always @(posedge <clock>)
    if (<reset>)
        begin
            <input_reg> <= 2'b00;
            <output_reg> <= 2'b00;
            <output_enable_reg> <= 2'b00;
    else
        begin
            <input_reg> <= <top_level_port>;
            <output_reg> <= <output_signal>;
            <output_enable_reg> <= {2{<output_enable_signal>}};
        end
```

## Xilinx Architecture Wizard

The Xilinx Architecture Wizard is an advanced software electronic design automation (EDA) tool that augments other Verilog HDL module creation tools such as the Xilinx CORE Generator. The only Architecture Wizard currently available for the Spartan-3 and Spartan-3E FPGA is the Digital Clock Manager (DCM). However, the DCM provides flexibility in the selection and generation of synchronous clocks in an embedded system design in an FPGA. The DCM is used in the Basys Board and the Spartan-3E Starter Board applications described in Chapter 3 Programmable Gate Array Hardware and in the digital signal processing, digital communication and digital control applications described in Chapter 4 Digital Signal Processing, Communications and Control.

The Architecture Wizard is a convenient way to customize the relatively complex parameters and to precisely define the operation of the DCM. A block definition file (*name.xaw*) is produced which can be translated to a Verilog HDL description and an instantiation template. The DCM Architecture Wizard has design rule checks to insure that the combination of parameters specified results in a valid configuration of the DCM.

The Architecture Wizard is installed as part of the Xilinx ISE WebPACK software electronic design automation (EDA) tools and is accessible from the *Project...New Source* menu command, which opens New Source Wizard – Select Source Type window, as shown in Figure 2.48. Select *IP (Coregen & Architecture Wizard)* and enter the file name (*dacs3edcm*) and project location for the Architecture Wizard Verilog HDL instantiation.

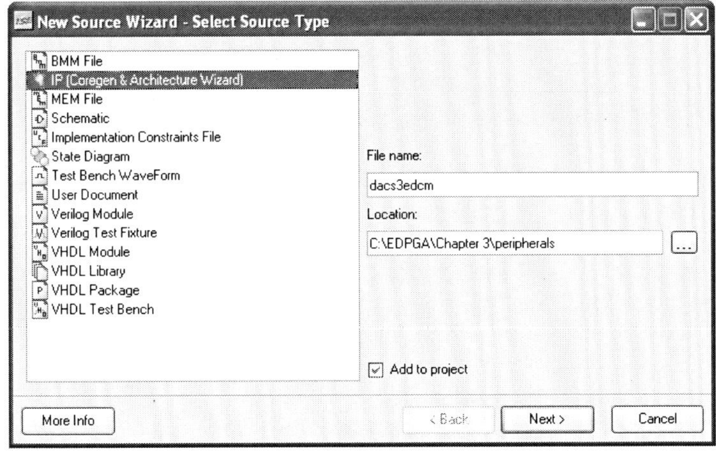


**Figure 2.48** New Source Window – Select Source Type window

The file name cannot contain uppercase letters or start with a digit. Here the location for the Architecture Wizard block definition file (dac3sedcm.xaw) is the *Chapter 3\peripherals* folder. *C:\EDPGA* is the installation directory for the Verilog modules and projects, as described in the Appendix. The resulting Verilog HDL module (dacs3edcm.v) is placed in the ISE project directory and the *Add to project* box should be checked.

Clicking *Next* opens the New Source Wizard – Select IP window, as shown in Figure 2.49. Opening the *FPGA Features and Design...Clocking...Spartan-3E, Spartan-3A* directory folder shows the five DCM Architecture Wizard products that are available.

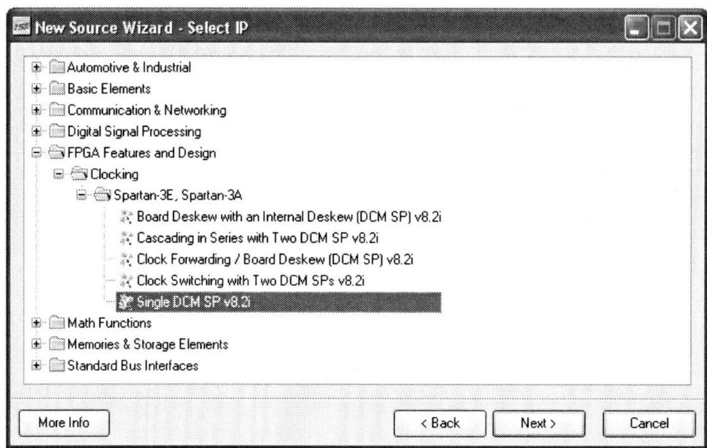


**Figure 2.49** New Source Wizard – Select IP window

Select the Single DCM product for the project here. The DCM here is a *frequency synthesizer* which increases the crystal clock frequency of the Spartan-3E Starter Board from 50 MHz to 83.33 MHz for an application utilizing the digital-to-analog converter (DAC), as described in Chapter 3 Programmable Gate Array Hardware.

Figure 2.50 is the Xilinx Clocking Wizard – General Setup window in which a reset (RST), a clock output (CLKO), synthesized clock output (CLKFX), a a delayed-locked loop (DLL) *lock-in* signal (LOCKED) and the external input clock frequency is specified. The default values of no phase shift (NONE) and an internal feedback source with a multiply by 1 (X1) feedback value is used. The DCM is described in a Xilinx application note (XAPP462, *www.xilinx.com*).

Clicking *Next* opens the Xilinx Clocking Wizard – Clock Buffers window, as shown in Figure 2.51. The default clock buffer setting using the global buffers for the clock output and the synthesized clock output is used.

Clicking *Next* opens the Xilinx Clocking Wizard – Clock Frequency Synthesizer window, as shown in Figure 2.52. The DCM frequency synthesizer uses an clock input frequency of 50 MHz and a specified multiplier (M) and divider (D) of 5 and 3 to produce a output frequency of 50 MHz × 5 / 3 = 83.333 MHz.

Clicking *Next* displays the Xilinx Clocking Wizard – Summary. Clicking *Finish* generates the Architecture Wizard block definition file dacs3edcm.xaw. The Verilog HDL module dacs3edcm.v is generated within the ISE project, as given in Listing 2.7. The module instantiations for the global clock buffer BUFG and the single-ended input global clock buffer IBUFG are generated automatically and are also available as a Xilinx Language Template.

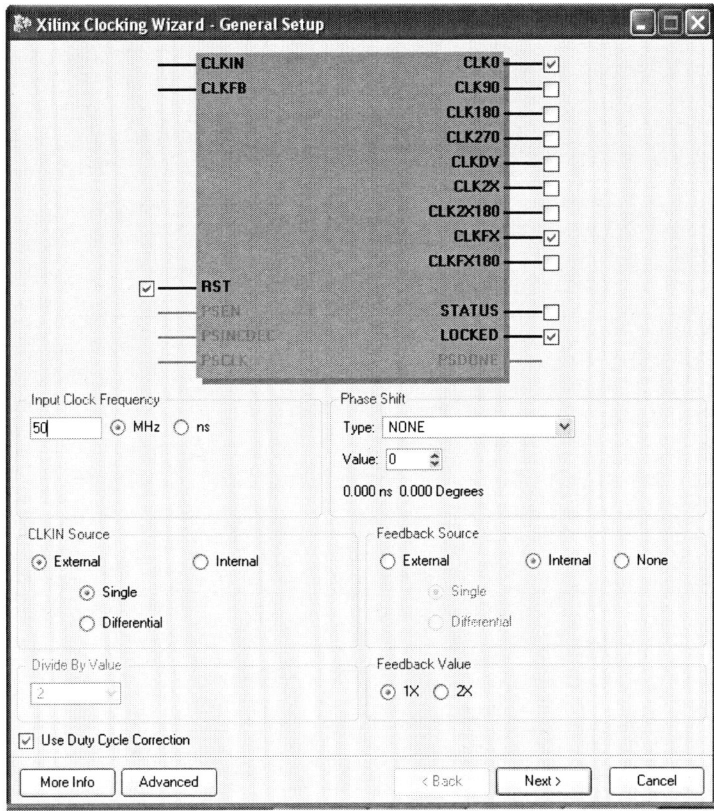


**Figure 2.50** Xilinx Clocking Wizard – General Setup window

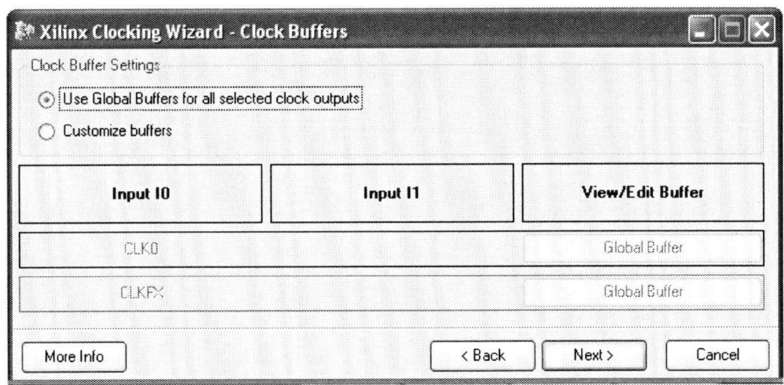


**Figure 2.51** Xilinx Clocking Wizard – Clock Buffers window

The module ports in Listing 2.7 utilize the *connection by position* option where the formal name in the declaration need not be the same as the actual name evoked in the hierarchical structure, as described in Chapter 1 Hardware Description Language. Note that the instantiation of the module uses the *connection by name* convention, also described in Chapter 1, and that unspecified signals in the DCM Architecture Wizard, as shown in Figure 2.50, are not implemented. Finally, the Verilog HDL keyword *defparam* sets the parameters of this instance of the nested module dacs3edcm.v, as described in Chapter 1 Hardware Description Language.

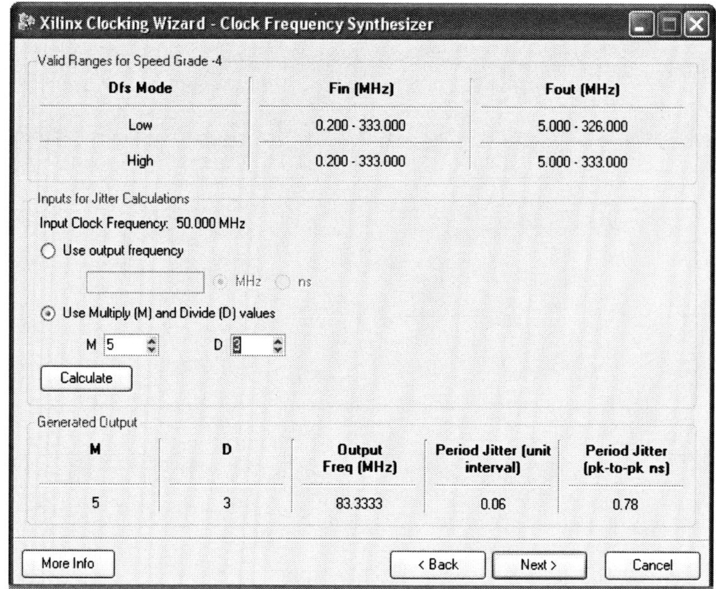


**Figure 2.52** Xilinx Clocking Wizard – Clock Frequency Synthesizer window

Clicking on the block definition file dacs3edcm.xaw in the *Sources* tab of the Xilinx ISE WebPACK Project Navigator design window opens the *Processes* window. Clicking on *View HDL Instantiation Template* opens the dacse3dcm.tif file as given in Listing 2.8.

*Copy and paste* the instantiation template file into the top module of the project and edit the connection by position variable names (within the parentheses) to those in use. The Verilog HDL module dacs3edcm.v in Listing 2.7 can be viewed in the Project Navigator design window by clicking on *View HDL Source* in the *Processes* window.

**Listing 2.7** Xilinx DCM Architecture Wizard Verilog HDL module dacs3edcm.v

```
////////////////////////////////////////////////////////////////////////////////
// Copyright (c) 1995-2006 Xilinx, Inc.  All rights reserved.
////////////////////////////////////////////////////////////////////////////////
//   ____  ____
//  /   /\/   /
// /___/  \  /    Vendor: Xilinx
// \   \   \/     Version : 9.2.01i
//  \   \         Application : xaw2verilog
//  /   /         Filename : dacs3edcm.v
// /___/   /\     Timestamp : 1/06/2008 16:25:51
// \   \  /  \
//  \___\/\___\
//
//Command: xaw2verilog -intstyle C:/EDPGA/Chapter 3/peripherals/dacs3e.xaw
           -st dacs3e.v
//Design Name: dacs3edcm
//Device: xc3s500e-4fg320
//
// Module dacs3e
// Generated by Xilinx Architecture Wizard
```

```
// Written for synthesis tool: XST
`timescale 1ns / 1ps

module dacs3edcm(CLKIN_IN,
            RST_IN,
            CLKFX_OUT,
            CLKIN_IBUFG_OUT,
            CLK0_OUT,
            LOCKED_OUT);

input CLKIN_IN;
input RST_IN;
output CLKFX_OUT;
output CLKIN_IBUFG_OUT;
output CLK0_OUT;
output LOCKED_OUT;
wire CLKFB_IN;
wire CLKFX_BUF;
wire CLKIN_IBUFG;
wire CLK0_BUF;
wire GND1;
assign GND1 = 0;
assign CLKIN_IBUFG_OUT = CLKIN_IBUFG;
assign CLK0_OUT = CLKFB_IN;

BUFG CLKFX_BUFG_INST (.I(CLKFX_BUF), .O(CLKFX_OUT));
IBUFG CLKIN_IBUFG_INST (.I(CLKIN_IN), .O(CLKIN_IBUFG));
BUFG CLK0_BUFG_INST (.I(CLK0_BUF), .O(CLKFB_IN));

// Period Jitter (unit interval) for block DCM_SP_INST = 0.06 UI
// Period Jitter (Peak-to-Peak) for block DCM_SP_INST = 0.78 ns

DCM_SP DCM_SP_INST (.CLKFB(CLKFB_IN),
                .CLKIN(CLKIN_IBUFG),
                .DSSEN(GND1),
                .PSCLK(GND1),
                .PSEN(GND1),
                .PSINCDEC(GND1),
                .RST(RST_IN),
                .CLKDV(),
                .CLKFX(CLKFX_BUF),
                .CLKFX180(),
                .CLK0(CLK0_BUF),
                .CLK2X(),
                .CLK2X180(),
                .CLK90(),
                .CLK180(),
                .CLK270(),
                .LOCKED(LOCKED_OUT),
                .PSDONE(),
                .STATUS());
```

```
defparam DCM_SP_INST.CLK_FEEDBACK = "1X";
defparam DCM_SP_INST.CLKDV_DIVIDE = 2.0;
defparam DCM_SP_INST.CLKFX_DIVIDE = 3;
defparam DCM_SP_INST.CLKFX_MULTIPLY = 5;
defparam DCM_SP_INST.CLKIN_DIVIDE_BY_2 = "FALSE";
defparam DCM_SP_INST.CLKIN_PERIOD = 20.0;
defparam DCM_SP_INST.CLKOUT_PHASE_SHIFT = "NONE";
defparam DCM_SP_INST.DESKEW_ADJUST = "SYSTEM_SYNCHRONOUS";
defparam DCM_SP_INST.DFS_FREQUENCY_MODE = "LOW";
defparam DCM_SP_INST.DLL_FREQUENCY_MODE = "LOW";
defparam DCM_SP_INST.DUTY_CYCLE_CORRECTION = "TRUE";
defparam DCM_SP_INST.FACTORY_JF = 16'hC080;
defparam DCM_SP_INST.PHASE_SHIFT = 0;
defparam DCM_SP_INST.STARTUP_WAIT = "FALSE";

endmodule
```

**Listing 2.8** Xilinx DCM Architecture Wizard Verilog HDL instantiation template file dacs3edcm.tif

```
// Instantiate the module
dacse3dcm instance_name (
   .CLKIN_IN(CLKIN_IN),
   .RST_IN(RST_IN),
   .CLKFX_OUT(CLKFX_OUT),
   .CLKIN_IBUFG_OUT(CLKIN_IBUFG_OUT),
   .CLK0_OUT(CLK0_OUT),
   .LOCKED_OUT(LOCKED_OUT)
   );
```

## Xilinx LogiCORE Blocks

The Xilinx CORE Generator™ System and the LogiCORE blocks facilitate embedded design in the Verilog HDL using the programmable gate array. Several of the LogiCORE products for the Spartan-3E FPGA are freely available and described in a Xilinx application note (XAPP474, *www.xilinx.com*) and subject to the LogiCORE site license agreement (www.xilinx.com/ipcenter /doc/xilinx_click_core_site_license.pdf).

These structural LogiCORE blocks are optimized and are used in Xilinx ISE WebPACK projects in Chapter 3 Field Programmable Gate Array Hardware and Chapter 4 Digital Signal Processing, Communications and Control. Each of the LogiCORE blocks is provided with design windows, as shown in Figure 2.30 and Figure 2.31 for the Divider Generator v1.0.

The LogiCORE basic element, memory, multiplexer and register blocks for the Spartan-3E FPGA provides comparators, counters, encoders and decoders, format conversions, logic gate and buffers, memory elements, multiplexers, registers, shifters, and pipelines, as shown in Figure 2.53 and as listed in Table 2.1. The LogiCORE block data sheets listed in Table 2.1 are available (*www.xilinx.com*) and provide more detailed information.

The Comparator LogiCORE block is used to create comparison logic that performs the equal, not equal, less than, less than or equal to, greater than, or greater than or equal to function for two N-bit data. The Comparator can operate with two's complement signed or unsigned data and has an option for clock enable, asynchronous and synchronous set and clear.

The Binary Counter LogiCORE block implements an up, down and up/down counter where the upper limit of the count can be set, a count threshold signal can be outputted, asynchronous and synchronous set, clear and initialize input signals are available. The Binary Decoder LogiCORE block provides M *one-hot* outputs from N binary inputs where $M = 2^N$. One-hot encoding of binary data is used in the address range decoding logic for device *chip select* signals or in the configuration of the controller of a finite state machine [Wakerly00].

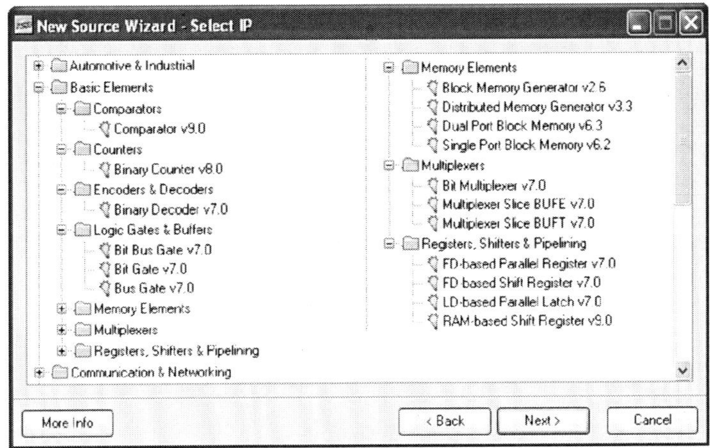


**Figure 2.53** LogiCORE basic elements, memory, multiplexer and register blocks

The Pipelined Divider LogiCORE block is described in the example of the Xilinx CORE Generator ISE project implementation in this Chapter. The Bit Bus Gate LogiCORE block outputs binary data each bit of which is either the logical AND, NAND, OR, NOR, XOR or NXOR (not exclusive OR) of the input bit and a control bit. The Bit Gate LogiCORE block outputs a single bit which is either the logical AND, NAND, OR, NOR, XOR or NXOR of the N-bit data input.

**Table 2.1** Basic element, memory, multiplexer and register LogiCORE blocks

| Function | Data Sheet | Application |
|---|---|---|
| Comparator | DS224 | Logical comparator for N-bit input data |
| Binary Counter | DS215 | Up, down and up/down counters |
| Binary Decoder | DS216 | M-bit one-hot output from N-bit input data |
| Bit Bus Gate | DS217 | Logical control bit processing of N-bit input data |
| Bit Gate | DS218 | Logical bit processing of N-bit input data |
| Bus Gate | DS222 | Logical bit processing of multiple bit buses |
| Block Memory Generator | DS512 | Single/dual port RAM/ROM using block memory |
| Distributed Memory Generator | DS322 | Single and dual port RAM and ROM from LUTs |
| Dual Port Block Memory | DS235 | Dual port RAM/ROM from block memory |
| Single Port Block Memory | DS234 | Single port RAM/ROM from block memory |
| Bit Multiplexer | DS219 | One bit output by N-bit selection M-bit input data |
| Multiplexer Slice BUFE | DS220 | N-bit bus buffer with output enable |
| Multiplexer Slice BUFT | DS221 | N-bit bus buffer with tristate control |
| FD-based Parallel Register | DS225 | N-bit type D flip-flop based parallel register |
| FD-based Shift Register | DS226 | N-bit type D flip-flop based shift register |
| LD-based Parallel Latch | DS227 | N-bit type D latch register |
| RAM-Based Shift Register | DS228 | Parallel input and output N-bit shift register |

The Bus Gate LogiCORE block inputs up to four, multiple bit data buses and outputs binary data each bit of which is the logical AND, NAND, OR, NOR, XOR or NXOR of the input buses. The Bus Gate also provides inversion (NOT) and buffering of a single N-bit data bus. The Bit Bus Gate, the

Bit Gate and the Bus Gate LogiCORE blocks all have clocked registered outputs with asynchronous and synchronous clear and initialize and asynchronous clear input signals.

The Dual Port Memory and Single Port Memory LogiCORE blocks shown in Figure 2.53 are superseded by the Block Memory Generator LogiCORE block for all new Spartan-3E FPGA designs in the Xilinx ISE. The Block Memory Generator uses the available block random access memory (RAM) of the fine grained architecture of the Spartan-3E FPGA, as described in Chapter 1 Hardware Description Language. The Block Memory Generator also uses the block memory primitives of the Verilog Language Templates of the Xilinx ISE Project Navigator, as shown in Figure 2.48, and extends their capabilities to random access memories (RAM) of arbitrary bit widths and word depths.

The Distributed Memory Generator LogiCORE block uses the look up tables (LUT) from the of the fine grained architecture of the Spartan-3E FPGA, as described in Chapter 1 Hardware Description Language. The Bit Multiplexer LogiCORE block utilizes an M-bit data to select one bit of an N-bit input data. The Multiplexer Slice BUFE and BUFT LogiCORE blocks provide output enable and *tristate* logic control of a M-bit input bus to an M-bit output bus.

The FD-based Parallel Register and the FD-based Shift Register LogiCORE blocks uses type D flip-flops to form these optimized functions with asynchronous and synchronous set, clear and initialize input signals and clock enable logic signals [Wakerly00]. The FD-based Shift Register can be loaded with both parallel or serial digital data and features parallel or serial digital data output and a bidirectional shift control.

The LD-based Parallel Latch LogiCORE block uses type D latches and has a gate enable logic signal with asynchronous and synchronous set, clear and initialize input signals. The RAM-based Shift Register LogiCORE block provides fast and efficient first-in-first-out (FIFO) N-bit parallel input and output shift registers, digital delay lines or time-skew buffers with asynchronous and synchronous set, clear and initialize input signals and clock enable logic signals.

The LogiCORE basic communications and networking blocks for the Spartan-3E FPGA provides encoders, decoders, Ethernet Media Access Control (MAC), direct digital synthesizer (DDS) and an interweaver and de-interweaver, as shown in Figure 2.54 and as listed in Table 2.2. The LogiCORE block data sheets listed in Table 2.2 are available (*www.xilinx.com*) and provide more detailed information.

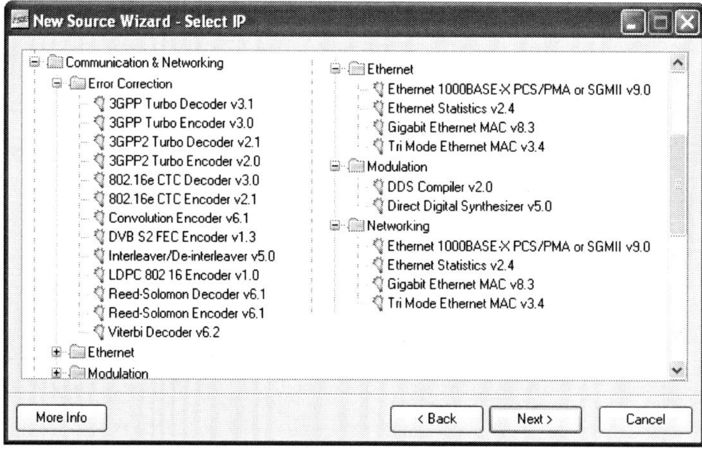


**Figure 2.54** LogiCORE communication and networking blocks

The 3GPP Turbo Encoder and the 3GPP Turbo Decoder implement the 3rd Generation Partnership Project (3GPP) mobile communication specification for transmission of data over noisy channels. The 802.16e CTC Encoder is a parallel implementation of the convolutional turbo code encoder as specified by the IEEE 802.16e standard for wireless metropolitan area network (MAN).

The Convolutional Encoder is a high speed implementation with parameterized *constrain lengths* and *puncturing* capability [Sklar01]. The DVB S2 FEC Encoder supports the second generation digital video broadcast (DVB) forward error correction (FEC) standard for satellite transmission. The FEC is implemented by a Bose-Chaudhuri-Hochquenghem (BCH) code and can be concatenated with a low density parity check (LDPC or Gallagher) code [Sklar01]. The Interweaver /De-interweaver randomizes data to mitigate the effect of burst errors in a data transmission using a convolutional or block structure.

**Table 2.2** Communication and networking LogiCORE blocks

| Function | Data Sheet | Application |
|---|---|---|
| 8b/10b Decoder | DS258 | Decoder for 8-bit to 10-bit protocol |
| 8b/10b Encoder | DS254 | Encoder of 8-bit to 10-bit protocol |
| 3GPP Turbo Decoder | DS318 | Turbo convolution code decoder |
| 3GPP Turbo Encoder | DS319 | Turbo convolution code encoder |
| 802.16e CTC Encoder | DS525 | Convolutional turbo code encoder |
| Convolutional Encoder | DS248 | Encoder for Viterbi decoder |
| DVB S2 FEC Encoder | DS505 | Forward error correction for DVB-S.2 |
| Interleaver/De-interleaver | DS250 | Data randomizer for burst errors |
| Reed-Solomon Decoder | DS252 | Forward error correction decoder |
| Reed-Solomon Encoder | DS251 | Forward error correction encoder |
| Viterbi Decoder | DS247 | Forward error correction decoder |
| Ethernet Statistics | DS323 | Network traffic statistics |
| Ethernet 1000Base-X | DS264 | PCS/PMA 1 Gb/sec controller |
| Gigabit Ethernet MAC | DS200 | 1 Gb/sec media access controller |
| Tri Mode Ethernet MAC | DS297 | 10/100/1000 Mb/sec media access controller |
| DDS Compiler | DS558 | Direct digital synthesizer common interface |
| Direct Digital Synthesizer | DS246 | Numerical controlled oscillator |

The Reed-Solomon Encoder and the Reed-Solomon Decoder implement FEC with the Reed-Solomon algorithm which uses a field polynomial and a generator polynomial to provide $n$ total number of symbols for $k$ data symbols (with $(n - k)$ check symbols in the $(n, k)$ code) [Sklar01]. The Viterbi Decoder processes convolutionally encoded data using the Viterbi algorithm with either a fast parallel implementation with a large FPGA resource requirement or a slower serial implementation that uses less FPGA resources.

The Ethernet Statistics LogiCORE block provides user-configurable statistical counters for the analysis of network traffic. The Ethernet 1000Base-X is a physical coding sublayer (PCS) for 1 gigabit/sec (Gb/sec) using a physical media attachment (PMA).

The Gigabit Ethernet Mac implements a 1 Gb/sec media access controller (MAC) compliant with the IEEE 802.3-2002 standard. The Tri Mode Ethernet MAC provides a 10 Mb/sec, 100 Mb/sec and 1 Gb/sec MAC that can be configured for half or full-duplex operation with flow control. However, the Ethernet 1000Base-X, the Gigabit Ethernet Mac and the Tri Mode Ethernet MAC can only be simulated in the Xilinx ISE and are not available for hardware synthesis without a license (see *www.xilinx.com/ipcenter*).

The Direct Digital Synthesizer (DDS) is a numerically controller oscillator (NCO) which provides a quadrature synthesizer for digital up and down converters, modulators and demodulators [Silage06]. The DDS (direct digital synthesizer) Compiler is a common user interface for the DSS and NCO which simplifies the LogiCORE block implementation and is utilized in Xilinx ISE Webpack projects in Chapter 4.

The LogiCORE digital signal processing (DSP) and clocking blocks for the Spartan-3E FPGA provide several FIR digital filters, a multiply accumulator, the Fast Fourier Transform (FFT), the

Coordinate Rotation Digital Computer (CORDIC) and sine-cosine look-up tables, as shown in Figure 2.55 and as listed in Table 2.3. The LogiCORE block data sheets listed in Table 2.3 are available (*www.xilinx.com*) and provide more detailed information.

The Cascaded Integrator-Comb (CIC) Compiler implements the multipierless, multirate Hogenauer filter LogiCORE block for digital up converters (DUC) and digital down converters (DDC) [Mitra06]. The Distributed Arithmetic (DA) FIR Filter implements the Hilbert transform and interpolated filters, including the polyphase and half-band decimator and interpolator [Mitra06]. The FIR Compiler is the common user interface for area efficient DA or multiply-and-accumulate (MAC) FIR digital filters and is utilized in Xilinx ISE Webpack projects in Chapter 4.

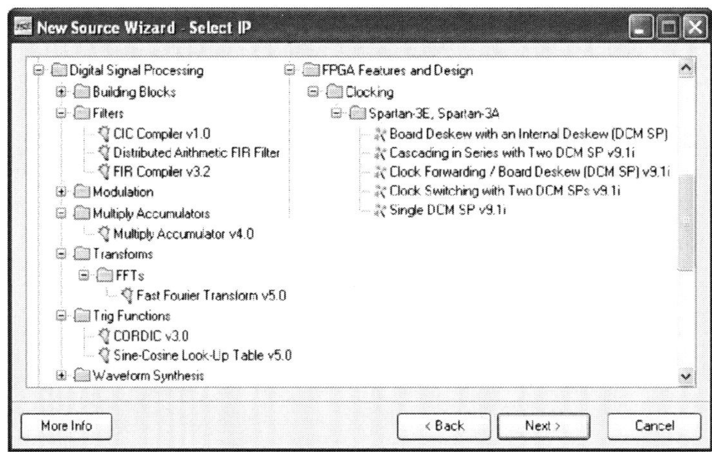


**Figure 2.55** LogiCORE digital signal processing and clocking blocks

The Multiply Accumulator (MAC) LogiCORE block is a parallel MAC with pipelining. The Fast Fourier Transform implements the discrete Fourier transform (DFT) and the FFT by the efficient Cooley-Tukey algorithm. The Coordinate Rotation Digital Computer (CORDIC) provides vector rotation and translation, sine, cosine and tangent, arctangent, hyperbolic sine and cosine and square root calculations.

The Sine-Cosine Look-Up Table can utilize distributed or block RAM in the Spartan-3E FPGA to produce the sine or cosine of an integer angle input and is utilized in Xilinx ISE Webpack projects in Chapter 4. The Xilinx Architecture Wizard (XAW) for the Spartan-3E FPGA implements the Digital Clock Manager (DCM), as described in this Chapter.

**Table 2.3** Digital signal processing and clocking blocks LogiCORE blocks

| Function | Data Sheet | Application |
|---|---|---|
| CIC Compiler | DS613 | Cascaded integrator-comb filter common interface |
| DA FIR Filter | DS240 | Serial and parallel FIR filter |
| FIR Compiler | DS534 | FIR filter common interface |
| Multiply Accumulator | DS336 | Parallel MAC |
| Fast Fourier Transform | DS260 | DFT and FFT |
| CORDIC | DS249 | Vector rotation, sine and cosine, square root |
| Sine-Cosine Look-Up Table | DS275 | Sine and cosine output |
| Digital Clock Manager | XAW | Clock generation |

The LogiCORE math functions, first-in-first-out (FIFO) and memory interface generator blocks for the Spartan-3E FPGA provides an accumulator, adder/subtracter, dividers, floating-point operations, multipliers, memory generators and packet and FIFO buffers, as shown in Figure 2.56 and as listed in Table 2.4. The LogiCORE block data sheets listed in Table 2.4 are available (*www.xilinx.com*) and provide more detailed information.

The Accumulator LogiCORE block implements an add, subtract or add/subtract accumulator with asynchronous and synchronous set, clear and initialize input signals. The Adder Subtractor has the same functionality but without an accumulator. The Divider Generator is a fixed-point divider based on radix-2 non-restoring division or a floating-point divider by repeated multiplications and is described in this Chapter. The Pipelined Divider is optimized for high-speed DSP applications where the divisor is not constant.

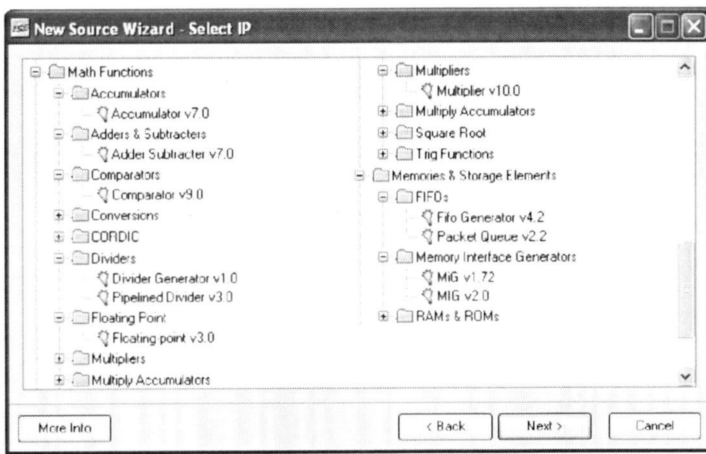


**Figure 2.56** LogiCORE math functions, FIFO and memory interface generator blocks

The Floating Point LogiCORE block provides the floating-point operations of add, subtract, divide, square root and comparison and fixed-point to floating-point and floating-point to fixed-point conversion with close compliance to the IEEE 754 standard for binary floating-point arithmetic, as described in Chapter 1 Hardware Description Language.

The Multiplier LogiCORE block accepts fixed-point data on two input busses and can utilize the 18-bit hardware multiplier of the Spartan-3E FPGA, as described in Chapter 1. The Multiplier can also accept data on a single bus and multiply it by either a fixed integer constant or an integer constant that can be reloaded from the other input bus.

**Table 2.4** Math functions, FIFO and memory interface generator blocks

| Function | Data Sheet | Application |
|---|---|---|
| Accumulator | DS213 | Add/subtract accumulator |
| Adder Subtracter | DS214 | Add, subtract or add and subtract |
| Divider Generator | DS530 | Fixed-point or floating-point division |
| Pipelined Divider | DS305 | Pipelined fixed-point division |
| Floating-Point | DS335 | Floating-point operations |
| Multiplier | DS255 | Parallel fixed-point multiplier |
| FIFO Generator | DS317 | First-in-first-out buffer |
| Packet Queue | DS509 | Multiple data stream buffer |
| Memory Interface Generator | UG086 | Static and dynamic RAM controller |

The FIFO Generator LogiCORE block is a first-in-first-out memory queue for application required in-order data storage and retrieval. The FIFO Generator can utilize block or distributed RAM or shift registers in the fine-grained architecture of the Spartan-3E FPGA, as described in Chapter 1 Hardware Description Language, and status and data transfer signals are available. The Packet Queue LogiCORE block buffers packet-based data from multiple data streams, however it can only be simulated in the Xilinx ISE and is not available for FPGA hardware synthesis without a license (see *www.xilinx.com/ipcenter*).

Finally, the Memory Interface Generator (MIG) is a LogiCORE block generator for static and dynamic RAM controllers and is described in an available User's Guide (UG) (*www.xilinx.com*). The Xilinx MIG can directly generate a design for the 512 Mb of double data rate (DDR) external SDRAM of the Spartan-3E Starter Board, which is described in Chapter 3 Programmable Gate Array Hardware.

The Xilinx LogiCORE blocks provide *drop-in module* support for embedded design in the Verilog HDL using programmable data arrays. The performance and applicability of the LogiCORE blocks are pre-verified and utilize optimal components of the fine-grained architecture of the Spartan-3E FPGA. The Pipelined Divider, the FIR Compiler, the Sine-Cosine Look-Up Table and the Direct Digital Synthesis Compiler LogiCORE blocks are used in the Xilinx ISE WebPACK projects described in Chapter 3 Field Programmable Gate Array Hardware and Chapter 4 Digital Signal Processing, Communications and Control.

## Warnings and Errors in Synthesis

The hardware synthesis of an embedded system application in a programmable gate array is subject to constraints which can generate implementation warnings and errors [Ciletti04]. The User Constraints File (UCF) not only assures that the input and output signals are routed to the proper pins of the FPGA but can include implementation and timing considerations. Warnings and errors appear in the Xilinx ISE WebPACK Project Navigator window, as shown in Figure 2.1.

Warning are non-fatal descriptions of the design implementation process which consists of synthesizing, translating, mapping and placing and routing of the interconnections of the configurable logic and input-output blocks of the programmable gate array, as described in Chapter 1 Verilog Hardware Description Language.

Although non-fatal, warnings indicate design changes that are made autonomously by the Xilinx ISE WebPACK during the implementation process. These warnings include signals that are inadvertently not used and other design implementation descriptions, as given in Listing 2.8.

The warning messages in Listing 2.9 are generated by the Xilinx ISE WebPACK synthesizer (XST) for the project ad1adctest.ise, as given in Listing 3.33. The warning messages include that the signal lcdhome is not used (Xst:647, Xst:1291 and Xst:1710). This warning is generated because the controller module genad1adc.v does not utilize the lcdhome signal to datapath module for the LCD lcd.v in this ISE WebPACK project. Rather than modify the datapath module lcd.v and complicate the design reuse of the Verilog HDL modules, the warning can be accepted and the signal lcdhome is set to logic 0.

**Listing 2.9** Xilinx XST synthesizer warning messages for the Xilinx ISE WebPACK project ad1adctest.ise

```
WARNING:Xst:905 - "../ad1adctest.v" line 68: The signals <adc4, adc3, adc2, adc1> are missing
    in the sensitivity list of always block.
WARNING:Xst:647 - Input <lcdhome> is never used.
WARNING:Xst:646 - Signal <value> is assigned but never used.
WARNING:Xst:1872 - Variable <i> is used but never assigned.
WARNING:Xst:1291 - FF/Latch <adcstate_34> is unconnected in block <ad1adc>.
WARNING:Xst:1426 - The value init of the FF/Latch gstate_0 hinder the constant cleaning in the
    block adclcd.
```

WARNING:Xst:1291 - FF/Latch <lcdstate_40> is unconnected in block <lcd>.
WARNING:Xst:1291 - FF/Latch <lcdhome> is unconnected in block <M2>.
WARNING:Xst:1710 - FF/Latch <adc1_3> (without init value) has a constant value of 0 in block <genad1adc>.
WARNING:Xst:1710 - FF/Latch <M2/lcdhome> (without init value) has a constant value of 0 in block <ad1adctest>.

Another warning message is that the signals adc4, adc3, adc2 and adc1 are not in the sensitivity list of the *always@* Verilog HDL statement, as described in Chapter 1 Verilog Hardware Description Language (Xst:905). This warning message can be ignored since the design of the controller and datapath modules here is that signal data is only to be updated when the signal digitmux changes, as given in Listing 3.33. Finally, the finite state machine (FSM) of the datapath modules ad1adc.v and lcd.v include an *idle state* which is entered when the FSM concludes its operation. The FSM then waits for the next controller module signal and the warning is generated for an unconnected signal (Xst:1291).

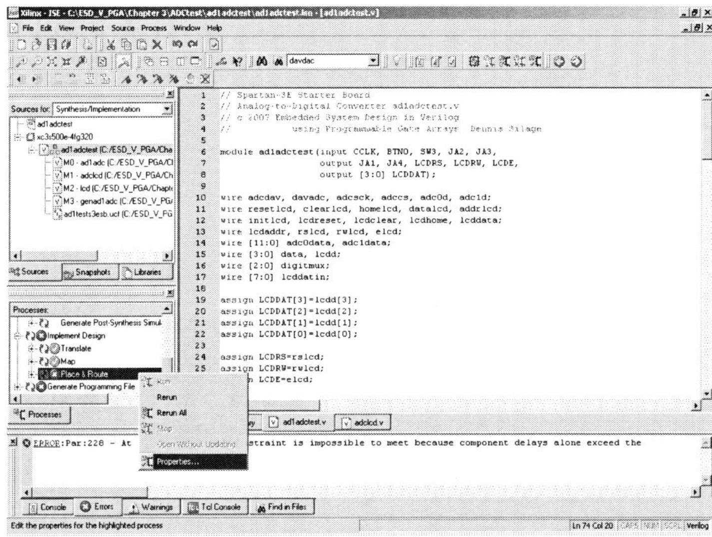


**Figure 2.57** Xilinx ISE WebPACK Project Navigator place and route error message

Error messages generated by the Xilinx ISE WebPACK project are fatal and interrupt the hardware synthesis of an embedded system design in a programmable gate array. Errors can occur for a variety of minor and severe causes. One minor cause of an error message is an active (uncommented) port in the User Constraints File (UCF). One severe error message occurs when timing constraints are not met in an implementation because of path delays in a design, as shown in Figure 2.57. Here the synthesis, translate and map processes of the Xilinx ISE WebPACK are successful but the place and route (Par:228) process fails.

Right mouse button clicking on the *Implement Design...Place & Route* in the Xilinx ISE WebPACK Project Navigator design window opens the functions menu, as shown in Figure 2.57. Clicking on *Properties* opens the Place & Route Properties window, as shown in Figure 2.58. The severe timing constraint error may usually be obviated by increasing the Place & Route Effort Level (Overall) to *High* and the rerunning the place and route process.

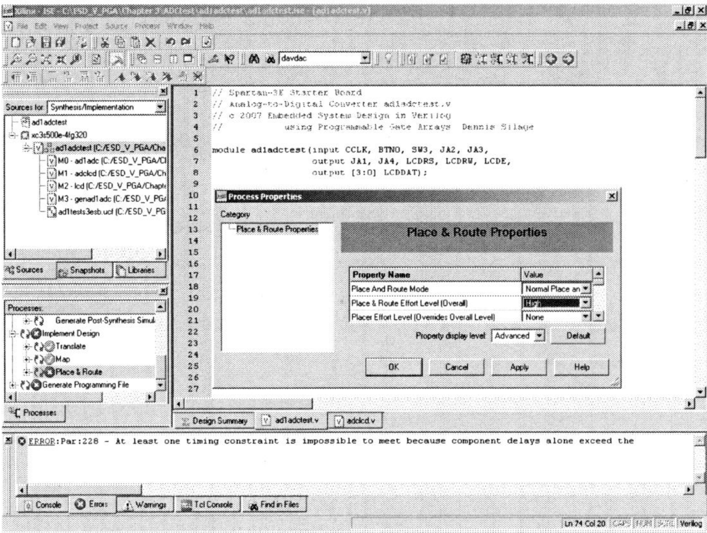

**Figure 2.58** Place & Route Properties window

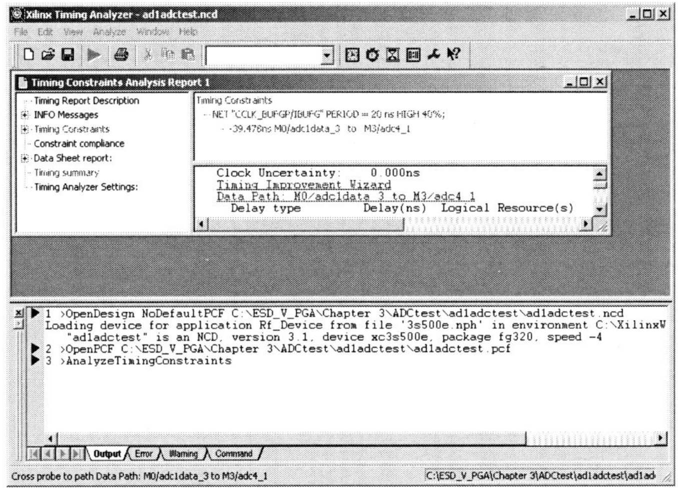

**Figure 2.59** Xilinx Timing Analyzer window

Timing constraint errors are often prevalent in Verilog HDL implementation designs using multiple modules that are placed and routed at distances across the fine-grained architecture of the Spartan-3E FPGA. The Xilinx ISE Floorplanner can assist in identifying the placement of the modules of the design and the interconnection paths. The Xilinx ISE WebPACK includes the Xilinx Timing Analyzer which provides a description of the timing constraint in error and design implementation suggestions to alleviate the error.

The Xilinx Timing Analyzer is evoked from the *Processes... Generate Post Place & Route Static Timing...Analyze Post Place & Route Static Time* (a clock icon) in the Xilinx ISE WebPACK Project Navigator design window. The Xilinx Timing Analyzer provides the Timing Improvement Wizard link, as shown in Figure 2.59.

Fortunately, a controller and datapath construct in the Verilog HDL is subjected less to actual timing constraint errors for an embedded design using a programmable gate array. The FSM of the controller and datapath, although event driven by a clock, essentially depends only upon the *data available* signals, as described in Chapter 1 Verilog Hardware Description Language. If the clock transition to the FSM is *late* then the state transition merely occurs on the next clock transition with a *design penalty* of only one clock cycle.

If a timing constraint error initially occurs for a controller and datapath construct, the *Use Timing Constraints* (scrolling down) in the Place & Route Properties window can be deselected, as shown in Figure 2.58. The place and route process is then rerun.

## Summary

In this Chapter embedded system design in the Verilog HDL using a field programmable gate array (FPGA) is shown to be facilitated by software electronic design automation (EDA) tools which provide both simulation and hardware synthesis. The Xilinx ISE WebPACK is presented using an embedded design stop watch project as a complete EDA tool in hardware synthesis for the Xilinx Spartan-3E FPGA and the Spartan-3E Starter Board. Several additional components of the Xilinx ISE WebPACK EDA are described, including the CORE Generator, the Floorplanner, the Simulator, the Verilog Language Templates, the Architecture Wizard and the Timing Analyzer.

Chapter 3 Programmable Gate Array Hardware describes two evaluation boards that utilize the Xilinx Spartan-3E FPGA. The integral components and available hardware modules of the evaluation boards are presented in operation by complete Xilinx ISE WebPACK projects. Chapter 4 Digital Signal Processing, Communications and Control presents projects in digital filtering, digital modulation in communication systems, digital data transmission and digital control system design. Chapter 5 Embedded Soft Core Processors describes the Xilinx PicoBlaze 8-bit soft-core processor with auxiliary EDA software and its comparison to the Verilog controller and datapath construct and Xilinx LogiCORE blocks in the Xilinx ISE WebPACK. These three Chapters survey the application of the Verilog HDL, the soft core processor and programmable gate arrays in embedded design.

## References


[Botros06] Botros, Nazeih M., *HDL Programming Fundamentals*. Thomson Delmar, 2006.

[Ciletti99] Cilletti, Michael D., *Modeling, Synthesis and Rapid Prototyping with the Verilog HDL*. Prentice Hall, 1999.

[Ciletti04] Cilletti, Michael D., *Starter's Guide to Verilog 2001*. Prentice Hall, 2004.

[Lee06] Lee, Sunggu, *Advanced Digital Logic Design*. Thomson, 2006.

[Mitra06] Mitra, Sanjit K., *Digital Signal Processing: A Computer Based Approach*. McGraw-Hill, 2006.

[Navabi06] Zainalabedin Navabi, *Verilog Digital System Design*. McGraw-Hill, 1999.

[Silage06] Silage, Dennis, *Digital Communication Systems using SystemVue*. Thomson Delmar, 2006

[Wakerly00] Wakerly, John F. *Digital Design Principles and Practice*, Prentice Hall, 2000.

# 3

# Programmable Gate Array Hardware

Embedded design in the Verilog hardware description language (HDL) using a field programmable gate array (FPGA) is investigated with an electronic design automation (EDA) software tool , a hardware development and evaluation board and peripherals. The evaluation boards feature a Xilinx Spartan™-3E FPGA supported by a Joint Test Action Group (JTAG) port (IEEE 1149.1) for remote programming of the bit file provided by the EDA and a variety of external hardware components and ports. The ports allow the interconnection of analog and digital input-output hardware peripherals for interfacing to external signals and devices. These components, ports and peripherals facilitate real-time processing with the FPGA in digital signal processing, digital communications, data communications and process control, as described in Chapter 4.

The Verilog source modules and project files are located in the *Chapter 3* folder as subfolders identified by the name of the appropriate module or project. The complete contents and the file download procedure are described in the Appendix.

The Verilog HDL source and project file structure for the Xilinx Integrated Synthesis Environment (ISE) WebPACK EDA software tool is described in Chapter 2 Verilog Design Automation. The projects in this Chapter illustrate not only the use of the components, ports and external hardware peripherals in applications, but the syntax, development and versatility of the Verilog HDL for embedded system design as an extension of the discussion in Chapter 1.

## Evaluation Boards

The Xilinx Spartan™-3E FPGA (data sheet DS312, *www.xilinx.com*) is a gate optimized device for lower cost with up to 376 single-ended I/O signals and 1600 K gates. The Xilinx Spartan-3E FPGAs are available as part of inexpensive development and evaluation boards and are used for the Verilog HDL hardware synthesis of an embedded design.

### Basys Board

The Digilent Basys (Basic Systems) Board (*www.digilentinc.com*) is an inexpensive Xilinx Spartan-3E FPGA development and evaluation boards, as shown in Figure 3.1. The Xilinx Spartan-3E XC3S100E-VQ100 FPGA on the evaluation board contains 100 K gates with the equivalent of 2160 logic cells. The fine grained architecture of the Xilinx Spartan-3E XC3S100E FPGA includes four 18 Kb blocks of random access memory (RAM), four 18 × 18 bit hardware multipliers, two digital clock managers (DCM) and up to 66 I/O signals. The complete description of the Spartan-3E FPGA is available (data sheet DS312, *www.xilinx.com*).

Available peripherals on the Basys Board are a Xilinx 2 megabit (Mb) XCF02S flash programmable read-only memory (PROM) and a selectable 25 megahertz (MHz), 50 MHz or 100 MHz crystal clock oscillator. Jumper JP3 selects the frequency of the clock. Unlike the Spartan-3E Starter Board, the Basys Board does not provide any external RAM. External ports include a video graphics array (VGA) port, a PS/2 mouse or keyboard port and a JTAG programming port.

The programming port is compatible with either the Digilent download PC parallel port cable or the Xilinx Platform Cable USB. The flash PROM can be programmed by the Xilinx IMPACT tool, as described in Chapter 2 Verilog Design Automation, and the FPGA can be configured from the PROM on power-up or by depressing the reset pushbutton (BTN-R). Jumper JP2 selects either the PROM or the host PC with the JTAG port for the configuration of the FPGA.

The Basys Board also has hard-wired accessories, as shown in Figure 3.1. Four pushbuttons, eight slide switches and eight light emitting diodes (LED) and four LED seven-segment displays provide FPGA application support. The Basys Board provides power and configuration LED indicators. Four 6-pin hardware module connectors (JA, JB, JC and JD) extend the capabilities of the Basys Board to additional external peripherals.

**Figure 3.1** Digilent Spartan-3E Basys Board

These hardware module connectors each provide four input-output (IO) signals from the FPGA, power (+VS) and ground. The +VS power source can be set by the JPA jumper to provide either +3.3 V DC or an external user input voltage (VU). The IO signals are short-circuit protected with a series resistor and have electrostatic discharge protection (ESD) diodes. A single on-board regulator provides 3.3 V, 2.5 V and 1.2 V digital logic power supplies.The Basys Board is described in detail in the publication basys_rm.pdf (*www.digilentinc.com*).

## Spartan-3E Starter Board

The Spartan-3E Starter Board (*www.xilinx.com/s3starter*) is a more versatile but more expensive development and evaluation board, as shown in Figure 3.2. The Xilinx Spartan-3E XC3S500E-FG320 FPGA on the evaluation board contains 500 K gates with the equivalent of 10 476 logic cells. The fine grained architecture of the Xilinx Spartan-3E XC3S500 FPGA includes twenty 18 Kb blocks of RAM, twenty 18 × 18 bit hardware multipliers, four digital clock managers (DCM) and up to 232 I/O signals. The complete description of the Spartan-3E FPGA is available (DS312, *www.xilinx.com*).

The Spartan-3E Starter Board with the standard XC3S500E is more than comparable in functionality to the XC3S100E for the Spartan-3E Basys Board. However for more available resources, the Spartan-3E-1600 Development Board (*www.digilentinc.com*) utilizes the Xilinx Spartan-3E XCS1600E-FG320 FPGA and contains 1600 K gates with the equivalent of 33 192 logic cells, thirty-six 18 Kb blocks of RAM, thirty-six 18 × 18 bit hardware multipliers, eight digital clock managers (DCM) and up to 376 I/O signals. The Spartan-3E Starter Board and the Spartan-3E-1600 Development Board are identical and functionally equivalent except for the available internal FPGA resources.

Available peripherals on the Spartan-3E Starter Board are an Intel 128 Mb parallel StrataFlash 28F256 PROM, 512 Mb of double data rate (DDR) SDRAM organized as 32 M of 16 bit words, a Xilinx XC2C64A CoolRunner-II complex programmable logic device (CPLD), a Xilinx 4 Mb XCF04S platform flash PROM, an STMicroelectronics (www.st.com) M25P16 16 Mb serial peripheral interface (SPI) flash PROM, a Standard Microsystems LAN83C185 10/100 Ethernet physical layer interface and a 50 MHz crystal clock oscillator.

External ports include two RS-232 serial ports configured as a data communication equipment (DCE) and a data terminal equipment (DTE) port, a VGA port, a PS/2 mouse or keyboard port, a 10/100 Mb/sec Ethernet port and a two programming ports. The primary programming port is an embedded USB controller but there is also a JTAG port which utilizes either the Digilent PC parallel port cable or the Xilinx Platform Cable USB.

The Spartan-3E Starter Board has hard-wired accessories, as shown in Figure 3.2. Four pushbuttons, four slide switches, a rotary shaft encoder and eight light emitting diodes (LED) and a 2 line by 16 character liquid crystal display (LCD) provide FPGA application support. The Spartan-3E Starter Board also provides power and configuration LED indicators. A 100-pin expansion connection and three 6-pin hardware module connectors (J1, J2 and J4 but for compatibility referred to as JA, JB and JC) extend the capabilities of the Spartan-3E Starter Board to additional external peripherals.

These module connectors each provide four input-output (IO) signals from the FPGA, power (+VS) and ground. The +V power source can be set by the JPA jumper to provide either +3.3 V DC or an external user input voltage (VU). The IO signals are short-circuit protected with a series resistor and have electrostatic discharge protection (ESD) diodes. A single on-board regulator provides 3.3 V, 2.5 V and 1.2 V digital logic power supplies.

Unlike the Basys Board which requires an external digital-to-analog converters (DAC) and analog-to-digital converters (ADC), the Spartan-3E Starter Board features an integral DAC and ADC. The four channel DAC is a Linear Technology LTC2624 with 12-bit unsigned resolution DAC. The two channel programmable scaling preamplifier and ADC are a Linear Technology LTC6912 and LTC1407A.

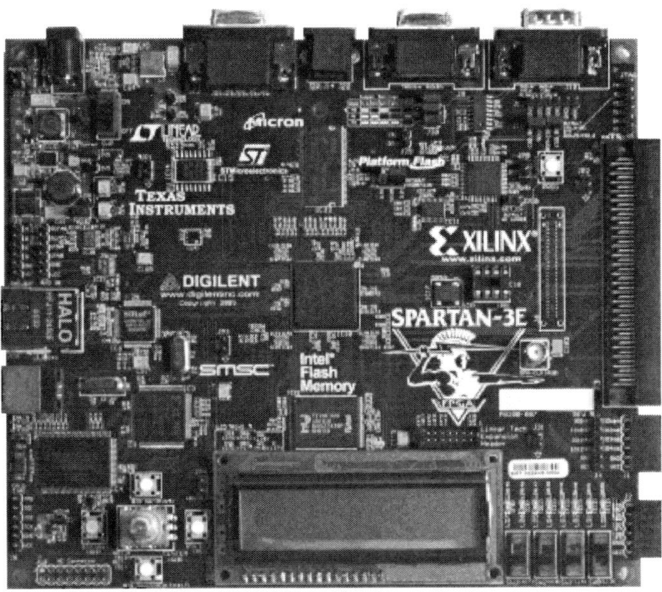


**Figure 3.2** Spartan-3E Starter Board

The Spartan-3E Starter Board is inherently more versatile than either the Basys Board because of the additional number, variety and type of peripherals. However, the LCD is slower and more complicated to use than the LED seven segment display and the DDR SDRAM requires a more elaborate memory controller than the SDRAM. The Spartan-3E Starter Board is described in detail in the publication UG230.pdf (*www.xilinx.com* or *www.digilentinc.com*).

## Selection of an Evaluation Board

Since the Spartan-3E Starter Board is apparently more versatile than the Basys Board, it will be used extensively here. Nevertheless, many of the FPGA embedded design projects execute on either of these Xilinx Spartan-3E or even other evaluation boards, as described in the Appendix. The Xilinx ISE WebPACK project files targeting either the Basys Board are configured from the Spartan-3E Starter Board project source files by substituting the appropriate User Constraints File and modules that support their unique peripherals, such as the LED seven-segment display and the external ADC and DAC modules that interface to the 6-pin header. Project files for the Basys Board are available as described in the Appendix.

The salient differences between the Spartan-3E Starter Board and the Basys Board include the size of the Spartan-3E FPGA, an integral ADC and DAC, a rotary shaft encoder and an LCD rather than an LED seven-segment display. The Basys Boards will be used for applications featuring the LED seven-segment display.

## User Constraints File

The User Constraints File (UCF) basys.ucf in Listing 3.1 provides the basic definitions for the hard-wired peripherals and accessories of the Basys Board and is located in the *Chapter 3\ucf* folder. The file download procedure is described in the Appendix.

The net labels in the UCF, such as CCLK for crystal clock oscillator, are arbitrary but are evocative of the peripheral or accessory function. Each net in this UCF is commented out with the asterisk ( # ) in Listing 3.1 and would be ignored. Active nets would be uncommented and the net label used in the UCF must also appear in the Verilog source modules of the project, as described in Chapter 2 Verilog Design Automation.

Note that register variables are written as q<0> in the UCF, while they are written as q[0] in the Verilog source modules. The function of the specific pin locations for the peripherals and accessories of the Basys Board are described in this Chapter by the development of modules in Verilog and example application project files.

**Listing 3.1** User Constraints File for the Basys Board basys.ucf

```
# Basys Evaluation Board
# User Constraints File (UCF) basys.ucf
# c 2008 Embedded Design using Programmable Gate Arrays  Dennis Silage

# Crystal Clock Oscillator
#NET    "CCLK"  LOC = "P36";

# Slide Switches
#NET    "SW0"   LOC = "P98";
#NET    "SW1"   LOC = "P95";
#NET    "SW2"   LOC = "P94";
#NET    "SW3"   LOC = "P92";
#NET    "SW4"   LOC = "P91";
#NET    "SW5"   LOC = "P90";
#NET    "SW6"   LOC = "P89";
#NET    "SW7"   LOC = "P88";
```

```
# Push Buttons
#NET    "BTN0"   LOC = "P69";
#NET    "BTN1"   LOC = "P30";
#NET    "BTN2"   LOC = "P13";
#NET    "BTN3"   LOC = "P11";

# LEDs
#NET    "LD0"    LOC = "P15";
#NET    "LD1"    LOC = "P12";
#NET    "LD2"    LOC = "P10";
#NET    "LD3"    LOC = "P9";
#NET    "LD4"    LOC = "P5"
#NET    "LD5"    LOC = "P4";
#NET    "LD6"    LOC = "P3";
#NET    "LD7"    LOC = "P2";

# Seven Segment Display
#NET    "AN0"    LOC = "P33";
#NET    "AN1"    LOC = "P32";
#NET    "AN2"    LOC = "P27";
#NET    "AN3"    LOC = "P26";
#NET    "CA"     LOC = "P42";
#NET    "CB"     LOC = "P24";
#NET    "CC"     LOC = "P22";
#NET    "CD"     LOC = "P17";
#NET    "CE"     LOC = "P16";
#NET    "CF"     LOC = "P43";
#NET    "CG"     LOC = "P23";
#NET    "CDP"    LOC = "P18";

# VGA Display
#NET    "RD"     LOC = "P49";
#NET    "GR"     LOC = "P47";
#NET    "BL"     LOC = "P48";
#NET    "HS"     LOC = "P41";
#NET    "VS"     LOC = "P40";

# PS/2 Mouse or Keyboard Port
#NET    "PS2DAT"     LOC = "P86";
#NET    "PS2CLK"     LOC = "P85";

# Peripheral Port A  (JA)
#NET "JA1"   LOC = "79";
#NET "JA2"   LOC = "78";
#NET "JA3"   LOC = "71";
#NET "JA4"   LOC = "70";

# Peripheral Port B  (JB)
#NET "JB1"   LOC = "68";
#NET "JB2"   LOC = "67";
#NET "JB3"   LOC = "66";
#NET "JB4"   LOC = "65";
```

```
# Peripheral Port C  (JC)
#NET "JC1"    LOC = "63";
#NET "JC2"    LOC = "62";
#NET "JC3"    LOC = "61";
#NET "JC4"    LOC = "60";

# Peripheral Port D  (JD)
#NET "JD1"    LOC = "58";
#NET "JD2"    LOC = "57";
#NET "JD3"    LOC = "54";
#NET "JD4"    LOC = "53";
```

The UCF s3esb.ucf in Listing 3.2 provides the basic definitions for the hard-wired peripherals and accessories of the Spartan-3E Starter Board which are somewhat compatible with the Basys Board, as given in Listing 3.1. The Spartan-3E Starter Board UCF s3esb.ucf is located in the *Chapter 3\ucf* folder. The file download procedure is described in the Appendix. The net labels in this UCF for the Spartan-3E Starter Board, such as CCLK, JA, JB and JC, are the same as the Basys Board wherever possible.

The initial *timing constraint* on CCLK for the Spartan-3E Starter Board can obviated by commenting or removing that line in the UCF s3esb.ucf. Alternatively, the *Use Timing Constraints* in the Place & Route Properties window can be deselected, as described in Chapter 2 Verilog Design Automation and shown in Figure 2.58.

However, the some of the peripherals and accessories are functionally quite different. Here there are only four slides switches and the four pushbuttons are arranged in a circle about the rotary shaft encoder rather than horizontally, as shown in Figure 3.2. The slide switches require a *pullup* resistor and the pushbuttons use a *pulldown* resistor which is accommodated in the Startan-3E Starter Board UCF. The 2 line by 16 character LCD replaces the four LED seven-segment displays and the control bus for the 512 Mb DDR SDRAM is more complicated than that for the SDRAM.

**Listing 3.2** User Constraints File for the Spartan-3E Starter Board  s3esb.ucf

```
# Xilinx Spartan-3E Starter Board
# User Constraints File (UCF) s3esb.ucf
# c 2008 Embedded Design using Programmable Gate Arrays  Dennis Silage

# Crystal Clock Oscillator
#NET    "CCLK"     LOC = "C9"    | IOSTANDARD = LVCMOS33;
#NET    "CCLK"     PERIOD = 20.0ns HIGH 40%;

# Slide Switches
#NET    "SW0"   LOC = "L13"  | IOSTANDARD = LVTTL | PULLUP;
#NET    "SW1"   LOC = "L14"  | IOSTANDARD = LVTTL | PULLUP;
#NET    "SW2"   LOC = "H18"  | IOSTANDARD = LVTTL | PULLUP;
#NET    "SW3"   LOC = "N17"  | IOSTANDARD = LVTTL | PULLUP;

# Push Buttons
#NET    "BTN0"  LOC = "V4"   | IOSTANDARD = LVTTL | PULLDOWN; # north
#NET    "BTN1"  LOC = "H13"  | IOSTANDARD = LVTTL | PULLDOWN; # east
#NET    "BTN2"  LOC = "K17"  | IOSTANDARD = LVTTL | PULLDOWN; # south
#NET    "BTN3"  LOC = "D18"  | IOSTANDARD = LVTTL | PULLDOWN; # west
```

```
# Rotary Shaft Encoder
#NET    "ROTA"      LOC = "K18"  | IOSTANDARD = LVTTL | PULLUP;
#NET    "ROTB"      LOC = "G18"  | IOSTANDARD = LVTTL | PULLUP;
#NET    "ROTCTR"    LOC = "V16"  | IOSTANDARD = LVTTL | PULLDOWN;

# LEDs
#NET    "LD0"   LOC = "F12"   | IOSTANDARD = LVTTL | DRIVE = 8 | SLEW = SLOW;
#NET    "LD1"   LOC = "E12"   | IOSTANDARD = LVTTL | DRIVE = 8 | SLEW = SLOW;
#NET    "LD2"   LOC = "E11"   | IOSTANDARD = LVTTL | DRIVE = 8 | SLEW = SLOW;
#NET    "LD3"   LOC = "F11"   | IOSTANDARD = LVTTL | DRIVE = 8 | SLEW = SLOW;
#NET    "LD4"   LOC = "C11"   | IOSTANDARD = LVTTL | DRIVE = 8 | SLEW = SLOW;
#NET    "LD5"   LOC = "D11"   | IOSTANDARD = LVTTL | DRIVE = 8 | SLEW = SLOW;
#NET    "LD6"   LOC = "E9"    | IOSTANDARD = LVTTL | DRIVE = 8 | SLEW = SLOW;
#NET    "LD7"   LOC = "F9"    | IOSTANDARD = LVTTL | DRIVE = 8 | SLEW = SLOW;

# Liquid Crystal Display
#NET "LCDE"  LOC = "M18"       | IOSTANDARD = LVCMOS33 | DRIVE = 4 | SLEW = SLOW;
#NET "LCDRS" LOC = "L18"       | IOSTANDARD = LVCMOS33 | DRIVE = 4 | SLEW = SLOW;
#NET "LCDRW" LOC = "L17"       | IOSTANDARD = LVCMOS33 | DRIVE = 4 | SLEW = SLOW;
#NET "LCDDAT<0>" LOC = "R15"       | IOSTANDARD = LVCMOS33 | DRIVE = 4 |
     SLEW = SLOW;
#NET "LCDDAT<1>" LOC = "R16"       | IOSTANDARD = LVCMOS33 | DRIVE = 4 |
     SLEW = SLOW;
#NET "LCDDAT<2>" LOC = "P17"       | IOSTANDARD = LVCMOS33 | DRIVE = 4 |
     SLEW = SLOW;
#NET "LCDDAT<3>" LOC = "M15"       | IOSTANDARD = LVCMOS33 | DRIVE = 4 |
     SLEW = SLOW;

# VGA Display
#NET    "RD" LOC = "H14"  | IOSTANDARD = LVTTL | DRIVE = 8 | SLEW = FAST;
#NET    "GR" LOC = "H15"  | IOSTANDARD = LVTTL | DRIVE = 8 | SLEW = FAST;
#NET    "BL" LOC = "G15"  | IOSTANDARD = LVTTL | DRIVE = 8 | SLEW = FAST;
#NET    "HS" LOC = "F15"  | IOSTANDARD = LVTTL | DRIVE = 8 | SLEW = FAST;
#NET    "VS" LOC = "F14"  | IOSTANDARD = LVTTL | DRIVE = 8 | SLEW = FAST;

# RS-232 Port
#NET "DCERXD"   LOC = "R7"    | IOSTANDARD = LVTTL;
#NET "DCETXD"   LOC = "M14"   | IOSTANDARD = LVTTL | DRIVE = 8 |
     SLEW = SLOW;
#NET "DTERXD"   LOC = "U8"    | IOSTANDARD = LVTTL;
#NET "DTETXD"   LOC = "M13"   | IOSTANDARD = LVTTL | DRIVE = 8 |
     SLEW = SLOW;

# PS/2 Mouse or Keyboard Port
#NET "PS2DAT"   LOC = "G13"   | IOSTANDARD = LVCMOS33 | DRIVE = 8 | SLEW = SLOW;
#NET "PS2CLK"   LOC = "G14"   | IOSTANDARD = LVCMOS33 | DRIVE = 8 | SLEW = SLOW;

# Peripheral Port A (J1)
#NET    "JA1"    LOC = "B4"    | IOSTANDARD = LVCMOS33;
#NET    "JA2"    LOC = "A4"    | IOSTANDARD = LVCMOS33;
#NET    "JA3"    LOC = "D5"    | IOSTANDARD = LVCMOS33;
#NET    "JA4"    LOC = "C5"    | IOSTANDARD = LVCMOS33;
```

```
# Peripheral Port B (J2)
#NET    "JB1"    LOC = "A6"    | IOSTANDARD = LVCMOS33;
#NET    "JB2"    LOC = "B6"    | IOSTANDARD = LVCMOS33;
#NET    "JB3"    LOC = "E7"    | IOSTANDARD = LVCMOS33;
#NET    "JB4"    LOC = "F7"    | IOSTANDARD = LVCMOS33;

# Peripheral Port C (J4)
#NET    "JC1"    LOC = "D7"    | IOSTANDARD = LVCMOS33;
#NET    "JC2"    LOC = "C7"    | IOSTANDARD = LVCMOS33;
#NET    "JC3"    LOC = "F8"    | IOSTANDARD = LVCMOS33;
#NET    "JC4"    LOC = "E8"    | IOSTANDARD = LVCMOS33;

# Programmable Gain Amplifier
#NET    "AMPCS"    LOC = "N7"    | IOSTANDARD = LVCMOS33 | DRIVE = 6 |
    SLEW = SLOW;
#NET    "AMPSD"    LOC = "P7"    | IOSTANDARD = LVCMOS33 | DRIVE = 6 |
    SLEW = SLOW;
#NET    "AMPDO"    LOC = "E18"   | IOSTANDARD = LVCMOS33;

# Digital to Analog Converter
#NET    "DACCLR"    LOC = "P8"    | IOSTANDARD = LVCMOS33 | DRIVE = 8 |
    SLEW = SLOW;
#NET    "DACCS"     LOC = "N8"    | IOSTANDARD = LVCMOS33 | DRIVE = 8 |
    SLEW = SLOW;

# Analog to Digital Converter
#NET    "ADCON"    LOC = "P11"    | IOSTANDARD = LVCMOS33 | DRIVE = 6 |
    SLEW = SLOW;

# SPI Bus Signals
#NET    "SPIMOSI"    LOC = "T4"    | IOSTANDARD = LVCMOS33 | DRIVE = 6 |
    SLEW = SLOW;
#NET    "SPIMISO"    LOC = "N10    | IOSTANDARD = LVCMOS33;
#NET    "SPISCK"     LOC = "U16"   | IOSTANDARD = LVCMOS33 | DRIVE = 6 |
    SLEW = SLOW;

# SPI BUS Disable
#NET    "SPISF"    LOC = "U3"    | IOSTANDARD = LVCMOS33 | DRIVE = 6 |
    SLEW = SLOW;
#NET    "SFCE"     LOC = "D16"   | IOSTANDARD = LVCMOS33 | DRIVE = 4 |
    SLEW = SLOW;

# FPGA Configuration
#NET "FPGAM0"    LOC="M10"    | IOSTANDARD = LVCMOS33 | DRIVE = 8 | SLEW = SLOW;
#NET "FPGAM1"    LOC="V11"    | IOSTANDARD = LVCMOS33 | DRIVE = 8 | SLEW = SLOW;
#NET "FPGAM2"    LOC="T10"    | IOSTANDARD = LVCMOS33 | DRIVE = 8 | SLEW = SLOW;
#NET "FPGAIB"    LOC="T3"     | IOSTANDARD = LVCMOS33 | DRIVE = 4 | SLEW = SLOW;
#NET "FPGARW"    LOC="M10"    | IOSTANDARD = LVCMOS33 | DRIVE = 4 | SLEW = SLOW;
#NET "FPGAHS"    LOC="M10"    | IOSTANDARD = LVCMOS33;
```

```
# DDR SDRAM Address Bus
#NET     "SDA<12>"    LOC = "P2"    | IOSTANDARD = SSTL2_I;
#NET     "SDA<11>"    LOC = "N5"    | IOSTANDARD = SSTL2_I;
#NET     "SDA<10>"    LOC = "T2"    | IOSTANDARD = SSTL2_I;
#NET     "SDA<9>"     LOC = "N4"    | IOSTANDARD = SSTL2_I;
#NET     "SDA<8>"     LOC = "H2"    | IOSTANDARD = SSTL2_I;
#NET     "SDA<7>"     LOC = "H1"    | IOSTANDARD = SSTL2_I;
#NET     "SDA<6>"     LOC = "H3"    | IOSTANDARD = SSTL2_I;
#NET     "SDA<5>"     LOC = "H4"    | IOSTANDARD = SSTL2_I;
#NET     "SDA<4>"     LOC = "F4"    | IOSTANDARD = SSTL2_I;
#NET     "SDA<3>"     LOC = "P1"    | IOSTANDARD = SSTL2_I;
#NET     "SDA<2>"     LOC = "R2"    | IOSTANDARD = SSTL2_I;
#NET     "SDA<1>"     LOC = "R3"    | IOSTANDARD = SSTL2_I;
#NET     "SDA<0>"     LOC = "T1"    | IOSTANDARD = SSTL2_I;

# DDR SDRAM Control Bus
#NET     "SDBA<0>"    LOC = "K5"    | IOSTANDARD = SSTL2_I;
#NET     "SDBA<1>"    LOC = "K6"    | IOSTANDARD = SSTL2_I;
#NET     "SDCAS"      LOC = "C2"    | IOSTANDARD = SSTL2_I;
#NET     "SDCKN"      LOC = "J4"    | IOSTANDARD = SSTL2_I;
#NET     "SDCKP"      LOC = "J5"    | IOSTANDARD = SSTL2_I;
#NET     "SDCKE"      LOC = "K3"    | IOSTANDARD = SSTL2_I;
#NET     "SDCS"       LOC = "K4"    | IOSTANDARD = SSTL2_I;
#NET     "SDLDM"      LOC = "J2"    | IOSTANDARD = SSTL2_I;
#NET     "SDLDQS"     LOC = "L6"    | IOSTANDARD = SSTL2_I;
#NET     "SDRAS"      LOC = "C1"    | IOSTANDARD = SSTL2_I;
#NET     "SDUDM"      LOC = "J1"    | IOSTANDARD = SSTL2_I;
#NET     "SDUDQS"     LOC = "G3"    | IOSTANDARD = SSTL2_I;
#NET     "SDWE"       LOC = "D1"    | IOSTANDARD = SSTL2_I;

# DDR SDRAM Data Bus
#NET     "SDD<0>"     LOC = "L2     | IOSTANDARD = SSTL2_I;
#NET     "SDD<1>"     LOC = "L1     | IOSTANDARD = SSTL2_I;
#NET     "SDD<2>"     LOC = "L3     | IOSTANDARD = SSTL2_I;
#NET     "SDD<3>"     LOC = "L4     | IOSTANDARD = SSTL2_I;
#NET     "SDD<4>"     LOC = "M3     | IOSTANDARD = SSTL2_I;
#NET     "SDD<5>"     LOC = "M4     | IOSTANDARD = SSTL2_I;
#NET     "SDD<6>"     LOC = "M5     | IOSTANDARD = SSTL2_I;
#NET     "SDD<7>"     LOC = "M6     | IOSTANDARD = SSTL2_I;
#NET     "SDD<8>"     LOC = "E2     | IOSTANDARD = SSTL2_I;
#NET     "SDD<9>"     LOC = "E1     | IOSTANDARD = SSTL2_I;
#NET     "SDD<10>"    LOC = "F1     | IOSTANDARD = SSTL2_I;
#NET     "SDD<11>"    LOC = "F2     | IOSTANDARD = SSTL2_I;
#NET     "SDD<12>"    LOC = "G6     | IOSTANDARD = SSTL2_I;
#NET     "SDD<13>"    LOC = "G5     | IOSTANDARD = SSTL2_I;
#NET     "SDD<14>"    LOC = "H6     | IOSTANDARD = SSTL2_I;
#NET     "SDD<15>"    LOC = "H5     | IOSTANDARD = SSTL2_I;

#DDR SDRAM DCM
#NET     "SDCKFB"     LOC = "B9"    | IOSTANDARD = LVCMOS33;
```

```
#Prohibit Use of Vref Pins
#CONFIG PROHIBIT = D2;
#CONFIG PROHIBIT = G4;
#CONFIG PROHIBIT = J6;
#CONFIG PROHIBIT = L5;
#CONFIG PROHIBIT = R4;
```

## Hardware Components and Peripherals

An embedded design using programmable gate arrays requires a variety of hardware components and peripherals which provide system support and facilitate the interface to the sensor and actuator environment and the operator. The available peripherals for the Basys Board and the Spartan-3E Starter Board is described here.

### Crystal Clock Oscillator

The 50 MHz crystal clock oscillator external peripheral on the Basys Board and the Spartan-3E Starter Board can function as the clock for synchronous logic operation of the FPGA. The clock oscillator has a 40% to 60% duty cycle and an accuracy of ± 2500 Hz or ± 50 parts per million (ppm). However, the minimum clock period of 20 nanoseconds (nsec) is exceedingly fast for projects that utilize the pushbuttons, slide switches, LEDs and the LED seven-segment display or LCD. The module clock.v in Listing 3.3 is located in the *Chapter 3\peripherals* folder. The file download procedure is described in the Appendix.

**Listing 3.3** Crystal clock oscillator module clock.v

```
// Basys Board and Spartan-3E Starter Board
// Crystal Clock Oscillator  clock.v
// c 2008 Embedded Design using Programmable Gate Arrays  Dennis Silage

module clock (input CCLK, input [31:0] clkscale, output reg clk);
                                  // CCLK master crystal clock oscillator 50 MHz
reg [31:0] clkq = 0;              // clock register, initial value of 0

always@(posedge CCLK)
    begin
        clkq = clkq + 1;                // increment clock register
            if (clkq >= clkscale)       // clock scaling
                begin
                    clk = ~clk;         // output clock
                    clkq = 0;           // reset clock register
                end
    end

endmodule
```

The module clock.v increases the period by comparing the 50 MHz master crystal clock input signal CCLK accumnulated in the register clkq to the input 32-bit variable clkscale to provide an external clk signal that is used in the application. The 32-bit register clkq is initialized to 0 here on global reset or power-up by a declaration, although the default value on power-up would also be 0 [Lee06].

The value of the clock scale factor net variable clkscale is determined by Equation 3.1, where *frequency* in Hertz (Hz) is the inverse of the period in seconds of the desired external clock signal. A square wave (50% duty cycle) clk signal with a frequency of 1 kHz requires that the input variable clkscale be 25 000.

$$\text{clkscale} = \frac{25\ 000\ 000}{\textit{frequency}} \qquad (3.1)$$

The statement output reg clk defines and maps the clock as a 1-bit register to the output net for use by other Verilog modules. The 32-bit clock register clkq can accommodate periods as long as 171 seconds with the 50 MHz master clock ($2^{32} - 1 = 4\ 294\ 967\ 296$) and matches the input 32-bit integer variable clkscale.

The clock oscillator module clock.v is verified by the Verilog top module clocktest.v in Listing 3.4, which is also located in the *Chapter 3\peripherals* folder. The file download procedure is described in the Appendix. The Xilinx ISE WebPACK project in the *Chapter 3\clocktest\s3eclocktest* folder s3eclocktest.ise (note that the ISE does not allow spaces in project names) uses a UCF clocktests3esb.ucf which uncomments the signals CCLK, LD0, LD1 and LD2 in the Spartan-3E Starter Board UCF of Listing 3.2.

The LEDs are simply mapped in clocktest.v module to the only output of each instance of the clock.v module and blink at rates of 1, 2, and 4 Hz. Unlike sequential processing, the three Verilog modules operate in parallel and some independently, as described in Chapter 1 Verilog Hardware Description Language.

As described in Chapter 2 Verilog Design Automation, the Design Utilization Summary for the top module clocktest.v shows the use of use of 41 slice flip-flops (1%), 18 4-input LUTs (1%), 46 occupied slices (1%) and a total of 944 equivalent gates in the XC3S500E Spartan-3E FPGA synthesis.

**Listing 3.4** Clock oscillator test top module clocktest.v

```
// Basys Board and Spartan-3E Starter Board
// Clock Oscillator Test clocktest.v
// c 2008 Embedded Design using Programmable Gate Arrays  Dennis Silage

module clocktest (input CCLK, output LD0, LD1, LD2);

clock M0 (CCLK, 25000000, LD0);  // 1 Hz clock
clock M1 (CCLK, 12500000, LD1);  // 2 Hz clock
clock M2 (CCLK, 6250000, LD2);   // 4 Hz clock

endmodule
```

## Light Emitting Diodes

The LED hard-wired accessory on the Basys Board or the Spartan-3E Starter Board can function as an indicator in embedded applications of the FPGA. The module bargraph.v in Listing 3.5, which is also located in the *Chapter 3\peripherals* folder, utilizes all eight available LEDs of the Spartan-3E Starter Board and provides a bar graph display useful for peak amplitude measurements in audio and communications signal processing.

The module demonstrates the simple mapping as a continuous assignment of an LED to a logic signal [Ciletti99]. The module is event driven on the positive edge of the input signal clk and uses the largest non-zero bit of the 8-bit input data to produce a bar graph display. The 8-bit LED data register leddata cannot be mapped directly to the eight LED output signals. In Listing 3.4 the 1-bit clock register could be mapped directly to a single LED. However, the eight continuous assignment Verilog statement assign performs the requisite 8-bit register mapping in Listing 3.5.

**Listing 3.5** LED bar graph module bargraph.v

```
// Basys Board and Spartan-3E Starter Board
// LED Bar Graph bargraph.v
// c 2008 Embedded Design using Programmable Gate Arrays  Dennis Silage

module bargraph (input clk, input [7:0] data, output LD7, LD6, LD5, LD4, LD3, LD2, LD1, LD0);

reg [7:0] leddata;            // LED data

assign LD7 = leddata[7];  // continuous assignment
assign LD6 = leddata[6];  // for LED data output
assign LD5 = leddata[5];
assign LD4 = leddata[4];
assign LD3 = leddata[3];
assign LD2 = leddata[2];
assign LD1 = leddata[1];
assign LD0 = leddata[0];

always@(posedge clk)      // local clock event driven
    begin
        leddata = 8'b00000000;          // bar graph pattern based
        if (data[0] == 1)               // based on the least
            leddata = 8'b00000001;      // bit set to logic 1
        if (data[1] == 1)
            leddata = 8'b00000011;
        if (data[2] == 1)
            leddata = 8'b00000111;
        if (data[3] == 1)
            leddata=8'b00001111;
        if (data[4] == 1)
            leddata=8'b00011111;
        if (data[5] == 1)
            leddata=8'b00111111;
        if (data[6] == 1)
            leddata=8'b01111111;
        if (data[7] == 1)
            leddata=8'b11111111;
    end

endmodule
```

The bar graph module has a resolution of 8-bits, but the most significant bit (MSB) is non-zero for 8-bit data greater than 128 or half the range (0 to 255). In an application the data might represent an analog signal that has been converted to a 12-bit digital signal by an analog-to-digital converter (ADC) peripheral. The most significant 8-bits would then be mapped to the 8-bits of the bar graph. To increase the resolution of the LED bar graph, a digital bias can be subtracted from the 12-bit data.

The bar graph module bargraph.v is verified by the Verilog top module bargraphtest.v in Listing 3.6, which is also in the *Chapter 3\bargraphtest\s3ebargraphtest* folder. The Xilinx ISE WebPACK project s3ebargraphtest.ise uses a UCF bargraphtests3esb.ucf which uncomments the signals CCLK, LD0 through LD7 in the Spartan-3E Starter Board UCF of Listing 3.2. The three Verilog modules operate in parallel and some independently. The file download procedure is described in the Appendix.

The data is generated as a simple ramp by the stimulus module gendata.v on the negative edge of the clock input signal. The wire net type establishes the 8-bit vector connectivity for data between the bargraph.v and gendata.v modules [Ciletti04]. The clock signal clk is outputted from the clock.v module and inputted to both the bargraph.v and gendata.v modules.

In Listing 3.6 the statement output reg [7:0] gdata defines and maps the generated data as an 8-bit register to the output vector net for use by other Verilog modules. Note that the name of the generated signal data in the top module bargraphtest.v (data) does not have to agree with the name of the generated register data in the gendata.v module (gdata) since the connection by position option for the ports of a module is used here, as described in Chapter 1 Verilog Hardware Description Language.

As described in Chapter 2 Verilog Design Automation, the Design Utilization Summary for the top module bargraphtest.v shows the use of use of 41 slice flip-flops (1%), 18 4-input LUTs (1%), 46 occupied slices (1%) and a total of 944 equivalent gates in the XC3S500E Spartan-3E FPGA synthesis.

**Listing 3.6** LED bar graph test top module bargraphtest.v

```
// Basys Board and Spartan-3E Starter Board
// LED Bar Graph Test bargraphtest.v
// c 2008 Embedded Design using Programmable Gate Arrays  Dennis Silage

module bargraphtest (input CCLK, output LD7, LD6, LD5, LD4, LD3, LD2, LD1, LD0);

wire [7:0] data;

clock M0 (CCLK, 250000, clk);        // 100 Hz clock
bargraph M1 (clk, data, LD7, LD6, LD5, LD4, LD3, LD2, LD1, LD0);
gendata M2 (clk, data);

endmodule

module gendata (input clock, output reg [7:0] gdata);        // generate bar graph test data

always@(negedge clock)          // local clock event driven
      gdata = gdata + 1;        // increment generated data

endmodule
```

## Push Buttons and Slide Switches

The hard-wired accessory push buttons and slide switches on the Basys Board or the Spartan-3E Starter Board function as asynchronous input signals in applications of the FPGA. Depressing the push buttons and setting the slide switches generate logic 1 on the associated FPGA pin, as given in the UCF of Listing 3.1 or Listing 3.2. They have no active debouncing circuitry.

The module pbsswtest.v in Listing 3.7, which is also in the *Chapter 3\pbsswtest\s3epbsswtest* folder, utilizes two of the four push buttons, the four slide switches and the four of the eight LEDs on the Spartan-3E Starter Board. The Basys Board has eight slide switches. The Xilinx ISE WebPACK project uses the UCF pbsswtests3esb.ucf which uncomments the signals BTN0 and BTN1, SW0 through SW3 and LD0 through LD3 in the Spartan-3E Starter Board UCF of Listing 3.2. The file download procedure is described in the Appendix.

The module is event driven on the asynchronous depression or release of either the push button 0 (BTN0) or push button 1 (BTN1). As in Listing 3.5, the continuous assignment Verilog statement assign performs the requisite mapping of leddata to the LED in Listing 3.7. Depressing BTN0 reads the push button switches individually and maps their output to the LEDs. Depressing BTN1 turns the LEDs off.

As described in Chapter 2 Verilog Design Automation, the Design Utilization Summary for the module pbsswtest.v shows the use of only 4 Input/Output Bus (IOB) latches (4%) and a total of 23 equivalent gates in the XC3S500E Spartan-3E FPGA synthesis.

**Listing 3.7** Push button and slide switch test modules pbsswtest.v

```
// Basys Board and Spartan-3E Starter Board
// Push Button and Slide Switch Test pbsswtest.v
// c 2008 Embedded Design using Programmable Gate Arrays  Dennis Silage
module pbsswtest (input BTN0, BTN1, SW0, SW1, SW2, SW3, output LD3, LD2, LD1, LD0);

reg [3:0] leddata;          // LED data

assign LD3 = leddata[3];  // continuous assignment
assign LD2 = leddata[2];  // for LED data output
assign LD1 = leddata[1];
assign LD0 = leddata[0];

always@(BTN0 or BTN1)          // pushbutton event driven
    begin
        if (BTN0 == 1)
            begin
                leddata[0] <= SW0;  // non-blocking
                leddata[1] <= SW1;  // assignment
                leddata[2] <= SW2;  // to read
                leddata[3] <= SW3;  // switches
            end
        if (BTN1 == 1)
            leddata <= 0;               // clear LEDs
    end

endmodule
```

The push buttons are often used for counting functions in FPGA applications where contact *bounce* can cause an aberrant result. The top module pbdebouncetest.v in Listing 3.8, which is also in the *Chapter 3\pbdebouncetest\s3epbdebouncetest* folder, illustrates the use of a serial shift register to insure that if a contact bounce occurs then only a single output pulse results. The module pbdebounce.v in Listing 3.9 is in the *Chapter 3\peripherals* folder and is utilized within the top module pbdebouncetest.v. The three Verilog modules operate in parallel and some independently in the top module [Botros06].

A 4-bit register pbshift is logically (no wrap-around) left shifted on the positive edge of a 10 Hz clock provided by the clock.v module. The least significant bit of the shift register is set equal to the push button (BTN0). If all four bits of pbshift contain logic 1 (1111 or 15 decimal), then the push button register output pbreg is set to logic 1. If all four bits contain logic 0, then the push button register output pbreg is set to logic 0. Any other 4-bit pattern indicates that a push button bounce has occurred and pbreg is unchanged.

**Listing 3.8** Push button contact debounce test top module pbdebouncetest.v

```
// Basys Board and Spartan-3E Starter Board
// Push Button Debounce Test pbdbtest.v
// c 2008 Embedded Design using Programmable Gate Arrays  Dennis Silage

module pbdebouncetest (input CCLK, input BTN0,output LD7, LD6, LD5, LD4, LD3,
                              LD2, LD1, LD0);

wire [3:0] leddata;
wire [3:0] dataled;

assign LD7 = dataled[3];  // continuous assignment
assign LD6 = dataled[2];  // for LED data output
assign LD5 = dataled[1];
assign LD4 = dataled[0];
assign LD3 = leddata[3];
assign LD2 = leddata[2];
assign LD1 = leddata[1];
assign LD0 = leddata[0];

clock M0 (CCLK, 2500000, clk);      // 10 Hz clock
pbdebounce M1 (clk, BTN0, pbreg);
ledtest M2 (pbreg, BTN0, leddata, dataled);

endmodule
module ledtest (input pbreg, input button, output reg [3:0] leddata, output reg [3:0] dataled);

always@(posedge pbreg)  // debounced pushbutton event
      begin                          // driven
            leddata = leddata + 1;   // increment counter
      end

always@(posedge button) // chattering pushbutton event
      begin                          // driven
            dataled = dataled + 1;   // increment counter
      end

endmodule
```

**Listing 3.9** Push button debounce pbdebounce.v

```
// Basys Board and Spartan-3E Starter Board
// Push Button Debounce pbdebounce.v
// c 2008 Embedded Design using Programmable Gate Arrays  Dennis Silage

module pbdebounce (input clk, input button, output reg pbreg);

reg [3:0] pbshift;
```

```
always@(posedge clk)          // local clock event driven
    begin
        pbshift = pbshift << 1;     // shift register
        pbshift[0] = button;        // read button
    if (pbshift == 0)               // if a bounce occurs
        pbreg = 0;      // clear the register
    if (pbshift == 15)  // 15 local clock tics without a bounce
        pbreg = 1;      //  sets the register
    end

endmodule
```

The module ledtest.v loads the rightmost four LEDs of the Basys Board or the Spartan-3E Starter Board with the 4-bit register leddata which increments on the positive edge of the register pbreg. The leftmost four LEDs are loaded with the 4-bit register dataled which increments on the positive edge of the same push button signal BTN0. A comparison of the output of these two registers clearly indicates the deleterious nature of push button bounce on performance. The push button bounce mitigation is based on a time window that is the product of the number of bits in the shift register and the clock period.

The Xilinx ISE WebPACK project is in the *Chapter 3\pbdebouncetest\s3epbdebouncetest* folder and uses the UCF pbdebouncetests3esb.ucf which uncomments the signals CCLK, LD0 through LD7 in the Spartan-3E Starter Board UCF of Listing 3.2. The file download procedure is described in the Appendix. The module is event driven on the asynchronous depression of push button 0 (BTN0) which toggles the LEDs.

As described in Chapter 2 Verilog Design Automation, the Design Utilization Summary for the top module pbdebouncetest.v describes the use of use of 45 slice flip-flops (1%), 18 4-input LUTs (1%), 46 occupied slices (1%) and a total of 873 equivalent gates in the XC3S500E Spartan-3E FPGA synthesis.

## Rotary Shaft Encoder

The hard-wired accessory rotary shaft encoder on the Spartan-3E Starter Board functions as asynchronous input signals in applications of the FPGA. The Basys Board does not have a rotary shaft encoder. Depressing the center shaft provides an additional push button switch which generates logic 1 on the ROTCTR signal. Rotating the shaft clockwise provides a logic 1 on the ROTA signal before a logic 1 appears on the ROTB signal. Rotating the shaft counter-clockwise provides a logic 1 on the ROTB signal before a logic 1 appears on the ROTA signal. When the shaft is stationary at the detent position both ROTA and ROTB are logic 0. The rotary shaft encoder signals ROTA, ROTB and ROTCTR have no active debouncing circuitry and are defined in the UCF of the Spartan-3E Starter Board in Listing 3.2.

The module rotary.v in Listing 3.10, which is also in the *Chapter 3\peripherals* folder, is event driven on the positive edge of the input signal clk provided by the clock.v module. Each of the rotary shaft encoder signals are debounced by the same process described in the module pbdebounce.v in Listing 3.9. The debounced shaft encoder signals here are the register variables rotAreg, rotBreg and rotCTRreg.

The rotary shaft encoder module rotary.v is verified by the Verilog top module rotarytest.v in Listing 3.11 which is in the *Chapter 3\rotarytest\s3erotarytest* folder. The complete Xilinx ISE WebPACK project uses the UCF rotarytests3esb.ucf which uncomments the signals CCLK, ROTA, ROTB, ROTCTR, and LD7 through LD0 in the Spartan-3E Starter Board UCF of Listing 3.2. The three Verilog modules operate in parallel and some independently in the top module. The file download procedure is described in the Appendix.

**Listing 3.10** Rotary shaft encoder module rotary.v

```
// Spartan-3E Starter Board
// Rotary Shaft Encoder rotary.v
// c 2008 Embedded Design using Programmable Gate Arrays  Dennis Silage
module rotary (input clk, ROTA, ROTB, ROTCTR, output reg rotAreg, output reg rotBreg,
                        output reg rotCTRreg);

reg [3:0] rotAshift;
reg [3:0] rotBshift;
reg [3:0] rotCTRshift;

always@(posedge clk)      // local clock event driven
    begin
        rotCTRshift = rotCTRshift << 1;         // debounce for ROTCTR
        rotCTRshift[0] = ROTCTR;
        if (rotCTRshift == 0)
            rotCTRreg = 0;
        if (rotCTRshift == 15)
            rotCTRreg = 1;

        rotAshift = rotAshift << 1;             // debounce for ROTA
        rotAshift[0] = ROTA;
        if (rotAshift == 15)
            rotAreg = 0;
        if (rotAshift == 0)
            rotAreg = 1;

        rotBshift = rotBshift << 1;             // debounce for ROTB
        rotBshift[0] = ROTB;
        if (rotBshift == 15)
            rotBreg = 0;
        if (rotBshift == 0)
            rotBreg = 1;
    end

endmodule
```

The wire net type establishes the 4-bit vector connectivity for leddata and dataled between the ledtest.v and rotarytest.v modules. The LED external signals LD7 through LD4 and LD7 through LD4 are assigned to register dataled[3]through dataled[0] and leddata[3]through leddata[0], respectively, since they are only referenced in the top module rotarytest.v.

The module ledtest.v loads the rightmost four LEDs of the Spartan-3E Starter Board with the 4-bit register leddata which increments or decrements on the logic of the debounced register variable rotBreg on the positive edge of the debounced register variable rotAreg. The leftmost four LEDs are loaded with the 4-bit register dataled which increments or decrements on the logic of the chattering register variable ROTB on the positive edge of the chattering register variable ROTA.

A clockwise rotation of the rotary shaft encoder increments the register, while a counterclockwise rotation decrements the register. These registers are cleared with the debounced register variable rotCTRreg or the chattering register variable ROTCTR which results from the depression of the shaft of the rotary encoder.

**Listing 3.11** Rotary shaft encoder top module rotarytest.v

```
// Spartan-3E Starter Board
// Rotary Switch Test rotarytest.v
// c 2008 Embedded Design using Programmable Gate Arrays  Dennis Silage
module rotarytest (input CCLK, ROTA, ROTB, ROTCTR, output LD7, LD6, LD5, LD4,
                        LD3, LD2, LD1, LD0);
wire rotAreg;
wire rotBreg;
wire [3:0] leddata;
wire [3:0] dataled;

assign LD7 = dataled[3];
assign LD6 = dataled[2];
assign LD5 = dataled[1];
assign LD4 = dataled[0];
assign LD3 = leddata[3];
assign LD2 = leddata[2];
assign LD1 = leddata[1];
assign LD0 = leddata[0];

clock M0 (CCLK, 25000, clk);          // 1 kHz clock
rotary M1 (clk, ROTA, ROTB, ROTCTR, rotAreg, rotBreg, rotCTRreg);
ledtest M2 (ROTA, ROTB, rotAreg, rotBreg, ROTCTR, rotCTRreg, leddata, dataled);

endmodule
module ledtest (input ROTA, ROTB, rotAreg, rotBreg, ROTCTR, rotCTRreg,
                output reg [3:0] leddata, output reg [3:0] dataled);

always@(posedge rotAreg)              // event driven on debounced
    begin                             // shaft encoder ROTA
        if (rotBreg == 0 && rotCTRreg == 0)
                leddata = leddata + 1;
            else
                leddata = leddata – 1;
            if (rotCTRreg == 1)
                leddata = 0;
    end

always@(posedge ROTA)                 // even driven on chattering
    begin                                 // shaft encoder ROTA
        if (ROTB == 0 && rotctr == 0)
            dataled = dataled + 1;
        else
            dataled = dataled – 1;
        if (ROTCTR == 1)
            dataled = 0;
        end

endmodule
```

A comparison of the output of these two registers clearly indicates the deleterious nature of rotary shaft encoder bounce on performance. The rotary shaft encoder bounce mitigation is based on a time window that is the product of the number of bits in the shift register and the clock period. As described in Chapter 2 Verilog Design Automation, the Design Utilization Summary the top module rotarytest.v shows the use of 158 slice flip-flops (1%), 789 4-input LUTs (8%), 417 occupied slices (8%) and a total of 6383 equivalent gates in the XC3S500E Spartan-3E FPGA synthesis.

## Seven Segment Display

The hard-wired accessory seven segment display on the Basys Board functions as an annunciator in applications of the FPGA. The seven segment display is commonly used to indicate a numerical output with the digitals 0 through 9 and a decimal point, but can be extended to show any pattern. For example, the additional characters that form the hexadecimal (base 16) number system can be displayed as A, b C, d, E and F. Other patterns can be formed which can be used in process control applications.

Any of the seven LED segments and the LED as a decimal point can be turned on separately by a logic 0 on the eight cathode signals CA, CB, CC, CD, CE, CF, CG and CDP. The Basys Board has four seven segment displays which are multiplexed by a logic 0 on the four anode signals AN0 (the rightmost seven segment display), AN1, AN2, and AN3 (the leftmost seven segment display). Multiplexing implies that the four seven segment displays must be scanned at a rate high enough to avoid flicker, but only 12 IO signals are used here rather than the 32 IO signals required if connected with the anodes grounded (logic 0).

The module sevensegment.v in Listing 3.12, which is in the *Chapter 3\peripherals* folder, is event driven on the negative edge of the input signal clk. A Verilog case statement assigns the 4-bit input signal data to the 8-bit cathode pattern signal cathodedata of the seven segment display for the normal 16 characters (the decimal numbers 0 through 9 and the extended hexadecimal digits A through F) and 15 special characters (including the minus sign and degree symbol). The decimal point as the least significant bit (LSB) is off (logic 1).

Another Verilog case statement assigns the 3-bit input signal digit to the 4-bit anode register anodedata to enable the seven segment display individually [Navabi06]. A 3-bit signal is used here because there are four digits and an all off condition (five states in all). The input signal setdp, if a logic 1, is used to logically AND (&) the 8-bit register cathodedata with FE hexadecimal (or 1111 1110 binary) to set the LSB of cathodedata to a logic 0 and turn the decimal point on. Finally, the continuous assignment Verilog statement assign performs requisite mapping of the cathode and anode signals to the output pins of the seven segment display, as given by the UCF of Listing 3.1.

**Listing 3.12** Seven segment display module sevensegment.v

```
// Basys Board
// Seven Segment Display sevensegment.v
// c 2008 Embedded Design using Programmable Gate Arrays  Dennis Silage

module sevensegment (input clock, input [4:0] data, input [2:0] digit, input setdp,
                     output AN0, AN1, AN2, AN3, CA, CB, CC, CD, CE, CF, CG, CDP);

reg [7:0] cathodedata;          // cathode data
reg [3:0] anodedata;            // anode data

assign CA = cathodedata[7];
assign CB = cathodedata[6];
assign CC = cathodedata[5];
assign CD = cathodedata[4];
```

```
assign CE = cathodedata[3];
assign CF = cathodedata[2];
assign CG = cathodedata[1];
assign CDP = cathodedata[0];

assign AN3 = anodedata[3];
assign AN2 = anodedata[2];
assign AN1 = anodedata[1];
assign AN0 = anodedata[0];

always@(negedge clock)  // local clock event driven
      begin
            case (data)
                  0: cathodedata = 8'b00000011;          // 0
                  1: cathodedata = 8'b10011111;          // 1
                  2: cathodedata = 8'b00100101;          // 2
                  3: cathodedata = 8'b00001101;          // 3
                  4: cathodedata = 8'b10011001;          // 4
                  5: cathodedata = 8'b01001001;          // 5
                  6: cathodedata = 8'b01000001;          // 6
                  7: cathodedata = 8'b00011111;          // 7
                  8: cathodedata = 8'b00000001;          // 8
                  9: cathodedata = 8'b00001001;          // 9
                  10: cathodedata = 8'b00010001;         // A
                  11: cathodedata = 8'b11000001;         // b
                  12: cathodedata = 8'b01100011;         // C
                  13: cathodedata = 8'b10000101;         // d
                  14: cathodedata = 8'b01100001;         // E
                  15: cathodedata = 8'b01110001;         // F
                  16: cathodedata = 8'b11111101;         // middle (minus sign)
                  17: cathodedata = 8'b01111111;         // top
                  18: cathodedata = 8'101111111;         // right top
                  19: cathodedata = 8'b11011111;         // right bottom
                  20: cathodedata = 8'b11101111;         // bottom
                  21: cathodedata = 8'b11110111;         // left bottom
                  22: cathodedata = 8'b11111011;         // left top
                  23: cathodedata = 8'b11011001;         // left top, middle, right bottom
                  24: cathodedata = 8'b10110101;         // left bottom, middle, right top
                  25: cathodedata = 8'b11000101;         // bottom small o
                  26: cathodedata = 8'b00111001;         // top small o (degree)
                  27: cathodedata = 8'b11010101;         // bottom, inverted small u
                  28: cathodedata = 8'b10111001;         // top, small u
                  29: cathodedata = 8'b11000111;         // bottom, small u
                  30: cathodedata = 8'b00111011;         // top, inverted small u
                  31: cathodedata = 8'b11111111;         // all OFF
            endcase

            if (setdp == 1)                              // decimal point
                  cathodedata = cathodedata & 8'hFE;
```

```
        case (digit)
            0:    anodedata = 4'b1111;          // all OFF
            1:    anodedata = 4'b1110;          // AN0
            2:    anodedata = 4'b1101;          // AN1
            3:    anodedata = 4'b1011;          // AN2
            4:    anodedata = 4'b0111;          // AN3
            default:
                  anodedata = 4'b1111;          // all OFF
        endcase
    end

endmodule
```

The seven segment display module sevensegment.v is verified by the Verilog top module sevensegtest.v in Listing 3.13 which is in the *Chapter 3\sevensegtest\basevensegtest* folder. The complete Xilinx ISE WebPACK project uses a UCF which uncomments the signals CCLK, BTN0, BTN1, BTN3, and the cathode and anode signals in the Basys Board UCF of Listing 3.1. The four Verilog modules operate in parallel and some independently in the top module. The file download procedure is described in the Appendix.

The wire net type establishes the 4-bit and 3-bit vector connectivity for data and digit between the elapsedtime.v and sevensegment.v modules. The clock signals clka and clkb are outputted from the two instances of the clock.v module at a frequency of 100 Hz and 1 kHz and inputted to the elapsedtime.v module. The signal clka is also inputted to sevensegment.v module.

The module elapsedtime.v is a stop watch application with a resolution of 10 msec. The four seven segment displays are multiplexed on the negative edge of the signal clkb at a 1 kHz rate to avoid flicker using the 3-bit signal digit and the 4-bit signal data. The signal setdp is set to logic 1 for the third digit to signify the decimal point for the elapsed time in seconds. The blocking Verilog assignment statement ( = ) is used to set the seven segment display to insure that both the data and digit register variables are correct before the display is updated [Ciletti99].

The elapsed time is an event triggered process on the positive edge of the signal clock at a 100 Hz rate for a resolution of 10 milliseconds (msec). A digit is incremented and set to 0 if greater than 9 and the next digit is incremented. An overflow sets the elapsed time to 0. The push buttons BTN0, BTN1 and BTN3 starts, stops and resets the elapsed time clock process. Contact debounce, as shown in Listing 3.9, is not required here since the depression of the push button is a positive latch and not a counting function.

As described in Chapter 2 Verilog Design Automation, the Design Utilization Summary for the top module sevensegtest.v shows the use of 93 slice flip-flops (4%), 113 4-input LUTs (4%), 126 occupied slices (10%) and a total of 2337 equivalent gates in the XC3S100 Spartan-3E FPGA synthesis.

**Listing 3.13** Seven segment display test top module for the Basys Board sevensegtest.v

```
// Basys Board
// Seven Segment Display Test sevensegtest.v
// c 2008 Embedded Design using Programmable Gate Arrays  Dennis Silage

module sevensegtest (input CCLK, BTN0, BTN1, BTN3, output AN0, AN1, AN2, AN3, CA, CB,
                     CC, CD, CE, CF, CG, CDP);

wire [4:0] data;
wire [2:0] digit;
```

```
clock M0 (CCLK, 250000, clka);                  // 100 Hz
clock M1 (CCLK, 25000, clkb);                   // 1 kHz
sevensegment M2 (clkb, data, digit, setdp, AN0, AN1, AN2, AN3, CA, CB, CC, CD, CE, CF, CG,
                 CDP);
elapsedtime M3 (clka, clkb, data, digit, setdp, BTN0, BTN1, BTN3);

endmodule

module elapsedtime(input clka, input clkb, output reg [3:0] data, output reg [2:0] digit,
                   output reg setdp, input BTN0, BTN1, BTN3);

reg [1:0] digitmux;
reg startstop;

reg [3:0] csec100;    // seconds 1/100s
reg [3:0] csec10;     // seconds 1/10s
reg [3:0] sec1;       // seconds 1s
reg [3:0] sec10;      // seconds 10s

always@(negedge clkb)   // local clock event driven
    begin
        digitmux = digitmux + 1;  // digit multiplexer
        setdp = 0;                // clear decimal point
        data[4] = 0;
        case (digitmux)
            0:  begin
                    data[3:0] = csec100;
                    digit =1;
                end
            1:  begin
                    data[3:0] = csec10;
                    digit = 2;
                end
            2:  begin
                    data[3:0] = sec1;
                    digit = 3;
                    setdp = 1;        // set decimal point
                end
            3:  begin
                    data[3:0] = sec10;
                    digit = 4;
                end
        endcase
    end

always@(posedge clka)          // local clock event driven
    begin
        if (BTN3 == 1)         // clear and stop
            begin
                startstop = 0;
                csec100 = 0;
                csec10 = 0;
```

```
                sec1 = 0;
                sec10 = 0;
            end

        if (BTN1==1)          // stop
            startstop = 0;
        if (BTN0 == 1)        // start
            startstop = 1;

        if (startstop == 1)
            begin
                csec100 = csec100+1;
                if (csec100 > 9)
                    begin
                        csec100 = 0;
                        csec10 = csec10+1;
                    end
                if (csec10 > 9)
                    begin
                        csec10 = 0;
                        sec1 = sec1 + 1;
                    end
                if (sec1 > 9)
                    begin
                        sec1 = 0;
                        sec10 = sec10+1;
                    end
                if (sec10 > 9)
                    sec10 = 0;
            end
        end

endmodule
```

## Liquid Crystal Display

The hard-wired accessory liquid crystal display (LCD) on the Spartan-3E Starter Board functions as an alphanumeric annunciator in applications of the FPGA. Although the functional interface is more complicated and inherently slower in response than the seven segment display of the Basys Board, the LCD features a 2-line by 16-character. Updating the LCD at even a 0.5 second interval produces diminished clarity. Although a standard LCD can support an 8-bit parallel data interface, the Spartan-3E Starter Board utilizes an alternative 4-bit parallel interface to minimize the total input/output (I/O) pin count.

The LCD interface signals consist of the four data bits LCDDAT[0], LCDDAT[1] LCDDAT[2]and LCDDAT[3], the read/write enable signal LCDE, the register select signal LCDRS and the read/write control signal LCDRW. Although the LCD uses a +5 V TTL logic supply and the Spartan-3E Starter Board FPGA I/O signals use a +3.3 V LVCMOS logic supply, the standard LCD controller recognizes the signals.

The four data bits are shared with the Intel 128 Mb StrataFlash 28F256 PROM on the Spartan-3E Starter Board. However, for these applications of the LCD in embedded system design the read/write control signal LCDRW is set to logic 0 which provides write access only to the LCD but full read/write access to the PROM.

The standard LCD has three internal memory regions. The display data (DD) RAM stores the reference to a specific character bitmap to be displayed on the screen. The character bitmaps are stored in either the character generator (CG) ROM region or the user-defined CG RAM region. For these applications of the LCD in embedded system design the CG ROM is used to provide the fixed font bitmaps referenced by their ASCII character code. The hexadecimal DD RAM addresses for the 2-line by 16-character display are listed in Table 3.1.

**Table 3.1** Hexadecimal DD RAM addresses for the 2-line by 16-character LCD

| Position | 1 | 2 | 3 | 4 | 5 | 6 | 7 | 8 | 9 | 10 | 11 | 12 | 13 | 14 | 15 | 16 |
|---|---|---|---|---|---|---|---|---|---|---|---|---|---|---|---|---|
| Line 1 | 00 | 01 | 02 | 03 | 04 | 05 | 06 | 07 | 08 | 09 | 0A | 0B | 0C | 0D | 0E | 0F |
| Line 2 | 40 | 41 | 42 | 43 | 44 | 45 | 46 | 47 | 48 | 49 | 4A | 4B | 4C | 4D | 4E | 4F |

The LCD 8-bit command is sent as two 4-bit nibbles with the most significant 4-bit nibble transferred first. The common LCD command set in binary is listed in Table 3.2. The entry X indicates a don't care condition. The LCD read/write enable signal LCDE must be logic 1 for the commands to be effective. The LCD register select signal LCDRS and the read/write control signal LCDRW are set, as listed in Table 3.2.

**Table 3.2** Common LCD command set

| Command | LCDRS | LCDRW | Upper Nibble | Lower Nibble |
|---|---|---|---|---|
| Clear Display | 0 | 0 | 0 0 0 0 | 0 0 0 1 |
| Cursor Home | 0 | 0 | 0 0 0 0 | 0 0 1 X |
| Entry Mode | 0 | 0 | 0 0 0 0 | 0 1 I/D S |
| Display On/Off | 0 | 0 | 0 0 0 0 | 1 D C B |
| Shift Mode | 0 | 0 | 0 0 0 1 | S/C R/L X X |
| Function Set | 0 | 0 | 0 0 1 0 | 1 0 X X |
| Set DD RAM Address | 0 | 0 | 1 A6 A5 A4 | A3 A2 A1 A0 |
| Read Busy Flag | 0 | 1 | BF A6 A5 A4 | A3 A2 A1 A0 |
| Write Data to DD RAM | 1 | 0 | D7 D6 D5 D4 | D3 D2 D1 D0 |

The entry mode command uses the increment/decrement (I/D) bit to auto-decrement with logic 0 or auto-increment with logic 1 the address counter. This appears as though the invisible or blinking cursor is moving either left or right. The clear display command writes a space or ASCII character code 20 hexadecimal (h) into all DD RAM locations, clears all the option settings and sets the I/D bit to logic 1 and the DD RAM address counter to 00h or the top-left corner of the LCD.

The cursor home command only sets the DD RAM address counter to 00h without clearing the LCD. The display on/off command display (D) bit turns the LCD on with logic 1 or off with logic 0, the cursor (C) displays an underscore cursor with logic 1 or no cursor with logic 0 and the blink (B) bit blinks the cursor at an interval of 0.5 seconds or no cursor blink with logic 0.

The shift mode command uses the shift/cursor (S/C) bit and right/left (R/L) bit to provide four functions which shift the cursor or the entire display without affecting the DD RAM contents. If these di-bits are 00 the cursor shifts to the left and the DD RAM address counter is decremented by one. If these di-bits are 01 the cursor shifts to the left and the DD RAM address counter is incremented by one. If these di-bits are 10 the entire display and cursor shifts to the left and the DD RAM address counter is unchanged. Finally, if these di-bits are 11 the entire display and cursor shift to the right and the DD RAM address counter is unchanged.

The function set command sets the LCD interface data length, number of display lines and character font. The Spartan-3E Starter Board supports only a single function code of 28h. The set DD RAM address command sets the initial value in the address counter and subsequent LCD interface commands are executed with the DD RAM.

The read busy flag/address command reads the busy flag (BF) bit can be used to determine if an internal LCD interface operation is in progress or returns the current DD RAM address if the set DD RAM address command was executed first. The BF can be used to test for completion of internal LCD operations which can require anywhere from 1 microsecond (μsec) to over 1.6 millisecond (msec).

However, since the LCD is available for write access only here, a fixed time delay for each operation can also be employed. Finally, the write data to DD RAM command writes the 8-bit data to the current DD RAM address location and either increments or decrements the address counter as set by the entry mode command.

The LCD interface requires a minimum setup time of 40 nanoseconds (nsec) for the signals LCDRS, LCDRW and the 4-bit data LCDDAT before the signal LCDE becomes active logic 1. LCD commands and data must be held for a minimum of 230 nsec and for at least 10 nsec after the signal LCDE becomes inactive logic 0. The 4-bit data LCDDAT nibbles must be spaced a minimum of 1 μsec apart with at least 40 μsec between 8-bit data transfers. Finally, the clear display and cursor home commands require an additional delay of at least 1.6 msec. These LCD interface operations are event driven by the Spartan-3E Starter Board crystal clock signal CCLK with a frequency of 50 MHz or a period of 20 nsec.

The module lcd.v in Listing 3.14, which is in the *Chapter 3\peripherals* folder, utilizes multiple finite state machines (FSM) as a datapath to perform the functions of the LCD on the positive edge of the crystal clock oscillator (CCLK) and provides signals to the controller, as described in Chapter 1 Verilog Hardware Description Language. LCD initialization occurs if the net variable initlcd is set to logic 1. The state register lcdstate sets each of the operations within a given function of the LCD and the register variable lcdcount determines the fixed delays required by the operation. The delays are rounded up to the next power of two ($2^n$) for efficiency in the behavioral synthesis of the FSM. The default value of the state registers lcdstate and lcdcount on power-up would be 0 [Lee06].

Although the datapath module lcd.v is lengthy and seemingly complicated, the multiple functions of the LCD are essentially separate FSMs. The external controller would not evoke another LCD function until the corresponding status function indicates that the function has completed. Only a partial listing of the module lcd.v is given in Listing 3.14. The listings of other lengthy modules are also to be truncated for brevity. The complete Verilog HDL source for lcd.v is available and the file download procedure is described in the Appendix.

**Listing 3.14** Liquid crystal display datapath module lcd.v

```
// Spartan-3E Starter Board
// Liquid Crystal Display lcd.v
// c 2008 Embedded Design using Programmable Gate Arrays  Dennis Silage

module lcd (input CCLK, resetlcd, clearlcd, homelcd, datalcd, addrlcd, output reg lcdreset,
            output reg lcdclear, output reg lcdhome, output reg lcddata, output reg lcdaddr,
            output reg rslcd, output reg rwlcd, output reg elcd, output reg [3:0] lcdd,
            input [7:0] lcddatin, input initlcd);

reg [18:0] lcdcount;        // LCD delay counter
reg [5:0] lcdstate;         // LCD state

always@(posedge CCLK)          // master clock event driven
    begin
        if (initlcd == 1)      // initialize LCD
            begin
                lcdstate = 0;     // LCD state register
                lcdcount = 0;     // LCD delay count
                lcdreset = 0;     // LCD reset response
```

```
                lcdclear = 0;     // LCD clear response
                lcdhome = 0;      // LCD home response
                lcdaddr = 0;      // LCD address response
                lcddata = 0;      // LCD data response
            end
        else
            lcdcount = lcdcount + 1;   // increment delay counter

// reset LCD
        if (resetlcd == 1 && lcdreset == 0)    // reset LCD
            begin
                rslcd = 0;        // register select for command
                rwlcd = 0;        // LCD read/write
                case (lcdstate)
                    0:   begin       // send '3'
                            lcdd = 3;
                            elcd = 0;
                            if (lcdcount == 16)
                                begin
                                    lcdcount = 0;
                                    lcdstate = 1;
                                end
                         end
{LCD reset continues similarly in states 1 through 38}
                    39:  begin            // wait 40 usec (2000 clocks)
                            if (lcdcount == 2048)
                                begin
                                    lcdcount = 0;
                                    lcdstate = 40;
                                    lcdreset = 1;
                                end
                         end
                    40: lcdstate = 40;
                    default: lcdstate = 40;
                endcase
            end

// send 8-bit data to LCD
        if (datalcd == 1 && lcddata == 0)     // LCD data
            begin
                rslcd = 1;  // register select for data
                rwlcd = 0; // LCD read/write
                case (lcdstate)
                    0:   begin            // send upper nibble
                            lcdd[3:0] = lcddatin[7:4];
                            elcd = 0;
                            if (lcdcount == 16)
                                begin
                                    lcdcount = 0;
                                    lcdstate = 1;
                                end
                         end
```

```
{LCD send 8-bit data continues in states 1 through 6}
                    7:    begin              // wait 40 usec (2000 clocks)
                              if (lcdcount == 2048)
                                  begin
                                      lcdcount = 0;
                                      lcdstate = 8;
                                      lcddata = 1;
                                  end
                          end
                    8:    lcdstate = 8;
                    default: lcdstate = 8;
                endcase
            end

// return cursor home
        if (homelcd == 1 && lcdhome == 0) // LCD home
            begin
                rslcd = 0;          // register select for command
                rwlcd = 0;          // LCD read/write
                case (lcdstate)
                    0:    begin              // send '0'
                              lcdd = 0;
                              elcd = 0;
                              if (lcdcount == 16)
                                  begin
                                      lcdcount = 0;
                                      lcdstate = 1;
                                  end
                          end
{LCD return cursor home continues in states 1 through 6}
                    7:    begin              // wait 1.6 usec (80000 clocks)
                              if (lcdcount == 131072)
                                  begin
                                      lcdcount = 0;
                                      lcdstate = 8;
                                      lcdhome = 1;
                                  end
                          end
                    8:    lcdstate = 8;
                    default: lcdstate = 8;
                endcase
            end

// clear display
        if (clearlcd == 1 && lcdclear == 0)  // LCD clear
            begin
                rslcd = 0;                   // register select for command
                rwlcd = 0;                   // LCD read/write
                case (lcdstate)
                    0:    begin              // send '0'
                              lcdd = 0;
                              elcd = 0;
```

```
                        if (lcdcount == 16)
                            begin
                                lcdcount = 0;
                                lcdstate = 1;
                            end
                    end
{LCD clear display continues in states 1 through 6}
                7:  begin               // wait 1.64 msec (82000 clocks)
                        if (lcdcount == 131072)
                            begin
                                lcdcount = 0;
                                lcdstate = 8;
                                lcdclear = 1;
                            end
                    end
                8:  lcdstate = 8;
                default: lcdstate = 8;
            endcase
        end

// set display address
    if (addrlcd == 1 && lcdaddr == 0)    // LCD address
        begin
            rslcd = 0;                   // register select for command
            rwlcd = 0;                   // LCD read/write
            case (lcdstate)
                0:  begin
                        lcdd[3] = 1;
                        lcdd[2:0] = lcddatin[6:4];
                        elcd = 0;
                        if (lcdcount == 16)
                            begin
                                lcdcount = 0;
                                lcdstate = 1;
                            end
                    end
{LCD set display address continues in states 1 through 6}
                7:  begin               // wait 40 usec (2000 clocks)
                        if (lcdcount == 2048)
                            begin
                                lcdcount = 0;
                                lcdstate = 8;
                                lcdaddr = 1;
                        end
                    end
                8:  lcdstate = 8;
                default: lcdstate = 8;
                endcase
        end
    end

endmodule
```

The initialize LCD function occurs if the net variable initlcd is set to logic 1. The FSM state variable lcdstate and the delay count variable lcdcount are set to 0. The lcd datapath status signals lcdreset, lcdclear, lcdhome, lcdaddr and lcddata are cleared. The initialize LCD function synchronizes and coordinates the multiple FSMs that are used for the various functions.

The reset LCD function occurs if the net variable resetlcd is set to logic 1 which changes the configuration of the LCD from the default 8-bit parallel interface to the required 4-bit parallel interface of the Spartan-3E Starter Board, sets the LCD to automatically increment the DD RAM address pointer, and turns the LCD on and disables the display of the cursor. After 39 active states, the reset LCD function enters state 40 and waits there until the controller responds to the datapath status signal lcdreset as logic 1 by resetting the net variable resetlcd to logic 0.

The clear LCD function occurs if the net variable clearlcd is set to logic 1 which writes ASCII character code 20h or space to all DD RAM address locations, effectively clearing the LCD. The DD RAM address is set to 00h or a cursor location of the upper left corner as listed in Table 3.1. The I/D bit is set to auto-increment the LCD. After 7 active states, the clear LCD function enters state 8 and waits there until the controller responds to the datapath status signal lcdclear as logic 1 by resetting the net variable clearlcd to logic 0.

The home LCD function occurs if the net variable homelcd is set to logic 1 which sets the DD RAM address for the location of the cursor to 00h or the upper left corner. After 7 active states, the home LCD function enters state 8 and waits there until the controller responds to the datapath status signal lcdhome as logic 1 by resetting the net variable homelcd to logic 0.

The set DD RAM address LCD function occurs if the net variable addrlcd is set to logic 1. The 8-bit register variable lcddatin contains the address to be set as the cursor location. The most significant bit (MSB) of the register variable lcddatin is set to logic 1 and the remaining 7 bits contain the DD RAM address. After 7 active states, the set DD RAM LCD function enters state 8 and waits there until the controller responds to the datapath status signal lcdaddr as logic 1 by resetting the net variable addrlcd to logic 0.

Finally, the LCD data function occurs if the net variable datalcd is set to logic 1. The 8-bit register variable lcddatin contains the ASCII character to be displayed at the current DD RAM address of the LCD. The upper 4-bits of lcddatin is sent first, followed by the lower 4-bits. After 7 active states, the LCD data function enters state 8 and waits there until the controller responds to the datapath status signal lcddata as logic 1 by resetting the net variable datalcd to logic 0. The 4-bit data as the register variable lcdd and the LCD control signals rslcd, rwlcd and elcd are outputted to the hierarchical Verilog top module.

The LCD datapath module lcd.v is verified by the Verilog top module lcdtest.v and controller module genlcd.v in Listing 3.15 which is in the *Chapter 3\lcdtest\s3elcdtest* folder. The complete Xilinx ISE WebPACK project uses the UCF lcdtests3esb.ucf which uncomments the signals CCLK, BTN0, LCDDAT<0>, LCDDAT<1>, LCDDAT<2>, LCDDAT<3>, LCDRS, LCDRW, and LCDE in the Spartan-3E Starter Board UCF of Listing 3.2. The four Verilog modules operate in parallel and some independently in the top module, as shown in Figure 3.3. The file download procedure is described in the Appendix.

The wire net type establishes the 8-bit and 4-bit vector connectivity for lcddatin and lcdd between the lcd.v and genlcd.v modules. The LCD external control signals LCDDAT[0], LCDDAT[1], LCDDAT[2], LCDDAT[3], LCDRS, LCDRW, and LCDE are assigned to internal datapath register signals lcdd[0], lcdd[1], lcdd[2], lcdd[3], rslcd, rwlcd, and elcd where they are then referenced in the top module lcdtest.v. Assignment of output signals to peripheral devices only in the top module provides a degree of flexibility in the FPGA hardware synthesis of an embedded system design [Navabi06].

The reset button BTN0 signal asynchronously initializes the LCD controller signals. The button is debounced by the pbdebounce.v module, as given in Listing 3.8. The genlcd.v module utilizes an FSM as the LCD controller to the LCD datapath module lcd.v to reset and clear the LCD, to set the DD RAM address to the second line at position 5 and to send the 11 character ASCII string *hello world*. The state register gstate sets each of the operations.

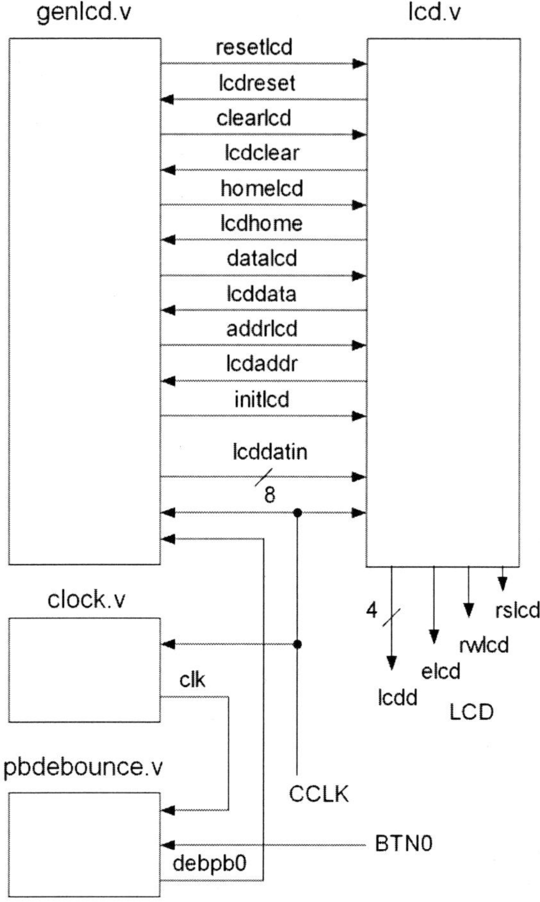


**Figure 3.3** Controller and datapath modules in the top module lcdtest.v

The LCD datapath is set to its initial state before each function by setting the register variable initlcd to logic 1 then resetting it to logic 0. Each LCD function in the datapath signals the controller when the function is completed. After 14 active states, the LCD test controller enters state 15 and waits there until a reset push button command is issued. The push button signal BTN0 is processed by the contact debounce module pbdebounce.v of Listing 3.8. The 50 MHz crystal clock signal CCLK is divided to a 1 KHz signal clk by the clock oscillator module clock of Listing 3.3.

The 11 character ASCII string *hello world* is stored in the 88-bit register variable strdata with the first character stored in the MSBs. The 8-bit character to be sent to the LCD is obtained by the Verilog decrementing, variable part select operator ( –: ), as described in Chapter 1 Verilog Hardware Description Language. This simple method of character storage in a behavioral synthesis register is not the most efficient in the fine-grained architecture of the Xilinx Spartan-3E FPGA, as described in Chapter 2 Verilog Design Automation. Either the internal FPGA block RAM or an external ROM could be used to provide more efficient data storage for embedded design applications.

As described in Chapter 2 Verilog Design Automation, the Design Utilization Summary for the top module lcdtest.v shows the use of 158 slice flip-flops (1%), 789 4-input LUTs (8%), 417 occupied slices (8%) and a total of 6383 equivalent gates in the XC3S500E Spartan-3E FPGA synthesis.

**Listing 3.15** LCD test top module lcdtest.v

```
// Spartan-3E Starter Board
// Liquid Crystal Display Test lcdtest.v
// c 2008 Embedded Design using Programmable Gate Arrays  Dennis Silage

module lcdtest (input CCLK, BTN0, output LCDRS, LCDRW, LCDE, output [3:0] LCDDAT);
wire [7:0] lcddatin;
wire [3:0] lcdd;
wire rslcd, rwlcd, elcd;

assign LCDDAT[3] = lcdd[3];
assign LCDDAT[2] = lcdd[2];
assign LCDDAT[1] = lcdd[1];
assign LCDDAT[0] = lcdd[0];

assign LCDRS = rslcd;
assign LCDRW = rwlcd;
assign LCDE = elcd;

lcd M0 (CCLK, resetlcd, clearlcd, homelcd, datalcd, addrlcd, lcdreset, lcdclear, lcdhome, lcddata,
        cdaddr, rslcd, rwlcd, elcd, lcdd, lcddatin, initlcd);
genlcd M1 (CCLK, debpb0, resetlcd, clearlcd, homelcd, datalcd, addrlcd, initlcd, lcdreset, lcdclear,
        lcdhome, lcddata, lcdaddr, lcddatin);
pbdebounce M2 (clk, BTN0, debpb0);
clock M3 (CCLK, 25000, clk);

endmodule

module genlcd(input CCLK, debpb0, output reg resetlcd, output reg clearlcd, output reg homelcd,
        output reg datalcd, output reg addrlcd, output reg initlcd, input lcdreset,
        lcdclear, input lcdhome, lcddata, lcdaddr, output reg [7:0] lcddatin);

reg [3:0] gstate;                       // state register
reg [87:0] strdata = "hello world";     // ASCII string data
integer i;

always@(posedge CCLK)       // master clock event driven
    begin
        if (debpb0 == 1)    // debounced push button reset
            begin
                resetlcd = 0;
                clearlcd = 0;
                homelcd = 0;
                datalcd = 0;
                gstate = 0;
            end
        else
```

```
case (gstate)
    0:  begin
            initlcd = 1;            // initialize LCD
            gstate = 1;
        end
    1:  begin
            initlcd = 0;
            gstate = 2;
        end
    2:  begin
            resetlcd = 1;           // reset LCD
            if (lcdreset == 1)
                begin
                    resetlcd = 0;
                    gstate = 3;
                end
        end
    3:  begin
            initlcd = 1;
            gstate = 4;
        end
    4:  begin
            initlcd = 0;
            gstate = 5;
        end
    5:  begin
            clearlcd = 1;           // clear LCD
            if (lcdclear == 1)
                begin
                    clearlcd = 0;
                    gstate = 6;
                end
        end
    6:  begin
            initlcd = 1;
            gstate = 7;
        end
    7:  begin
            initlcd = 0;
            gstate = 8;
        end
    8:  begin                       // DD RAM address 44h
            lcddatin[7:0] = 8'b01000100;
            addrlcd = 1;
            if (lcdaddr == 1)
                begin
                    addrlcd = 0;
                    gstate = 9;
                end
        end
```

```
            9:  begin
                    initlcd = 1;
                    gstate = 10;
                end
            10: begin
                    initlcd = 0;
                    i = 87;                 // character bit count
                    gstate = 11;
                end
            11: begin                       // display string
                    lcddatin[7:0] = strdata[i-:8];
                    datalcd = 1;
                    if (lcddata == 1)
                        begin
                            datalcd = 0;
                            gstate = 12;
                        end
                end
            12: begin
                    initlcd = 1;
                    gstate = 13;
                end
            13: begin
                    initlcd = 0;
                    gstate = 14;
                end
            14: begin                           // loop until finished
                    i = i – 8;
                    if (i < 0)
                        gstate = 15;
                    else
                        gstate = 11;
                end
            15: gstate = 15;
            default: gstate = 15;
        endcase
    end

endmodule
```

The elapsed time project designed for the seven segment display of the Basys Board in Listing 3.13 can be modified to execute on the LCD of the Spartan-3E Starter Board. This project is the Verilog behavioral synthesis example that is used to illustrate the Xilinx ISE WebPACK EDA software tool in Chapter 2 Verilog Design Automation.

The top module s3eelapsedtime.v in Listing 3.16 is also in the *Chapter 3\elapsedtime \s3eelapsedtime* folder. The complete Xilinx ISE WebPACK project uses the UCF elapsedtimes3esb.ucf which uncomments the signals CCLK, BTN0, LCDDAT<0>, LCDDAT<1>, LCDDAT<2>, LCDDAT<3>, LCDRS, LCDRW, and ELCD in the Spartan-3E Starter Board UCF of Listing 3.2. The four Verilog modules operate in parallel and some independently in the top module, as shown in Figure 3.4. The file download procedure is described in the Appendix.

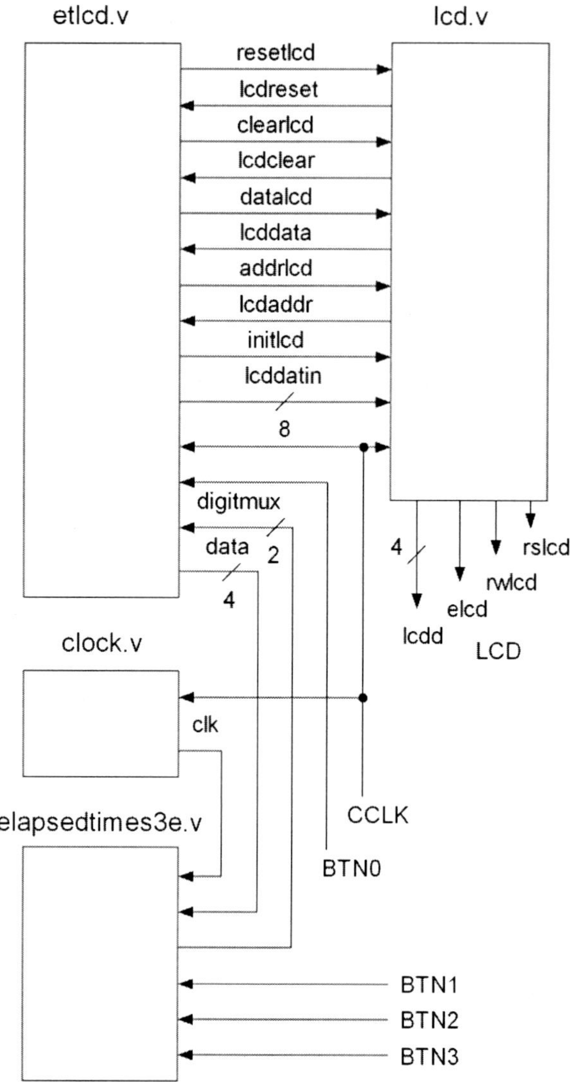


**Figure 3.4** Controller and datapath modules in the top module s3eelapsedtime.v

The wire net type establishes the 8-bit , 4-bit and 2-bit vector connectivity for lcddatin, data, lcdd and digitmux between the lcd.v, elapsedtime.v and etlcd.v modules. The LCD external signals LCDDAT[0], LCDDAT[1], LCDDAT[2], LCDDAT[3], LCDRS, LCDRW, and LCDE are assigned to internal controller signals lcdd[0], lcdd[1], lcdd[2], lcdd[3], rslcd, rwlcd, and elcd since they are only referenced in the top module s3eelapsedtime.v. The reset button BTN0 signal asynchronously initializes the LCD controller signals [Ciletti04]. The etlcd.v module is similar to the genlcd.v of Listing 3.15 and only a partial listing is provided here. The complete Verilog HDL source for etlcd.v is available and the file download procedure is described in the Appendix.

The etlcd.v utilizes an FSM as the LCD controller to the LCD datapath module lcd.v to reset and clear the LCD and set the DD RAM address to 46h or the second line at position 7 as listed in Table 3.1. The state register gstate sets each of the operations.

As described in Chapter 2 Verilog Design Automation Verilog Design Automation, the Design Utilization Summary for the top module s3eelapsedtime.v shows the use of 160 slice flip-flops

(1%), 609 4-input LUTs (6%), 381 occupied slices (8%) and a total of 5623 equivalent gates in the XC3S500E Spartan-3E FPGA synthesis.

The module elaspedtimes3e.v in Listing 3.17 is in the *Chapter 3\elapsedtime \s3eelapsedtime* folder. This module for the LCD on the Spartan-3E Starter Board is similar to the elapsedtime.v module in Listing 3.13 for the seven segment display on the Basys Board. The salient differences in the two modules are that only one 100 Hz clock signal clk from the clock.v module for the elapsed time process is required and there is no required set decimal point signal setdp or signal digit. However, an input 2-bit register signal digitmux is used here to select which digit is returned to the controller module etlcd.v. This elapsed time project illustrates the relative ease of design reuse of modules in Verilog behavioral synthesis for embedded design [Botros06].

**Listing 3.16** Elapsed time top module for the Spartan-3E Starter Board s3eelapsedtime.v

```
// Spartan-3E Starter Board
// Elapsed Time Test s3eelapsedtime.v
// c 2008 Embedded Design using Programmable Gate Arrays  Dennis Silage

module s3eelapsedtime (input CCLK, BTN0, BTN1, BTN2, BTN3, output LCDRS, LCDRW,
                        LCDE, output [3:0] LCDDAT);

wire [3:0] data;
wire [1:0] digitmux;
wire [7:0] lcddatin;
wire [3:0] lcdd;
wire rslcd, rwlcd, elcd;

assign LCDDAT[3] = lcdd[3];
assign LCDDAT[2] = lcdd[2];
assign LCDDAT[1] = lcdd[1];
assign LCDDAT[0] = lcdd[0];

assign LCDRS = rslcd;
assign LCDRW = rwlcd;
assign LCDE = elcd;

clock M0 (CCLK, 250000, clk);        // 100 Hz
lcd M1 (CCLK, resetlcd, clearlcd, homelcd, datalcd, addrlcd, lcdreset, lcdclear, lcdhome, lcddata,
        lcdaddr, rslcd, rwlcd, elcd, lcdd, lcddatin, initlcd);
elapsedtimes3e M2 (clk, BTN1, BTN2, BTN3, digitmux, data);
etlcd M3 (CCLK, BTN0, resetlcd, clearlcd, homelcd, datalcd, addrlcd, initlcd, lcdreset, lcdclear,
        lcdhome, lcddata, lcdaddr, lcddatin, digitmux, data);

endmodule

module etlcd(input CCLK, BTN0, output reg resetlcd, output reg clearlcd, output reg homelcd,
            output reg datalcd, output reg addrlcd, output reg initlcd, input lcdreset, lcdclear,
            input lcdhome, lcddata, lcdaddr, output reg [7:0] lcddatin,
            output reg [1:0] digitmux, input [3:0] data);

reg [4:0] gstate;        // state register
```

```
always@(posedge CCLK)
      begin
            if (BTN0 == 1)
                  begin
                        resetlcd = 0;
                        clearlcd = 0;
                        homelcd = 0;
                        datalcd = 0;
                        gstate = 0;
                  end
            else
                  case (gstate)
                        0:    begin
                                    initlcd = 1;
                                    gstate = 1;
                              end
{LCD initialize, clear display and set display address continues in states 1 through 10}
                        11:   begin
                                    lcddatin[7:4] = 3;      // 30h
                                    digitmux = 3;           // sec10
                                    lcddatin[3:0] = data[3:0];
                                    datalcd = 1;
                                    if (lcddata == 1)
                                          begin
                                                datalcd = 0;
                                                gstate = 12;
                                          end
                              end
                        12:   begin
                                    initlcd = 1;
                                    gstate = 13;
                              end
                        13:   begin
                                    initlcd = 0;
                                    gstate = 14;
                              end
                        14:   begin
                                    digitmux = 2;           // sec1
                                    lcddatin[3:0] = data[3:0];
                                    datalcd = 1;
                                    if (lcddata == 1)
                                          begin
                                                datalcd = 0;
                                                gstate = 15;
                                          end
                              end
                        15:   begin
                                    initlcd = 1;
                                    gstate = 16;
                              end
```

```
16:  begin
         initlcd = 0;
         gstate = 17;
     end
17:  begin
         lcddatin[7:0] = 58;    // ASCII :
         datalcd = 1;
         if (lcddata == 1)
             begin
                 datalcd = 0;
                 gstate = 18;
             end
     end
18:  begin
         initlcd = 1;
         gstate = 19;
     end
19:  begin
         initlcd = 0;
         gstate = 20;
     end
20:  begin
         lcddatin[7:4] = 3;     // 30h
         digitmux = 1;          // csec10
         lcddatin[3:0] = data[3:0];
         datalcd = 1;
         if (lcddata == 1)
             begin
                 datalcd = 0;
                 gstate = 21;
             end
     end
21:  begin
         initlcd = 1;
         gstate = 22;
     end
22:  begin
         initlcd = 0;
         gstate = 23;
     end
23:  begin
         digitmux = 0;          // csec100
         lcddatin[3:0] = data[3:0];
         datalcd = 1;
         if (lcddata == 1)
             begin
                 datalcd = 0;
                 gstate = 6;
             end
     end
```

```
                default: gstate = 0;
            endcase
    end

endmodule
```

**Listing 3.17** Elapsed time module for the Spartan-3E Starter Board elapsedtimes3e.v

```
// Spartan-3E Starter Board
// Elapsed Time Module elpasedtimes3e.v
// c 2008 Embedded Design using Programmable Gate Arrays  Dennis Silage

module elapsedtimes3e (input clk, BTN1, BTN2, BTN3, input [1:0] digitmux, output reg [3:0] data);

reg startstop;              // start or stop

reg [3:0] csec100;          // seconds 1/100s
reg [3:0] csec10;           // seconds 1/10s
reg [3:0] sec1;             // seconds 1s
reg [3:0] sec10;            // seconds 10s

always@(digitmux)           // digit multiplex even driven
    begin
        case (digitmux)
            0:  data[3:0] = csec100;
            1:  data[3:0] = csec10;
            2:  data[3:0] = sec1;
            3:  data[3:0] = sec10;
        endcase
    end

always@(posedge clk)        // local clock event driven
    begin
        if (BTN2 == 1) // clear and stop
            begin
                startstop = 0;
                csec100 = 0;
                csec10 = 0;
                sec1 = 0;
                sec10 = 0;
            end

        if (BTN1 == 1) // stop
            startstop=0;
        if (BTN3 == 1) // start
            startstop=1;

        if (startstop == 1)
            begin
                csec100 = csec100 + 1;
```

```
                    if (csec100 > 9)
                        begin
                            csec100 = 0;
                            csec10 = csec10 + 1;
                        end
                    if (csec10 > 9)
                        begin
                            csec10 = 0;
                            sec1 = sec1 + 1;
                        end
                    if (sec1 > 9)
                        begin
                            sec1 = 0;
                            sec10 = sec10 + 1;
                        end
                    if (sec10>9)
                        sec10=0;
                end
    end

endmodule
```

## PS/2 Keyboard Port

The hard-wired accessory PS/2 keyboard port on the Basys Board or the Spartan-3E Starter Board provides ASCII character input in applications of the FPGA. A single PS/2 6-pin mini-DIN connector functions either as the input for a standard PC keyboard or mouse (but not at the same time).

The keyboard (or mouse) uses the two-wire (data and clock) PS/2 serial bus protocol to communicate with the host processor. The keyboard sends 11-bit data (PS2DAT) on the synchronous negative edge of the PS/2 clock (PS2CLK). The data is initially held as logic 1 and the first bit at the negative edge of PS2CLK is the logic 0 start bit. Next, the 8-bit data keyboard scan code is sent with the LSB sent first followed by the odd parity bit and a stop bit that is logic 1. The number of logic 1s in the 8-bit data and the parity bit must be an odd number if the transmission is correct.

The module keyboard.v in Listing 3.18, which is also in the *Chapter 3\peripherals* folder, is event driven on the negative edge of the input signal PS2CLK. A Verilog case statement uses the internal 4-bit register count as the state register of a finite state machine (FSM) to assess which of the 11 data bits are sent. The register count is initially set to 0 in a declaration [Lee06]. The output data available register dav is initially set to logic 0 by the start bit and a logic 1 by the stop bit. On the eleventh data bit the 4-bit state register count is reset to zero. The scan code of the depressed key is outputted in the 8-bit register kbddata with the separate register parity for the parity bit.

**List 3.18** PS/2 keyboard module keyboard.v

```
// Basys Board and Spartan-3E Starter Board
// PS/2 Keyboard keyboard.v
// c 2008 Embedded Design using Programmable Gate Arrays  Dennis Silage

module keyboard (input PS2CLK, input PS2DAT, output reg dav, output reg [7:0] kbddata,
                 output reg parity);

reg [3:0] count = 0;            // keyboard data bit count
```

```
always@(negedge PS2CLK)    // PS2 clock event driven
    begin
        count = count + 1;
        case (count)
            1:  dav = PS2DAT;
            2:  kbddata[0] = PS2DAT;
            3:  kbddata[1] = PS2DAT;
            4:  kbddata[2] = PS2DAT;
            5:  kbddata[3] = PS2DAT;
            6:  kbddata[4] = PS2DAT;
            7:  kbddata[5] = PS2DAT;
            8:  kbddata[6] = PS2DAT;
            9:  kbddata[7] = PS2DAT;
            10: parity = PS2DAT;
            11: begin
                    count = 0;
                    dav =PS2DAT;        // data available
                end
        endcase
    end

endmodule
```

The PS/2 keyboard scan code of a key is not the ASCII character code and Table 3.3 lists their relationship in hexadecimal. If a key is depressed and held the PS/2 keyboard repeats the scan code approximately every 100 msec. Upon release of a key the keyboard first sends F0 hexadecimal (F0h) then a repeat of the scan code.

The left and right side shift keys actually have different scan codes and the application must keep track of the depression and release of the shift key for the upper case alphabetical characters and alternate symbols. The extended keys also send E0h before their scan code, even though the scan code for all of the keys are unique. If depressed and held the extended keys sends E0h and F0h then a repeat of the scan code.

The PS/2 keyboard module keyboard.v is verified by the Verilog top module bakeyboardtest.v for the Basys Board in Listing 3.19, which is also in the *Chapter 3\keyboardtest \bakeyboardtest* folder, and the top module s3ekeyboardtest.v for the Spartan-3E Starter Board in Listing 3.20, which is also in the *Chapter 3\keyboardtest\s3ekeyboardtest* folder.

The complete Xilinx ISE WebPACK projects use UCFs keyboardtestbasys.ucf and keyboardtests3esb.ucf which uncomment the signals CCLK, PS2CLK, PS2DAT, BTN3, and the cathode and anode signals in the Basys Board UCF of Listing 3.1 or the LCD signals in the Spartan-3E Starter Board UCF of Listing 3.2. The four Verilog modules operate in parallel and some independently in the top module, as shown in Figure 3.5 for the Basys Board and Figure 3.6 for the Spartan-3E Starter Board. The file download procedure is described in the Appendix.

The wire net type in the bakeyboardtest.v module for the Basys Board establishes the 8-bit, 3-bit and 2-bit vector connectivity for kbddata, digit, and data between the keyboard.v, sevensegment.v and checkkbd.v modules, as given in Listing 3.19. The clock signal clk is outputted from the clock.v module at a frequency of 1 kHz and inputted to the checkkbd.v and sevensegment.v modules.

The module checkkbd.v is *event driven* on the positive edge of the data available signal dav from the keyboard.v module and the posedge edge of 1 kHz signal clk from the clock.v module. These events are independent and operate in parallel but are coordinated by the digit available register digitav, as described in Chapter 1 Verilog Hardware Description Language.

The data available event selects the most significant digit register msdigit and least significant digit register lsdigit from the 8-bit register kbddata and checks the parity of the keyboard data with the

Verilog exclusive or (xor) bit-wise operation ( ^ ) on the 8-bit keyboard data and the parity data. The Verilog integer register variable i is used to sequentially index through the register variable kbddata to form the check parity register variable chkparity, which is a logic 1 for a correct data transmission.

**Table 3.3** Hexadecimal scan code and ASCII code for the PS/2 keyboard characters

| Character | Scan Code | ASCII Code | Character | Scan Code | ASCII Code |
|---|---|---|---|---|---|
| a A | 1C | 61 41 | – _ | 4E | 2D 5F |
| b B | 32 | 62 42 | = + | 55 | 3D 2B |
| c C | 21 | 63 43 | BS | 66 | 08 |
| d D | 23 | 64 44 | TAB | 0D | 09 |
| e E | 24 | 65 45 | [ { | 54 | 5B 7B |
| f F | 2B | 66 46 | ] } | 5B | 5D 7D |
| g G | 34 | 67 47 | \ \| | 5D | 5C 7C |
| h H | 33 | 68 48 | CAPS LOCK | 58 | |
| i I | 43 | 69 49 | ; : | 4C | 3B 3A |
| j J | 3B | 6A 4A | ' " | 52 | 27 22 |
| k K | 42 | 6B 4B | ENTER | 5A | |
| l L | 4B | 6C 4C | Left SHIFT | 12 | |
| m M | 3A | 6D 4D | , < | 41 | 2C 3C |
| *n N* | *31* | *6E 4E* | . > | 49 | 2E 3E |
| *o* O | 44 | 6F 4F | / ? | 4A | 2F 3F |
| *p* P | 4D | 70 50 | Right SHIFT | 59 | |
| q Q | 15 | 71 51 | Left CTRL | 14 | |
| r R | 2D | 72 52 | Left ALT | 11 | |
| s S | 1B | 73 53 | SPACE | 29 | 20 |
| t T | 2C | 74 54 | Right ALT | E0 11 | |
| u U | 3C | 75 55 | Right CTRL | E0 14 | |
| v V | 2A | 76 56 | ESC | 76 | 1B |
| w W | 1D | 77 57 | F1 | 05 | |
| x X | 22 | 78 58 | F2 | 06 | |
| y Y | 35 | 79 59 | F3 | 04 | |
| z Z | 1A | 7A 5A | F4 | 0C | |
| ` ~ | 0E | 60 7E | F5 | 03 | |
| 1 ! | 16 | 31 21 | F6 | 0B | |
| 2 @ | 1E | 32 22 | F7 | 83 | |
| 3 # | 26 | 33 23 | F8 | 0A | |
| 4 $ | 25 | 34 24 | F9 | 01 | |
| 5 % | 2E | 35 25 | F10 | 09 | |
| 6 ^ | 36 | 36 5E | F11 | 78 | |
| 7 & | 3D | 37 26 | F12 | 07 | |
| 8 * | 3E | 38 2A | ↑ | E0 75 | |
| 9 ( | 46 | 39 28 | → | E0 74 | |
| 0 ) | 45 | 30 29 | ← | E0 6B | |
| | | | ↓ | E0 72 | |

The clock event displays the keyboard hexadecimal scan code on the two rightmost and the parity and the check parity bits with the decimal point enabled on the leftmost of the seven segment display digits on the Basys Board. The set decimal point register setdp of the seven segment display is the most significant bit of the 2-bit state register digitmux. The 1-bit check parity and parity registers chkparity and parity are the least significant bits of the 4-bit segment display register data and the remaining three bits are set to logic 0.

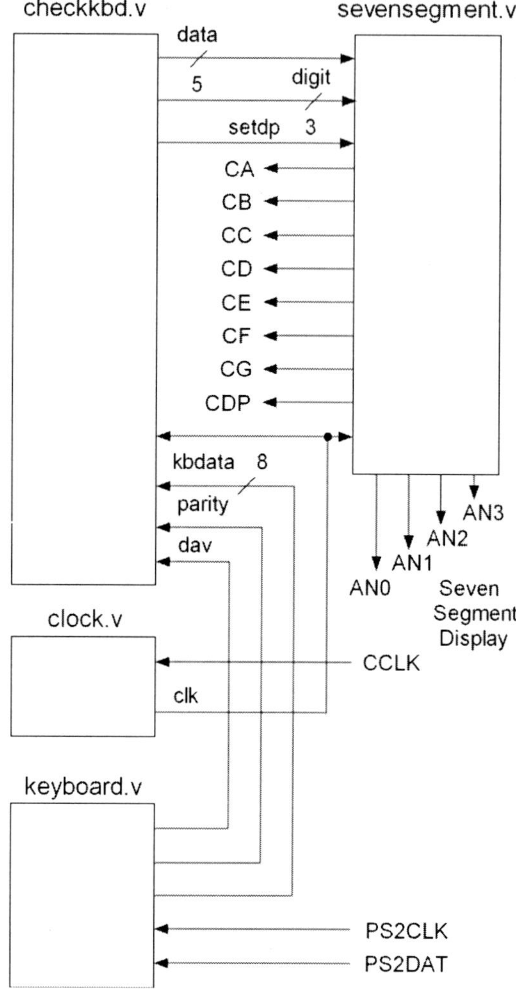


**Figure 3.5** Controller and datapath modules in the top module bakeyboardtest.v

The PS/2 keyboard hexadecimal scan code, the parity and the check parity bits are displayed on the LCD for the Spartan-3E Starter Board. Assignment of output signals to peripheral devices only in the top module provides a degree of flexibility in the FPGA hardware synthesis of an embedded design.

This Spartan-3E Starter Board project demonstrates the *design reuse* concept with the inclusion of the lcd.v module, as given in Listing 3.14, with the keyboard.v module. Also the module genlcd.v in Listing 3.20 is similar to the genlcd.v of Listing 3.15 and only a partial listing is provided here. The genlcd.v module in design reuse is expanded to check the parity of the keyboard data and replaces the checkkbd.v module of Listing 3.19.

**Listing 3.19** Keyboard test top module for the Basys Board bakeyboardtest.v

```
// Basys Evaluation Board
// PS/2 Keyboard Test bakeyboardtest.v
// c 2008 Embedded Design using Programmable Gate Arrays  Dennis Silage

module bakeyboardtest (input CCLK, PS2CLK, PS2DAT, output AN0, AN1, AN2, AN3,
                        output CA, CB, CC, CD, CE, CF, CG, CDP);

wire [4:0] data;
wire [2:0] digit;
wire [7:0] kbddata;

clock M0 (CCLK, 25000, clk);          // 1 kHz
keyboard M1 (PS2CLK, PS2DAT, dav, kbddata, parity);
sevensegment M2(clk, data, digit, setdp, AN0, AN1, AN2, AN3, CA, CB, CC, CD, CE, CF, CG,
                    CDP);
checkkbd M3 (clk, data, digit, setdp, dav, kbddata, parity);

endmodule

module checkkbd (input clk, output reg [4:0] data, output reg [2:0] digit, output reg setdp, input dav,
                    input [7:0] kbddata, input parity);

reg [1:0] digitmux = 0;          // digit multiplexer, initialize
reg [3:0] msdigit;               // most significant digit
reg [3:0] lsdigit;               // least significant digit
reg chkparity, dataav;           // check parity, data available

integer i;

always@(posedge dav)             // data available event driven
    begin                        // calculate parity check
        dataav = 0;
        msdigit[3:0] <= kbddata[7:4];
        lsdigit[3:0] <= kbddata[3:0];
        chkparity = kbddata[0] ^ kbddata[1];
        for (i = 2; i <= 7; i = i + 1)
            chkparity = kbddata[i] ^ chkparity;
        chkparity = parity ^ chkparity;
        datav = 1;               // data available
    end

always@(posedge clock)           // local clock event driven
    begin
        if (digitav)             // data available
            begin
                digitmux = digitmux + 1;
                setdp = digitmux[1];
                data[4] <= 0;
```

```
                    case (digitmux)
                         0:   begin
                                   data[3:0] <= lsdigit;
                                   digit <= 1;
                              end
                         1:   begin
                                   data[3:0] <= msdigit;
                                   digit <= 2;
                              end
                         2:   begin
                                   data[0] <= chkparity;
                                   data[3:1] <= 3'b000;
                                   digit <= 3;
                               end
                         3:   begin
                                   data[0] <= parity;
                                   data[3:1] <= 3'b000;
                                   digit <= 4;
                                end
                    endcase
               end
     end

endmodule
```

As described in Chapter 2 Verilog Design Automation Verilog Design Automation, the Design Utilization Summary for the top module bakeyboardtest.v shows the use of 136 slice flip-flops (3%), 542 4-input LUTs (12%), 328 occupied slices (12%) and a total of 4624 equivalent gates in the XC3S100 Spartan-3E FPGA synthesis.

**Listing 3.20** Keyboard test top and parity check modules for the Spartan-3E Starter Board s3ekeyboardtest.v

```
// Spartan-3E Starter Board
// PS/2 Keyboard Test s3ekeyboardtest.v
// c 2008 Embedded Design using Programmable Gate Arrays  Dennis Silage

module s3ekeyboardtest (input CCLK, PS2CLK, PS2DAT, BTN0, output LCDRS, LCDRW, LCDE,
                         output [3:0] LCDDAT);
wire [7:0] kbddata;
wire [3:0] lcdd;
wire [7:0] lcddatin;
wire rslcd, rwlcd, elcd;

assign LCDDAT[3] = lcdd[3];
assign LCDDAT[2] = lcdd[2];
assign LCDDAT[1] = lcdd[1];
assign LCDDAT[0] = lcdd[0];

assign LCDRS = rslcd;
assign LCDRW = rwlcd;
assign LCDE = elcd;
```

```
keyboard M0 (PS2CLK, PS2DAT, dav, kbddata, parity);
lcd M1 (CCLK, resetlcd, clearlcd, homelcd, datalcd, addrlcd, lcdreset, lcdclear, lcdhome, lcddata,
        lcdaddr, rslcd, rwlcd, elcd, lcdd, lcddatin, initlcd);
genlcd M2 (CCLK, BTN0, resetlcd, clearlcd, homelcd, datalcd, addrlcd, initlcd, lcdreset, lcdclear,
        lcdhome, lcddata, lcdaddr, lcddatin, dav, parity, kbddata);

endmodule

module genlcd (input CCLK, BTN0, output reg resetlcd, output reg clearlcd, output reg homelcd,
        output reg datalcd, output reg addrlcd, output reg initlcd, input lcdreset, lcdclear,
        input lcdhome, lcddata, lcdaddr, output reg [7:0] lcddatin, input dav, parity,
        input [7:0] kbddata);

reg [4:0] gstate;       // state register
reg chkparity;          // parity check
integer i;

always@(posedge dav)            // data available event driven
    begin                       // calculate parity check
        chkparity = kbddata[0] ^ kbddata[1];
        for (i = 2; i <= 7; i = i + 1)
            chkparity = kbddata[i] ^ chkparity;
        chkparity = parity ^ chkparity;
    end

always@(posedge CCLK)           // master clock event driven
    begin
        if (BTN0 == 1)          // reset
            begin
                resetlcd = 0;
                clearlcd = 0;
                homelcd = 0;
                datalcd = 0;
                gstate = 0;
            end
        else
            case (gstate)
                0:  begin
                        initlcd = 1;        // initialize LCD
                        gstate = 1;
                    end
{LCD initialize, clear display and set display address continues in states 1 through 10}
                11: begin
                        lcddatin[7:4] = 3;          // ASCII digit 3Xh
                        lcddatin[3:0] = kbddata[7:4];
                        datalcd = 1;
                        if (lcddata == 1)
                            begin
                                datalcd = 0;
                                gstate = 12;
                            end
                    end
```

```
12: begin
        initlcd = 1;
        gstate = 13;
    end
13: begin
        initlcd = 0;
        gstate = 14;
    end
14: begin
        lcddatin[7:4] = 0;
        lcddatin[3:0] = kbddata[3:0];
        if (lcddatin < = 9)
            lcddatin[7:4] = 3;      // ASCII digit 3Xh
        else
            lcddatin = lcddatin + 55;   // ASCII A, B...
        datalcd = 1;
        if (lcddata == 1)
            begin
                datalcd = 0;
                gstate = 15;
            end
    end
15: begin
        initlcd = 1;
        gstate = 16;
    end
16: begin
        initlcd = 0;
        gstate = 17;
    end
17: begin
        lcddatin[7:0] = 58;         // ASCII :
        datalcd = 1;
        if (lcddata == 1)
            begin
                datalcd = 0;
                gstate = 18;
            end
    end
18: begin
        initlcd = 1;
        gstate = 19;
    end
19: begin
        initlcd = 0;
        gstate = 20;
    end
20: begin
        lcddatin[7:4] = 3;          // 30h
        lcddatin[3:1] = 3'b000;
        lcddatin[0] = chkparity;
        datalcd = 1;
```

```
                        if (lcddata == 1)
                            begin
                                datalcd = 0;
                                gstate = 21;
                            end
                    end
                21: begin
                        initlcd = 1;
                        gstate = 22;
                    end
                22: begin
                        initlcd = 0;
                        gstate = 23;
                    end
                23: begin
                        lcddatin[7:0] = 58;         // ASCII :
                        datalcd = 1;
                        if (lcddata == 1)
                            begin
                                datalcd = 0;
                                gstate = 24;
                            end
                    end
                24: begin
                        initlcd = 1;
                        gstate = 25;
                    end
                25: begin
                        initlcd = 0;
                        gstate = 26;
                    end
                26: begin
                        lcddatin[7:4] = 3;          // 30h
                        lcddatin[3:1] = 3'b000;
                        lcddatin[0] = parity;
                        datalcd = 1;
                        if (lcddata == 1)
                            begin
                                datalcd = 0;
                                gstate = 6;
                            end
                    end
                default: gstate = 0;
            endcase
    end

endmodule
```

As described in Chapter 2 Verilog Design Automation Verilog Design Automation, the Design Utilization Summary for the top module s3ekeyboardtest.v shows the use of 127 slice flip-flops (1%), 544 4-input LUTs (5%), 325 occupied slices (6%) and a total of 4563 equivalent gates in the XC3S500E Spartan-3E FPGA synthesis. The PS/2 keyboard scan codes are converted to ASCII characters for serial data transmission in a complete Xilinx ISE WebPACK project in Chapter 4 Digital Signal Processing, Communications and Control.

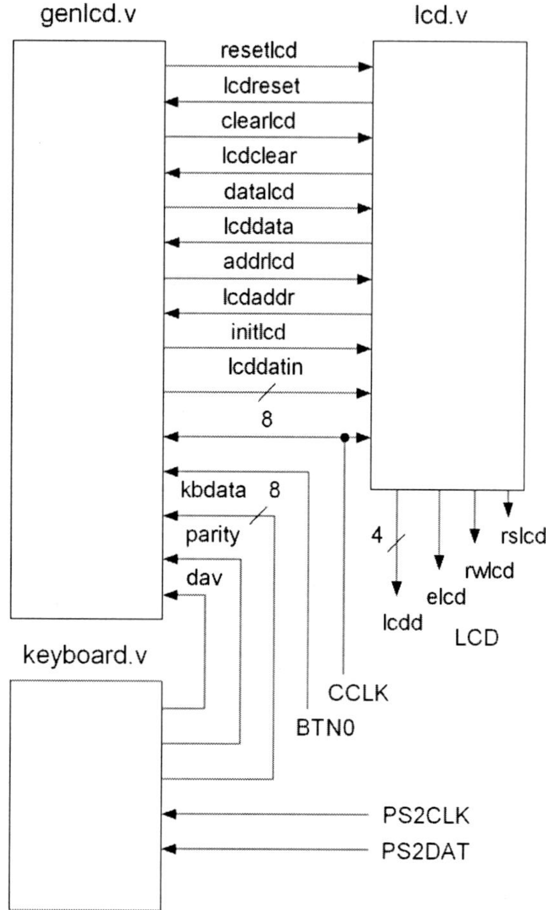


**Figure 3.6** Controller and datapath modules in the top module s3ekeyboardtest.v

## PS/2 Mouse Port

The hard-wired accessory PS/2 mouse port on the Basys Board or the Spartan-3E Starter Board provides positional information and button input in applications of the FPGA. A single PS/2 6-pin mini-DIN connector functions either as the input for a mouse or a standard PC keyboard (but not at the same time). The mouse (or keyboard) uses the two-wire (data and clock) PS/2 serial bus protocol to communicate *bidirectionally* with the Basys Board or the Spartan-3E Starter Board host processor and the clock (PS2CLK) and data (PS2DAT) signal is therefore defined with the Verilog type inout. The PS/2 mouse always generates the clock but, if the host is to send data, it must first inhibit communication by setting the clock to logic 0.

The PS/2 mouse enters the reset mode at initialization (power-up) and, unlike the PS/2 keyboard, data reporting is disabled. Although there are a variety of commands and resulting modes of operation for the PS/2 mouse, at the minimum the host should set the mouse to Enable Data Reporting with the 8-bit command code F4h.

The host first issues a Request to Send to the PS/2 mouse by setting the clock to logic 0 for at least 100 microseconds (μsec), setting the data to logic 0 and then setting the clock to logic 1. The host then sends an 11-bit data packet on the negative edge of the synchronous PS/2 clock consisting of a start bit (logic 0), 8-bit data command (F4h), an odd parity bit (logic 0) and a stop bit (logic 1). The PS/2 mouse responses with Acknowledge (FA H) as data, but this response can be ignored.

The PS/2 mouse then sends three 11-bit data packets (PS2DAT) on the synchronous negative edge of the PS/2 clock (PS2CLK) when either movement or button depression is sensed. The data is initially held as logic 1 and the first bit at the negative edge of PS2CLK is the start bit (logic 0). Next, the PS/2 mouse 8-bit status data is sent followed by the odd parity bit and a stop bit (logic 1). The number of logic 1s in the 8-bit data and the parity bit must be an odd number if the transmission is correct. The X direction and Y direction 11-bit data packets are sent next.

Table 3.4 shows the contents of the 8-bit status and the X directional and Y directional data without the accompanying start, parity and stop bits. The L and R bits of the status are logic 1 when the left or right mouse button is depressed. The XS and YS bits indicate the sign of the mouse movement with respect to the last quiescent position and logic 0 indicates a positive sign. The XV and YV bits are logic 1 if an overflow in the movement in the x or y direction occurs. Finally, the X0 through X7 and Y0 through Y7 bits are the 8-bit mouse movement data.

**Table 3.4** PS/2 mouse 8-bit data packets

| | | | | | | | | |
|---|---|---|---|---|---|---|---|---|
| Status | L | R | 0 | 1 | XS | YS | XV | YV |
| X Direction | X0 | X1 | X2 | X3 | X4 | X5 | X6 | X7 |
| Y Direction | Y0 | Y1 | Y2 | Y3 | Y4 | Y5 | Y6 | Y7 |

The module mouse.v, which is in the *Chapter 3\peripherals* folder, is given in Listing 3.21 and utilizes three finite state machines (FSM). The register reset is set to logic 1 and cleared to logic 0 on the positive edge of the crystal clock oscillator (CCLK), but only initially by the first FSM using the 2-bit state register resstate.

After reset has occurred, the second FSM uses the 5-bit sent data state register sstate to send the command Enable Data Reporting (F4h) from the host to the PS/2 mouse. Finally, the third FSM uses the 6-bit received data state register rstate to report the movement of the mouse or button depression. The data is read on the negative edge of the PS/2 clock input signal PS2CLK.

The 13-bit register count is used to set the PS/2 clock to logic 0 for at least 100 μsec. The 50 MHz crystal clock has a period of 0.02 μsec and a count of 5500 (5500 × 0.02 μsec = 110 μsec) assures that this occurs. The host then releases the PS/2 clock by setting it to a high impedance state (z).

A Verilog case statement uses the 5-bit state register sstate to send the Enable Data Reporting command from the host and to recognize the Acknowledge response from the PS/2 mouse. The 5-bit register sstate utilizes 22 states for the transmission and reception of these two 11-bit Enable Data Reporting and Acknowledge data packets.

A Verilog case statement uses the 6-bit state register rstate to assess which of the 33 data bits are sent. The output data available register dav is initially set to logic 0 by the start bit of the first data packet and to logic 1 by the stop bit of the third data packet. The PS/2 mouse X and Y directional data are outputted as a 9-bit signed (two's complement) binary number distributed in the 8-bit X direction data register mousexdata and the 8-bit Y direction data register mouseydata and the XS and YS sign bits, as listed in Table 3.4.

The sign bits are outputted in the 2-bit register sign. The 2-bit overflow register ovf and the two bit odd parity register parity are also provided for the PS/2 mouse X and Y directional data. Finally, the depression of the two PS/2 mouse buttons is outputted in the 2-bit register button.

**Listing 3.21** Mouse module mouse.v

```
// Basys Board and Spartan-3E Starter Board
// PS/2 Mouse mouse.v
// c 2008 Embedded Design using Programmable Gate Arrays  Dennis Silage

module mouse (input CCLK, inout PS2CLK, inout PS2DAT, output reg dav, output reg [2:0] parity,
                        output reg [1:0] button, output reg [1:0] ovf, output reg [1:0] sign,
                        output reg [7:0] mousexdata, output reg [7:0] mouseydata);

reg [12:0] count;          // delay counter
reg [1:0] resstate;        // reset state
reg [4:0] sstate;          // sent data state
reg [5:0] rstate;          // received data state
reg clkps2;
reg datps2;
reg reset;

assign PS2CLK = clkps2; // PS2 clock
assign PS2DAT = datps2; // PS2 data

always@(posedge CCLK)        // master clock event driven
    begin
        resstate <= resstate + 1;
        if (resstate == 1)
            reset = 1;
        if (resstate == 3)
            begin
                resstate <= 2;
                reset = 0;
            end
    end

always@(count or reset)   // count or reset event driven
    begin
        if (reset == 1)
            clkps2 = 1'bz;
        else if (count <= 5500 && reset == 0)
            clkps2 = 0;
        else
            clkps2 = 1'bz;
    end

always@(posedge CCLK or posedge reset)         // master clock or
    begin                                      // reset event driven
        if (reset == 1)
            count <= 0;
        else if (count == 6000)
            count <= count;
        else
            count <= count + 1;
    end
```

```
always@(count or reset)   // count or reset event driven
    begin
        if (reset == 1)
            datps2 = 1'bz;
        else if (count > 5000 && sstate == 0)
            datps2 = 0;              // start bit
        else if (sstate == 1)        // F4h
            datps2 = 0;
        else if (sstate == 2)
            datps2 = 0;
        else if (sstate == 3)
            datps2 = 1;
        else if (sstate == 4)
            datps2 = 0;
        else if (sstate == 5)
            datps2 = 1;
        else if (sstate == 6)
            datps2 = 1;
        else if (sstate == 7)
            datps2 = 1;
        else if (sstate == 8)
            datps2 = 1;
        else if (sstate == 9)        // parity
            datps2 = 0;
        else if (sstate == 10)       // stop bit
            datps2 = 1;
        else
            datps2 = 1'bz;
    end

always@(negedge PS2CLK or posedge reset)
    begin
        if (reset == 1)
            sstate <= 0;
        else if (sstate <= 21)
            sstate <= sstate + 1;
        else if (sstate == 22)
            sstate <= 22;
        else
            sstate <= 0;
    end

always@(negedge PS2CLK)    // read mouse data
    begin
        if (reset == 1)
            rstate = 0;
        else if (sstate == 22)
            begin
                rstate = rstate + 1;
                case (rstate)
                    1:   dav=0;
                    2:   button[1] = PS2DAT;          // x button
```

```
                    3:  button[0] = PS2DAT;         // y button
                    4:  dav = 0;
                    5:  dav = 0;
                    6:  sign[1] = PS2DAT;           // x sign
                    7:  sign[0] = PS2DAT;           // y sign
                    8:  ovf[1] = PS2DAT;            // x overflow
                    9:  ovf[0] = PS2DAT;            // y overflow
                    10: parity[0] = PS2DAT;
                    11: dav = 0;
                    12: dav = 0;
                    13: mousexdata[0] = PS2DAT;     // x data
                    14: mousexdata[1] = PS2DAT;
                    15: mousexdata[2] = PS2DAT;
                    16: mousexdata[3] = PS2DAT;
                    17: mousexdata[4] = PS2DAT;
                    18: mousexdata[5] = PS2DAT;
                    19: mousexdata[6] = PS2DAT;
                    20: mousexdata[7] = PS2DAT;
                    21: parity[1] = PS2DAT;         // x data parity
                    22: dav = 0;
                    23: dav = 0;
                    24: mouseydata[0] = PS2DAT;     // y data
                    25: mouseydata[1] = PS2DAT;
                    26: mouseydata[2] = PS2DAT;
                    27: mouseydata[3] = PS2DAT;
                    28: mouseydata[4] = PS2DAT;
                    29: mouseydata[5] = PS2DAT;
                    30: mouseydata[6] = PS2DAT;
                    31: mouseydata[7] = PS2DAT;
                    32: parity[2] = PS2DAT;         // y data parity
                    33: begin
                            rstate = 0;
                            dav = 1;                // data available
                        end
                endcase
            end
    end

endmodule
```

The PS/2 mouse module mouse.v is verified by the Verilog top module for the Basys Board bamousetest.v, which is in the *Chapter 3\mousetes\bamousetest* folder, is given in Listing 3.22. The complete Xilinx ISE WebPACK project uses UCF mousetestbasys.ucf which uncomments the signals CCLK, PS2CLK, PS2DAT, SW0, four LEDs and the cathode and anode signals in the UCF of Listing 3.1. The five Verilog modules operate in parallel and some independently in the top module. The file download procedure is described in the Appendix.

The wire net type establishes the 8-bit, 5-bit, 3-bit and 2-bit vector connectivity for mousexdata, mouseydata, data, digit, parity, button, sign and ovf between the mouse.v, sevensegment.v and checkmouse.v modules for the Spartan-3 Starter Board as given in Listing 3.10 [Ciletti04]. The clock signal clock is outputted from the clock.v module at a frequency of 1 kHz and inputted to the checkmouse.v and sevensegment.v modules. The 2-bit PS/2 mouse button register button and the 2-bit data overflow register ovf are continuously assigned to the four LEDs (LD0, LD1, LD2 and LD3) for display.

The module checkmouse.v is event driven on the positive edge of the data available signal dav from the mouse.v module and the posedge edge of 1 kHz signal clock from the clock.v module. These events are independent and operate in parallel but are coordinated by the digit available register digitav, as described in Chapter 1 Verilog Hardware Description Language.

The data available event extracts the PS/2 mouse X and Y directional data and calculates the absolute X and Y position for a standard Video Graphics Array (VGA) grid of 640 elements horizontally and 480 elements vertically. Depressing the two PS/2 mouse buttons simultaneously homes the absolute position to the center of the grid (X = 320, Y = 240).

The 9-bit X or Y directional data is converted to a magnitude in the 8-bit registers xvalue and yvalue with the original 2-bit register sign. If the X or Y directional data is negative, then the magnitude is calculated by the complement-and-increment Verilog statement mousedata = ~mousedata + 1; which converts the two's-complement binary number. The original 8-bit X and Y direction data registers mousexdata and mouseydata from the module mouse.v are inputted to the module checkmouse.v sequentially as the 8-bit register mousedata.

Unlike parameter passing in a conversational language such as C, mousexdata and mouseydata are signals and not registers and can not be manipulated. The X and Y directional data is the either added or subtracted from the 11-bit absolute position registers xpos and ypos, as determined by the sign bits.

An 11-bit absolute position register is required to check for underflow (X < 0 or Y < 0), which can occur after subtraction, by the value of the most significant bit (MSB). The MSB is a logic 1 for a negative two's complement binary number. The overflow value (X > 639 or Y > 479), which can occur after addition, is checked by the Verilog if statement. The Y absolute position register ypos need only be 10 bits but is 11 bits here for computational compatibility with the X absolute position register xpos.

Next, the absolute X and Y position registers xpos and ypos as binary numbers are converted to a three digit binary coded decimal (BCD) number. One structural model to do this is the serial binary to BCD conversion algorithm, as described in Chapter 2 Verilog Design Automation (Xilinx application note XAPP029, *www.xilinx.com*). However, this method is not intuitive and requires an understanding of the relationship of the binary and the BCD number systems.

Alternatively, Verilog can utilize a conversational language algorithm, as in the C language, to produce a behavioral model for the same result. This illustrates the salient concept that such algorithms can be used in Verilog for embedded design, as described in Chapter 1 Verilog Hardware Description Language.

The behavioral model iteratively subtracts 100 then 10 from the binary number to form the most significant (msdigit), middle (middigit) and least significant (lsdigit) BCD digit. This behavioral model iterative subtraction algorithm is described in Chapter 2 Verilog Design Automation Verilog Design Automation.

The Y position of the PS/2 mouse is selected for the three digit BCD display when slide switch 0 (SW0) is logic 1. The X position is selected when SW0 is logic 0. This is required because there are only four seven segment displays available on the Basys Board.

The clock event displays the PS/2 mouse X or Y position on the three seven segment display digits on the Basys Board. The set decimal point register setdp of the seven segment display is logic 0 and no decimal point is displayed. The three digit display clock event uses the same configuration as that in Listing 3.12 but blanks the unused fourth digit with a digit data code of 31, as shown in sevensegment.v in Listing 3.9.

As described in Chapter 2 Verilog Design Automation Verilog Design Automation, the Design Utilization Summary for the top module bamousetest.v shows the use of 134 slice flip-flops (3%), 88 4-input LUTs (4%), 302 occupied slices (12%) and a total of 3106 equivalent gates in the XC3S500E Spartan-3E FPGA synthesis.

**Listing 3.22** Mouse test top module for the Basys Board bamousetest.v

```
// Basys Board
// PS/2 Mouse Test bamousetest.v
// c 2008 Embedded Design in Verilog using Programmable Gate Arrays  Dennis Silage

module bamousetest (input CCLK, SW0, inout PS2CLK, inout PS2DAT, output AN0, AN1, AN2,
                    AN3, LD0, LD1, LD2, LD3, CA, CB, CC, CD, CE, CF, CG, CDP);

wire [4:0] data;
wire [2:0] digit;
wire [1:0] button;
wire [1:0] sign;
wire [1:0] ovf;
wire [2:0] parity;
wire [7:0] mousexdata;
wire [7:0] mouseydata;
assign LD1 = button[1];
assign LD0 = button[0];
assign LD3 = ovf[1];
assign LD2 = ovf[0];

clock M0 (CCLK, 25000, clk);          // 1 KHz
mouse M1 (CCLK, PS2CLK, PS2DAT, dav, parity, button, ovf, sign, mousexdata, mouseydata);
sevensegment M2 (clock, data, digit, setdp, AN0, AN1, AN2, AN3, CA, CB, CC, CD, CE, CF, CG,
                 CDP);
checkmouse M3 (clk, SW0, button, data, digit, setdp, dav, sign, mousexdata, mouseydata);

endmodule

module checkmouse (input clk, input SW0, input [1:0] button, output reg [4:0] data,
                   output reg [2:0] digit, output reg setdp, input dav, input [1:0] sign,
                   input [7:0] mousexdata, input [7:0] mouseydata);

reg [1:0] digitmux;   // digit multiplexer
reg [3:0] msdigit;    // most significant digit
reg [3:0] middigit;   // middle digit
reg [3:0] lsdigit;    // least significant digit
reg [7:0] mousedata;  // local mouse x or y data
reg dataav;           // data available

reg [10:0] xvalue;    // mouse x value
reg [10:0] yvalue;    // mouse y value
reg [10:0] value;     // local x or y value
reg [10:0] xpos;      // x position
reg [10:0] ypos;      // y position

integer i;

always@(posedge dav)     // data available even driven
    begin
        dataav = 0;
```

```
if (button[0] && button[1])       // reset x and y position
      begin
            xpos = 320;           // center
            ypos = 240;
      end
yvalue[10:8] = 0;
xvalue[10:8] = 0;

mousedata = mouseydata;

if (sign[0] == 1)
      mousedata = ~mousedata + 1;
yvalue[7:0] = mousedata[7:0];
mousedata = mousexdata;
if (sign[1] == 1)
      mousedata =~ mousedata + 1;
xvalue[7:0] = mousedata[7:0];

if (sign[0] == 1)
      ypos = ypos – yvalue;
else
      ypos=ypos + yvalue;

if (ypos[10] == 1)
      ypos = 0;
if (ypos > 479)
      ypos = 479;

if (sign[1] == 1)
      xpos = xpos – xvalue;
else
      xpos = xpos + xvalue;
if (xpos[10] == 1)
      xpos = 0;
if (xpos > 639)
      xpos = 639;

if (SW0)                      // select x or y position to display
      value = ypos;
else
      value = xpos;
                              // binary to BCD Conversion
msdigit = 0;                  // most significant digit
for (i = 1; i <= 9; i = i + 1)
      begin
            if (value >= 100)
                  begin
                        msdigit = msdigit + 1;
                        value = value – 100;
                  end
      end
```

```
            middigit = 0;                    // middle digit
            for (i= 1; i<= 9; i= i+1)
                begin
                    if (value >= 10)
                        begin
                            middigit = middigit + 1;
                            value = value – 10;
                        end
                end

            lsdigit = value;                 // least significant digit
            dataav = 1;                      // data available
        end

always@(posedge clk)                         // local clock event driven
        begin                                // display mouse data
            if (dataav)
                begin
                    digitmux = digitmux + 1;
                    setdp = 0;
                    data[4] = 0;
                    case (digitmux)
                        0:  begin
                                data[3:0] = lsdigit;
                                digit = 1;
                            end
                        1:  begin
                                data[3:0] = middigit;
                                digit = 2;
                            end
                        2:  begin
                                data[3:0] = msdigit;
                                digit = 3;
                            end
                        3:  begin
                                data = 31;       // blank
                                digit = 4;
                            end
                    endcase
                end
        end

endmodule
```

For the Spartan-3E Starter Board the PS/2 mouse the X and Y positions, the status of the buttons and data overflow are displayed on the LCD. The PS/2 mouse module mouse.v is verified by the Verilog top module for the Spartan-3E Starter Board s3emousetest.v, which is in the *Chapter 3 \mousetest\s3emousetest* folder and is also given in Listing 3.23.

The complete Xilinx ISE WebPACK project in the *Chapter 3\s3emousetest* folder uses the UCF mousetests3esb.ucf which uncomments the signals CCLK, PS2CLK, PS2DAT and the LCD signals in the Spartan-3E Starter Board UCF of Listing 3.2. The four Verilog modules operate in parallel and some independently in the top module, as shown partially in Figure 3.7. However, to reduce the complexity of the displayed configuration of the controllers and datapaths in Figure 3.7, the lcd.v module and its interconnections are not shown.

The genlcd.v module used here is the controller for the lcd.v datapath module, as described in Chapter 1 Verilog Hardware Description Language, and is similar to the genlcd.v module used for keyboard data for the Spartan-3E Starter Board, as given in Listing 3.20. The checkmouse.v module used here for the Spartan-3E Starter Board is similar to the checkmouse.v module given in Listing 3.22 for use with the seven segment display of the Basys Board.

However, there is no need to select which mouse cursor data is being displayed when using the LCD display and both the X and Y most significant, middle and least significant digits are outputted. The genlcd.v and checkmouse.v modules demonstrate the design reuse concept for embedded design [Botros06]. As described in Chapter 2 Verilog Design Automation Verilog Design Automation, the Design Utilization Summary for the top module s3emousetest.v shows the use of 203 slice flip-flops (2%), 920 4-input LUTs (9%), 553 occupied slices (11%) and a total of 7858 equivalent gates in the XC3S500E Spartan-3E FPGA synthesis.

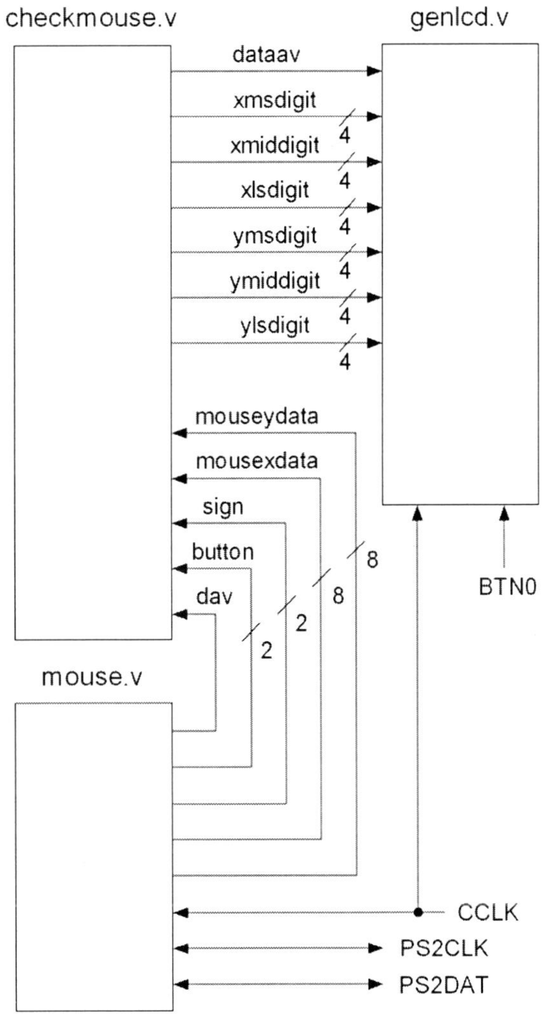


**Figure 3.7** Partial controller and datapath modules in the top module s3emousetest.v

**Listing 3.23** Mouse and LCD controller modules for the Spartan-3E Starter Board s3emousetest.v

```
// Spartan-3E Starter Board
// PS/2 Mouse Test s3emousetest.v
// c 2007 Embedded Design in Verilog using Programmable Gate Arrays  Dennis Silage

module s3emousetest (input CCLK, BTN0, inout PS2CLK, inout PS2DAT, output LD0, LD1, LD2,
                     LD3, LCDRS, LCDRW, LCDE, output [3:0] LCDDAT);

wire [3:0] data;
wire [1:0] digitmux;
wire [7:0] lcddatin;
wire [3:0] lcdd;
wire rslcd, rwlcd, elcd;

assign LCDDAT[3] = lcdd[3];
assign LCDDAT[2] = lcdd[2];
assign LCDDAT[1] = lcdd[1];
assign LCDDAT[0] = lcdd[0];

assign LCDRS = rslcd;
assign LCDRW = rwlcd;
assign LCDE = elcd;

wire [1:0] button;
wire [1:0] sign;
wire [1:0] ovf;
wire [2:0] parity;
wire [7:0] mousexdata;
wire [7:0] mouseydata;
wire [3:0] xmsdigit;
wire [3:0] xmiddigit;
wire [3:0] xlsdigit;
wire [3:0] ymsdigit;
wire [3:0] ymiddigit;
wire [3:0] ylsdigit;

assign LD1 = button[1];
assign LD0 = button[0];
assign LD3 = ovf[1];
assign LD2 = ovf[0];

mouse M0 (CCLK, PS2CLK, PS2DAT, dav, parity, button, ovf, sign, mousexdata, mouseydata);
lcd M1 (CCLK, resetlcd, clearlcd, homelcd, datalcd, addrlcd, lcdreset, lcdclear, lcdhome, lcddata,
        lcdaddr, rslcd, rwlcd, elcd, lcdd, lcddatin, initlcd);
checkmouse M2 (button, dav, sign, mousexdata, mouseydata, dataav, xmsdigit, xmiddigit,
        xlsdigit, ymsdigit, ymiddigit, ylsdigit);
genlcd M3 (CCLK, BTN0, dataav, resetlcd, clearlcd, homelcd, datalcd, addrlcd, initlcd, lcdreset,
        lcdclear, lcdhome, lcddata, lcdaddr, lcddatin, xmsdigit, xmiddigit, xlsdigit,
        ymsdigit, ymiddigit, ylsdigit);

endmodule
```

```
module genlcd (input CCLK, BTN0, dataav, output reg resetlcd, output reg clearlcd,
        output reg homelcd, output reg datalcd, output reg addrlcd, output reg initlcd,
        input lcdreset, lcdclear, lcdhome, lcddata, lcdaddr, output reg [7:0] lcddatin,
        input [3:0] xmsdigit, input [3:0] xmiddigit, input [3:0] xlsdigit, input [3:0] ymsdigit,
        input [3:0] ymiddigit, input [3:0] ylsdigit);
reg [4:0] gstate;

always@(posedge CCLK)       // master clock event driven
    begin
        case (gstate)
            0:  begin
                    initlcd = 1;            // initialize LCD
                    gstate = 1;
                end
{LCD initialize and clear display continues in states 1 through 5}
            6:  begin
                    if (dataav == 0)        // data available?
                        gstate = 6;
                    else
                        begin
                            initlcd = 1;
                            gstate = 7;
                        end
                end
{LCD set display address continues in states 7 through 10}
            11: begin
                    lcddatin[7:4] = 3;          // ASCII digit 3Xh
                    lcddatin[3:0] = ymsdigit[3:0];
                    datalcd = 1;
                    if (lcddata == 1)
                        begin
                            datalcd = 0;
                            gstate = 12;
                        end
                end
            12: begin
                    initlcd = 1;
                    gstate = 13;
                end
            13: begin
                    initlcd = 0;
                    gstate = 14;
                end
            14: begin
                    lcddatin[3:0] = ymiddigit[3:0];
                    datalcd = 1;
                    if (lcddata == 1)
                        begin
                            datalcd = 0;
                            gstate = 15;
                        end
                end
```

```
15: begin
        initlcd = 1;
        gstate = 16;
    end
16: begin
        initlcd = 0;
        gstate = 17;
    end
17: begin
        lcddatin[3:0] = ylsdigit[3:0];
        datalcd = 1;
        if (lcddata == 1)
            begin
                datalcd = 0;
                gstate = 18;
            end
    end
18: begin
        initlcd = 1;
        gstate = 19;
    end
19: begin
        initlcd = 0;
        gstate=20;
    end
20: begin
        lcddatin[7:0] = 8'h20;       // ASCII space
        datalcd = 1;
        if (lcddata == 1)
            begin
                datalcd = 0;
                gstate = 21;
            end
    end
21: begin
        initlcd = 1;
        gstate = 22;
    end
22: begin
        initlcd = 0;
        gstate = 23;
    end
23: begin
        lcddatin[7:4] = 3;           // ASCII digit 3Xh
        lcddatin[3:0] = xmsdigit[3:0];
        datalcd = 1;
        if (lcddata == 1)
            begin
                datalcd = 0;
                gstate = 24;
            end
    end
```

```
            24: begin
                    initlcd = 1;
                    gstate = 25;
                end
            25: begin
                    initlcd = 0;
                    gstate = 26;
                end
            26: begin
                    lcddatin[3:0] = xmiddigit[3:0];
                    datalcd = 1;
                    if (lcddata == 1)
                        begin
                            datalcd = 0;
                            gstate = 27;
                        end
                end
            27: begin
                    initlcd = 1;
                    gstate = 28;
                end
            28: begin
                    initlcd = 0;
                    gstate = 29;
                end
            29: begin
                    lcddatin[3:0] = xlsdigit[3:0];
                    datalcd = 1;
                    if (lcddata == 1)
                        begin
                            datalcd = 0;
                            gstate = 6;
                        end
                end
            default: gstate = 0;
        endcase
    end

endmodule
```

## Digital-to-Analog Converter

A digital-to-analog converter (DAC) provides a step-wise analog output signal from an n-bit binary data input signal for embedded system design in audio processing, analog and digital baseband and bandpass communication and digital process control. The analog output voltage resolution (or step size) $\Delta V$ for a DAC that inputs an unsigned n-bit binary values is determined by Equation 3.2.

$$\Delta V = \frac{V_{REF}}{2^n} \qquad (3.2)$$

$V_{REF}$ is the precision DC voltage used as a reference. The maximum rate at which the DAC can accept binary data and update its analog voltage output is its throughput rate $R_{DAC}$ in samples per second.

## Spartan-3E Starter Board DAC

The Spartan-3E Starter Board provides a serial peripheral interface (SPI) compatible, four channel, 12-bit unsigned DAC utilizing the Linear Technology LTC2624 device (*www.linear.com*). The four analog signal outputs from the four channels of the DAC are on the 6-pin header J5. The SPI bus is a full-duplex, synchronous serial data communication channel. The FPGA operates as the SPI bus master and generates the clock signal SPISCK to synchronously transmit the serial data signal SPIMOSI and to receive the serial data signal SPIMISO.

The SPISCK, SPIMOSI and SPIMISO signals are shared with the serial flash PROM, the parallel flash PROM, the platform flash PROM, the programmable analog pre-amplifier and the analog-to-digital converter (ADC) SPI peripherals on the Spartan-3E Starter Board and these must be disabled. Although the Strataflash PROM is a parallel device, its least significant bit is shared with the SPIMISO signal and it must also be disabled. However, the SPIMISO signal merely echoes the binary data sent to the DAC and is ignored here.

**Table 3.5** Spartan-3E Starter Board Linear Technology LTC2624 DAC 24-bit data packet

| Bits | | Contents (MSB...LSB) |
|---|---|---|
| 23-20 | Command: | 0000 Write to DAC input register |
| | | 0001 Update (power-up) DAC register |
| | | 0010 Write to DAC input register and update (power up) all DACs |
| | | 0011 Write to and update (power up) DAC |
| | | 0100 Power down DAC |
| 19-16 | Address: | 0000 DAC A |
| | | 0001 DAC B |
| | | 0010 DAC C |
| | | 0011 DAC D |
| | | 1111 DACs A, B, C and D |
| 15-4 | Data: | 12-bit unsigned DAC data, MSB first |
| 3-0 | X (don't care) | |

After these SPI devices are disabled, the DAC SPI data communication protocol begins with the chip select signal DACCS set to logic 0. The rising edge of the SPI bus clock signal SPISCK is used to transmit a 24-bit data packet, with the most significant bit (MSB) of the command, address and data sent first, as listed in Table 3.5. The DAC SPI data communication protocol also optionally supports a 32-bit data packet, but the additional bits are not required and only lowers the throughput frequency. After all 24 bits have been sent, the chip select signal DACCS is set to logic 1 and whose rising edge starts the actual DAC process.

Although the DAC transmits the previous 24-bit data packet as the SPIMISO signal, it can be ignored. The asynchronous clear signal DACCLR when logic 0 clears all the DAC registers and sets the analog output voltage to 0 V. The analog output voltage for the 12-bit DAC is determined by Equation 3.3.

$$V_{OUT} = \frac{D[11:0] \times V_{REF}}{4096} \text{ V} \qquad (3.3)$$

The DAC reference voltage $V_{REF}$ is 3.3 V for DAC A and DAC B and 2.5 V for DAC C and D. The four unipolar DAC analog output voltages (A, B, C, D) appear on the 6-pin header J5 (adjacent to the Ethernet RJ-45 connector and the 6-pin analog preamplifier header J7) with ground (GND) and a 3.3 V DC signal (VCC).

Although the Linear Technology LTC2624 DAC has a maximum SPI clock frequency of 50 MHz, the crystal clock oscillator of the Spartan-3E Starter Board is also 50 MHz. To be conservative, the actual SPI clock frequency should be less than 50 MHz. The Spartan-3E FPGA Digital Clock Manager (DCM) has a frequency synthesizer function where the input clock frequency can be multiplied and divided by two integers, as described in Chapter 2 Verilog Design Automation. The DCM Xilinx Architecture Wizard provides a block definition file dacs3edcm.xaw and a Verilog HDL instantiation template dacs3edcm.tif for implementation in the ISE project in the *Chapter 3\peripherals* folder, as described in Chapter 2 Verilog Design Automation.

The module s3edac.v is in the *Chapter 3\peripherals* folder and given in abbreviated form in Listing 3.24. This module is a datapath utilizing a finite state machine (FSM) with the 6-bit state register dacstate and provides signals to the controller and the DAC device, as described in Chapter 1 Verilog Hardware Description Language.

The FSM has 51 states that sequentially inputs the 4-bit DAC command daccmd, the 4-bit DAC address dacaddress, the 12-bit DAC data datadata and 4 don't care bits. Because of this, the module s3edac.v is lengthy and repetitive and is truncated for brevity. The complete Verilog HDL source for any module is available and the file download procedure is described in the Appendix.

**Listing 3.24** DAC datapath module for the Spartan-3E Starter Board s3edac.v

```
// Spartan-3E Starter Board
// Digital-to-Analog Converter s3edac.v
// c 2008 Embedded Design using Programmable Gate Arrays  Dennis Silage

module s3edac (input dacclk, input dacdav, output reg davdac, input [11:0] dacdata,
            input [3:0] dacaddr, input [3:0] daccmd, output reg dacsck, output reg dacspid,
            output reg csdac, output reg clrdac = 1);
reg [5:0] dacstate = 0;       // state register

always@(posedge dacclk)
    begin
        if (dacdav == 0)              // DAC data?
            begin
                dacstate = 0;
                davdac = 0;       // DAC data NAK
            end

        if (dacdav == 1 && davdac == 0)
            begin
                case (dacstate)
                0:    begin
                            csdac = 1;
                            dacsck = 0;
                            dacstate = 1;
                        end
                1:    begin
                            csdac = 0;          // select DAC
                            dacstate = 2;
                        end
                2:    begin
                            dacspid = daccmd[3];        // DAC command
                            dacstate = 3;
                        end
```

```
                3:  begin
                        dacsck = 1;
                        dacstate = 4;
                    end
{DAC command bits 2 and 1 are sent similarly in states 4 through 7}
                8:  begin
                        dacsck = 0;
                        dacspid = daccmd[0];
                        dacstate = 9;
                    end
                9:  begin
                        dacsck = 1;
                        dacstate = 10;
                    end
                10: begin
                        dacsck = 0;
                        dacspid = dacaddr[3];       //DAC address
                        dacstate = 11;
                    end
                11: begin
                        dacsck = 1;
                        dacstate = 12;
                    end
{DAC address bits 2 and 1 are sent similarly in states 12 through 15}
                16: begin
                        dacsck = 0;
                        dacspid = dacaddr[0];
                        dacstate = 17;
                    end
                17: begin
                        dacsck = 1;
                        dacstate = 18;
                    end
                18: begin
                        dacsck = 0;
                        dacspid = dacdata[11];      //DAC data
                        dacstate = 19;
                    end
                19: begin
                        dacsck = 1;
                        dacstate = 20;
                    end
{DAC data bits 10 through 1 are sent similarly in states 20 through 39}
                40: begin
                        dacsck = 0;
                        dacspid = dacdata[0];
                        dacstate = 41;
                    end
                41: begin
                        dacsck = 1;
                        dacstate = 42;
                    end
```

```
                    42: begin
                            dacsck = 0;
                            dacspid = 0;        // X (don't care)
                            dacstate = 43;
                        end
                    43: begin
                            dacsck = 1;         // X3
                            dacstate = 44;
                        end
{DAC don't care bits 2 through 1 are sent similarly in states 44 through 47}
                    48: begin
                            dacsck = 0;
                            dacstate = 49;
                        end
                    49: begin
                            dacsck = 1;         // X0
                            dacstate=50;
                        end
                    50: begin
                            dacsck = 0;
                            csdac = 1;
                            davdac = 1;         // DAC data ACK
                            dacstate = 51;
                        end
                    51: dacstate=51;
                    default: dacstate=51;
                endcase
            end
        end

endmodule
```

The DCM frequency synthesizer inputs the 50 MHz crystal clock oscillator and outputs a 50 MHz × 5 / 3 = 83.333 MHz clock. This DCM clock signal is inputted as the DAC clock dacclk. The actual SPI bus clock in the DAC datapath module s3edac.v is 83.333 MHz / 2 = 41.666 MHz. This is because the FSM transitions in the module s3edac.v generate the SPI bus clock.

The DAC requires a chip select signal csdac for a duration of two FSM transitions (2/2) and a 24-bit data packet to update a single analog output signal with 12 bits of data, so the approximate maximum throughput data rate $R_{DAC}$ = 41.666 MHz/ (2/2 +24 bits/packet) = 1.666 Msamples/sec.

The data available signal dacdav from the controller sets the dacstate register to 0 if it is a logic 0 or activates the FSM on the positive edge of the frequency synthesizer clock signal CLKFX if it is a logic 1 and if the datapath signal davdac is logic 0. The datapath returns the status signal davdac as a logic 1 to the controller when the process of loading the DAC with a command, address and data is completed. The DAC chip select signal DACCS is controlled by the module output register variable csdac.

The SPI bus clock SPISCK is set by the output register variable dacsck and on its positive edge clocks the 4-bit DAC command daccmd, the 4-bit DAC address dacaddr, the 12-bit DAC data dacdata and 4 don't care bits (set to logic 0) onto the SPI bus. The SPI bus serial data signal SPIMOSI is set by the output register variable dacspid. The asynchronous DAC clear input signal DACCLR is set by the output register clrdac but is disabled by setting it to a logic 1 in the declaration [Lee06].

The Spartan-3E Starter Board DAC datapath module s3edac.v is verified by the Verilog top module s3edactest.v, which is in the *Chapter 3\DACtest\s3edactest* folder, is given in Listing 3.25. The top module file s3edactest.v also includes the controller module gendac.v. The complete Xilinx ISE WebPACK project uses the UCF dactests3esb.ucf which uncomments the signals CCLK, DACCS, DACCLR, SPISF, AMPCS, ADCON, SFCE and FPGAIB in the Spartan-3E Starter Board UCF of Listing 3.2. The three Verilog modules operate in parallel and some independently in the top module. The file download procedure is described in the Appendix.

The device signals for the serial flash PROM SPISF, the programmable analog pre-amplifier AMPCS, the analog-to-digital converter ADCON, the parallel flash PROM SFCE and the platform flash PROM FPGAIB, as given in the Spartan-3E Starter Board UCF in Listing 3.2, are disabled in the top module with continuous assignment statements.

The DAC controller module gendac.v provides a signal and data to the DAC datapath using an FSM with the 2-bit state register gstate and generates a 12-bit (4096 levels) linear ramp. The direct measurement of the period of the 4096 samples in the linear ramp voltage output at the DAC header J5-A is 2.70 msec, which indicates that the actual data rate $R_{DAC} = 4096 / 2.70 \times 10^{-3} \approx 1.517$ Msamples/sec (the maximum throughput data rate is 1.666 Msamples/sec). The DAC command register daccmd is 0011 binary (3 decimal) and the DAC address register dacaddr is 0 which allows data to be written to DAC A.

As described in Chapter 2 Verilog Design Automation, the Design Utilization Summary for the top module s3edactest.v shows the use of 72 slice flip-flops (1%), 88 4-input LUTs (1%), 55 occupied slices (1%) and a total of 8194 equivalent gates in the XC3S500E Spartan-3E FPGA synthesis. The large number of equivalent gates is due to the inclusion of the DCM module dacs3edcm.v.

**Listing 3.25** DAC test top module for the Spartan-3E Starter Board s3edactest.v

```
// Spartan-3E Starter Board
// Digital-to-Analog Converter s3edactest.v
// c 2008 Embedded Design using Programmable Gate Arrays  Dennis Silage

module s3edactest (input CCLK, output SPIMOSI, SPISCK, DACCS, DACCLR, SPISF, AMPCS,
                   output ADCON, SFCE, FPGAIB);

wire dacdav, davdac, dacsck, dacspid, csdac, clrdac;
wire CLKFX, CLKOUT;
wire [11:0] dacdata;
wire [3:0] dacaddr, daccmd;

assign SPISCK = dacsck; // SPI clock
assign SPIMOS I = dacspid;    // SPI input data
assign DACCS = csdac;         // select DAC
assign DACCLR = clrdac;       // clear DAC
assign SPISF = 1;             // disable serial Flash
assign AMPCS = 1;             // disable prog amp
assign ADCON = 0;             // disable ADC
assign SFCE = 1;              // disable StrataFlash
assign FPGAIB = 1;            // disable Platform Flash

s3edac M0 (CLKFX, dacdav, davdac, dacdata, dacaddr, daccmd, dacsck, dacspid, csdac, clrdac);
gendac M1 (CLKOUT, dacdav, davdac, dacdata, dacaddr, daccmd);
dacs3edcm M2 (.CLKIN_IN(CCLK), .RST_IN(0), .CLKFX_OUT(CLKFX),
              .CLKIN_IBUFG_OUT(CLKINIBO), .CLK0_OUT(CLKOUT),
              .LOCKED_OUT(LOCK));
endmodule
```

```
module gendac (input genclk, output reg dacdav, input davdac, output reg [11:0] dacdata,
                output reg [3:0] dacaddr = 0, output reg [3:0] daccmd = 3);

reg [1:0] gstate = 0;        // state register

always@(posedge genclk)
    begin
        case (gstate)
            0:  begin               // generate linear ramp
                    dacdata = dacdata + 1;
                    dacdav = 0;
                    gstate = 1;
                end
            1:  begin
                    dacdav = 1;
                    gstate = 2;
                end
            2:  begin
                    if (davdac == 0)
                        gstate = 2;
                    else
                        gstate = 0;
                end
            default: gstate = 0;
        endcase
    end

endmodule
```

## Basys Board DAC

The Basys Board does not have an integral digital-to-analog converter (DAC). However, this Spartan-3E FPGA evaluation board has four 6-pin peripheral hardware module connectors and two accessory DACs, the Digilent Pmod DA1 and Pmod DA2, are available (*www.digilentinc.com*), as shown in Figure 3.8. The Spartan-3E Starter Board also has two 6-pin peripheral hardware module connectors and these DACs can be used with this evaluation board also.

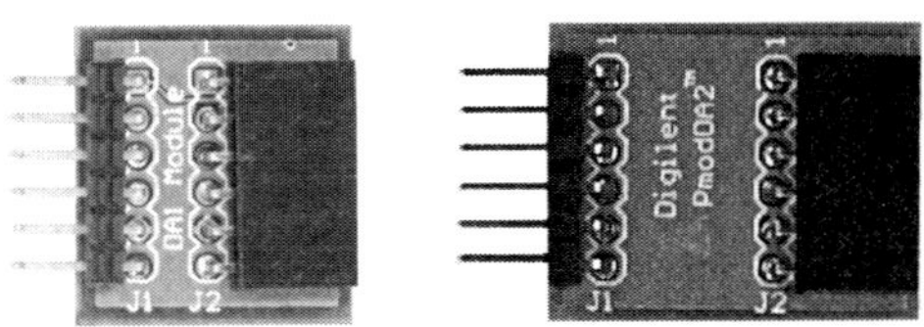


**Figure 3.8** Digilent Pmod DA1 module with two, two channel 8-bit DACs and the Pmod DA2 module with two 12 bit DACs

The Pmod DA1 hardware module features two Analog Devices (*www.analog.com*) AD7303 two channel, 8-bit DACs. The Pmod DA2 hardware module features two National Semiconductor (*www.national.com*) DAC121S101 12-bit DACs. Both of these accessory DACs also utilize the serial peripheral interface (SPI) bus protocol. Both the AD7303 and DAC121S101 DAC devices utilize an active low SYNC signal (similar to the DACCS signal for the Linear Technology LTC2624 DAC on the Spartan-3E Starter Board), an SPI bus clock signal SCLK, and two data signals DACD0 and DACD1. The DAC SPI data communication protocol begins with the sync signal SYNC set to logic 0.

For the Pmod DA1 AD7303 DAC the rising edge of the SPI bus clock signal SCLK is used to transmit a 16-bit data packet, with the most significant bit (MSB) of the command, address and data sent first, as listed in Table 3.6. After all 16 bits have been sent, the sync signal SYNC is set to logic 1 and whose rising edge starts the actual DAC process.

The Pmod DA1 DAC is connected to a 6-pin peripheral hardware module connector. The SYNC signal is pin 1, the two input data signals DACD0 and DACD1 are pins 2 and 3 and the SPI bus clock signal SCLK is pin 4. Pins 5 and 6 are DC ground and power (VDD = +3.3 V DC). The analog output voltage for the 8-bit DAC is determined by Equation 3.4.

$$V_{OUT} = \frac{D[7:0] \times 2 \times V_{REF}}{256} \text{ V} \qquad (3.4)$$

**Table 3.6** Pmod DA1 Analog Devices AD7303 DAC 16-bit data packet

| Bits | Contents (MSB...LSB) | | | |
|---|---|---|---|---|
| 15 | Internal (0) or external (1) reference voltage | | | |
| 14 | X (don't care) | | | |
| 13 | Load DAC | | | LDAC (see Control below) |
| 12-11 | Power-down: | | 00 | Both DAC A and DAC B active |
| | | | 01 | DAC A powered-down, DAC B active |
| | | | 10 | DAC A active, DAC B powered-down |
| | | | 11 | Both DAC A and DAC B powered-down |
| 10 | Address A/B: | | 0 | Select DAC A |
| | | | 1 | Select DAC B |
| 9-10 | Control | LDAC | A/B | |
| | 00 | 0 | X | DAC A and DAC B loaded from shift register |
| | 01 | 0 | 0 | Update DAC A input register from shift register |
| | 01 | 0 | 1 | Update DAC A input register from shift register |
| | 10 | 0 | 0 | Update DAC A from input register |
| | 10 | 0 | 1 | Update DAC B from input register |
| | 11 | 0 | 0 | Update DAC A from shift register |
| | 11 | 0 | 1 | Update DAC A from input register |
| | XX | 1 | 0 | Load DAC A input register from shift register and update both DAC A and DAC B |
| | XX | 1 | 1 | Load DAC B input register from shift register and update both DAC A and DAC B |
| 7-0 | Data: | | | 8-bit unsigned DAC data, MSB first |

If the Pmod DA1 DAC internal reference voltage is selected with a logic 0 in bit 15 of the data packet, VREF in Equation 3.4 is VDD/2 = 1.65 V DC. The four Pmod DA1 DAC unipolar analog output voltages A1, B1, A2 and B2 appear on a 6-pin header at pins 1, 2, 3 and 4.

The module da1dac.v for the Pmod DA1 DAC, which is in the *Chapter 3\peripherals* folder, is given in abbreviated form in Listing 3.26. This module is a datapath utilizing a finite state machines (FSM) with the 6-bit state register dacstate and provides signals to the controller and the DAC device, as described in Chapter 1 Verilog Hardware Description Language.

The FSM has 34 states that sequentially inputs the 8-bit DAC command daccmd and the 8-bit DAC data dacdata. Because of this, the module da1dac.v is repetitive and the listing is truncated for brevity. The complete Verilog HDL source for any module is available and the file download procedure is described in the Appendix.

The Analog Devices AD7303 DAC of the Pmod DA1 hardware module has a maximum SPI clock frequency of 30 MHz, while the crystal clock oscillator of the Basys Board or the Spartan-3E Starter Board is 50 MHz. Rather than using the DCM frequency synthesizer module to output a 30 MHz clock signal, the 50 MHz crystal clock oscillator can be divided by two by the transitions of the FSM to produce an effective 25 MHz SPI bus clock signal.

The 8-bit AD7303 DAC requires a sync signal dacsync for a duration on one FSM transition and a 16-bit data packet to update a single analog output signal, so the approximate maximum throughput data rate $R_{DAC}$ = 25 MHz / (1/2 +16 bits/packet) = 1.515 Msamples/sec. Although the 14-bit DAC of the Spartan-3E Starter Board requires a 24-bit data packet, it uses a 41.666 MHz SPI bus clock signal and its maximum throughput data rate $R_{DAC}$ = 1.666 Msamples/sec.

The data available signal dacdav from the controller sets the dacstate register to 0 if it is a logic 0 or activates the FSM on the positive edge of the clock signal (dacclk) if it is a logic 1. The datapath returns the status signal davdac as a logic 1 to the controller when the process of loading the DAC with a command and data is completed [Ciletti99]. The DAC SPI bus clock is set by the output register variable dacsck and on its positive edge clocks the 8-bit DAC command daccmd and the 8-bit DAC data dacdata as the DAC output signal dacout to either of the two data signals DACD0 and DACD1.

**Listing 3.26** Datapath module for the Pmod DA1 DAC da1dac.v

```
// Basys Board and Spartan-3E Starter Board
// Digital-to-Analog Converter da1dac.v
// c 2008 Embedded Design using Programmable Gate Arrays  Dennis Silage

module da1dac (input dacclk, input dacdav, output reg davdac, output reg dacout,
            output reg dacsck = 0, output reg dacsync = 0, input [7:0] daccmd,
            input [7:0] dacdata);

reg [5:0] dacstate = 0;

always@(posedge dacclk)
   begin
      if (dacdav == 0)           // DAC data?
            dacstate = 0;
      else
         begin
            davdac = 0;     // DAC data NAK
            case (dacstate)
               0:    begin
                        dacsync = 1;
                        dacsck = 0;
                        dacstate = 1;
                     end
               1:    begin      // DAC command
                        dacsync = 0;
                        dacout = daccmd[7];
                        dacstate = 2;
                     end
               2:    begin
                        dacsck = 1;
                        dacstate = 3;
                     end
```

```
{DAC command bits 6 through 1 are sent similarly in states 3 through 14}
                        15: begin
                                dacsck = 0;
                                dacout = daccmd[0];
                                dacstate = 16;
                            end
                        16: begin
                                dacsck = 1;
                                dacstate = 17;
                            end
                        17: begin       // DAC data
                                dacsck = 0;
                                dacout = dacdata[7];
                                dacstate = 18;
                            end
                        18: begin
                                dacsck = 1;
                                dacstate = 19;
                            end
{DAC data bits 6 through 1 are sent similarly in states 19 through 30}
                        31: begin
                                dacsck = 0;
                                dacout = dacdata[0];
                                dacstate = 32;
                            end
                        32: begin
                                dacsck = 1;
                                dacstate = 33;
                            end
                        33: begin
                                dacsync = 1;
                                dacsck = 0;
                                davdac = 1;     // DAC data ACK
                                dacstate = 34;
                            end
                        34: dacstate = 34;
                        default: dacstate = 34;
                    endcase
                end
    end

endmodule
```

The Pmod DA1 DAC datapath module da1dac.v is verified by the Verilog top module da1test.v, which is in the *Chapter 3\da1test\s3eda1test* folder and is also given in Listing 3.27. The Pmod DA1 DAC datapath module da1dac.v is also verified for the Basys Board in the *Chapter 3 \da1test\bada1test* folder.

The top module file also includes the controller module gendac.v. The complete Xilinx ISE WebPACK project uses the UCF da1tests3esb.ucf that uncomments the signals CCLK, JA1, JA2, JA3 and JA4 in the UCF of Listing 3.2 for the Spartan-3E Starter Board. The Pmod DA1 DAC is connected to the 6-pin peripheral connector J1. The Digilent Pmod TPH test point header, as shown in Figure

3.13, provides a convenient connection for the Pmod DA1 DAC at the 6-pin peripheral connector J1 of the Sparatn-3E Starter Board.

The two Verilog modules operate in parallel and some independently in the top module. The second Analog Devices AD7303 DAC on the Pmod DA1 hardware module can be processed by another da1dac.v module in the top module in Lisiting 3.27 but is not implemented here. A complete Pmod DA1 DAC Xilinx ISE WebPACK project for the Basys Board is in the *Chapter 3\bada1test* folder. The file download procedure is described in the Appendix.

The DAC 8-bit command register daccmd is 0001 0011 binary or 19 decimal which selects the internal reference, updates DAC A from the input shift register and activates DAC A. The DAC controller module gendac.v provides a signal and data to the DAC datapath using an FSM with the 2-bit state register gstate and generates an 8-bit (256 levels) linear ramp. The direct measurement of the period of the 256 samples in the linear ramp voltage output at the Pmod DA1 DAC header A1 is 0.189 msec, which indicates that the actual data rate $R_{DAC} = 256 / 0.189 \times 10^{-3} \approx 1.352$ Msamples/sec (the maximum throughput data rate is 1.515 Msamples/sec).

**Listing 3.27** Top module for the Pmod DA1 DAC da1test.v

```
// Spartan-3E Starter Board
// Digital-to-Analog Converter da1test.v
// c 2008 Embedded Design using Programmable Gate Arrays  Dennis Silage

module da1test (input CCLK, output JA1, JA2, JA3, JA4);

wire dacdav, davdac, dacout, dacsck, dacsync;
wire [7:0] dacdata;
wire [7:0] daccmd;

assign JA1 = dacsync;
assign JA2 = dacout;
assign JA3 = 0;
assign JA4 = dacsck;

da1dac M0 (CCLK, dacdav, davdac, dacout, dacsck, dacsync, daccmd, dacdata);
gendac M1 (CCLK, dacdav, davdac, daccmd, dacdata);

endmodule

module gendac (input genclk, output reg dacdav, input davdac, output reg [7:0] daccmd = 19,
            output reg [7:0] dacdata);

reg [1:0] gstate = 0;        // state register

always@(posedge genclk)
    begin
        case (gstate)
            0:   begin        // generate linear ramp
                     dacdata = dacdata + 1;
                     dacdav = 0;
                     gstate = 1;
                 end
```

```
                1:    begin
                          dacdav = 1;
                          gstate = 2;
                      end
                2:    begin
                          if (davdac == 0)
                          gstate = 2;
                      else
                          gstate = 0;
                      end
                default: gstate = 0;
            endcase
    end

endmodule
```

As described in Chapter 2 Verilog Design Automation, the Design Utilization Summary for the top module da1test.v shows the use of 33 slice flip-flops (1%), 46 4-input LUTs (1%), 37 occupied slices (1%) and a total of 591 equivalent gates in the XC3S500E Spartan-3E FPGA synthesis.

For the Pmod DA2 National Semiconductor DAC121S101 12-bit DAC the SPI data communication protocol begins with the sync signal SYNC set to logic 0. The falling edge of the SPI bus clock signal SCLK is used to transmit a 16-bit data packet, with the most significant bit (MSB) of the command and data sent first, as listed in Table 3.7. After all 16 bits have been sent, the sync signal SYNC is set to logic 1, whose rising edge starts the actual DAC process.

**Table 3.7** Pmod DA2 National Semiconductor DAC121S101 DAC 16-bit data packet

| Bits | Contents (MSB...LSB) | | |
|---|---|---|---|
| 15-14 | X (don't care) | | |
| 13-12 | Command: | 00 | Normal operation |
| | | 01 | Power-down with 1 kΩ to ground |
| | | 10 | Power-down with 100 kΩ to ground |
| | | 11 | Power-down with high impedance to ground |
| 11-0 | Data: | | 12-bit unsigned DAC data, MSB first |

The Pmod DA2 DAC is connected to a 6-pin peripheral hardware module connector. The SYNC signal is pin 1, the two input data signals DACD0 and DACD1 are pins 2 and 3 and the SPI bus clock signal SCLK is pin 4. Pins 5 and 6 are DC ground and power (VDD = +3.3 V DC). The analog output voltage for the 12-bit DAC is determined by Equation 3.3. The Pmod DA2 DAC reference voltage is the power supply VDD and the two unipolar analog output voltages A and B appear on a 6-pin header at pins 1 and 3.

The module da2dac.v for the Pmod DA2 DAC, which is in the *Chapter 3\peripherals* folder and is given in abbreviated form in Listing 3.28. The file download procedure is described in the Appendix. This module is a datapath utilizing a finite state machines (FSM) with the 6-bit state register dacstate and provides signals to the controller and the DAC device, as described in Chapter 1 Verilog Hardware Description Language.

The National Semiconductor DAC121S101 of the Pmod DA2 hardware module also has a maximum SPI bus clock frequency of 30 MHz, while the crystal clock oscillator of the Basys Board or the Spartan-3E Starter Board is 50 MHz. As for the Pmod DA1 DAC hardware module, rather than using the DCM frequency synthesizer module to output a 30 MHz clock signal, the 50 MHz crystal clock oscillator can be divided by two by the transitions of the FSM to produce a conservative 25 MHz SPI bus clock signal.

The 16-bit DAC121S101 also requires a a sync signal dacsync for a duration on one FSM transition (1/2) and a 16-bit data packet for 12-bit binary data to update a single analog output signal. The approximate maximum throughput data rate is then the same as the Pmod DA1 8-bit DAC with $R_{DAC}$ = 25 MHz / (1/2 + 16 bits/packet) = 1.515 Msamples/sec.

The data available signal dacdav from the controller sets the dacstate register to 0 if it is a logic 0 or activates the FSM on the positive edge of the clock signal (dacclk) if it is a logic 1. The datapath returns the status signal davdac as a logic 1 to the controller when the process of loading the DAC with a command and data is completed. The DAC SPI bus clock is set by the output register variable dacsck and on its negative edge clocks the 2-bit DAC command daccmd and the 12-bit DAC data dacdata as the DAC output signal dacout to either of the two data signals DACD0 and DACD1.

The Pmod DA2 DAC datapath module da2dac.v is verified by the Verilog top module da2test.v, which is in the *Chapter 3\da2test\s3eda2test* folder. The top module da2test.v is similar to the da1test.v top module for the Pmod DA1 in Listing 3.27 and is not given here. The top module file also includes the controller module gendac.v. The complete Xilinx ISE WebPACK project uses the UCF da2tests3esb.ucf that uncomments the signals CCLK, JA1, JA2, JA3 and JA4 in the UCF of Listing 3.2 for the Spartan-3E Starter Board.

**Listing 3.28** Datapath module for the Pmod DA2 DAC da2dac.v

```
// Spartan-3E Starter Board
// Digital-to-Analog Converter da2dac.v
// c 2008 Embedded Design using Programmable Gate Arrays  Dennis Silage

module da2dac (input dacclk, input dacdav, output reg davdac, output reg dacout, output reg dacsck,
            output reg dacsync, input [1:0] daccmd, input [11:0] dacdata);

reg [5:0] dacstate = 0;

always@(posedge dacclk)
    begin
        if (dacdav == 0)            // DAC data?
            dacstate = 0;
        else
            begin
                davdac = 0;     // DAC data NAK

                case (dacstate)
                    0:  begin
                            dacsync = 1;
                            dacsck = 1;
                            dacstate = 1;
                        end
                    1:  begin
                            dacsync = 0;
                            dacout = 0;       // X don't care
                            dacstate = 2;
                        end
```

```
        2:  begin
                dacsck = 0;
                dacstate = 3;
            end
        3:  begin
                dacsck = 1;
                dacout = 0;        // X don't care
                dacstate = 4;
            end
        4:  begin
                dacsck = 0;
                dacstate = 5;
            end
        5:  begin
                dacsck = 1;
                dacout = daccmd[1];
                dacstate = 6;
            end
        6:  begin
                dacsck = 0;
                dacstate = 7;
            end
        7:  begin
                dacsck = 1;
                dacout = daccmd[0];
                dacstate = 8;
            end
        8:  begin
                dacsck = 0;
                dacstate = 9;
            end
        9:  begin
                dacsck = 1;
                dacout = dacdata[11];
                dacstate = 10;
            end
        10: begin
                dacsck = 0;
                dacstate = 11;
            end
{DAC data bits 10 through 1 are sent similarly in states 11 through 30}
        31: begin
                dacsck = 1;
                dacout = dacdata[0];
                dacstate = 32;
            end
        32: begin
                dacsck = 0;
                dacstate = 33;
            end
```

```
                33:  begin
                        dacsync = 1;
                        dacsck = 1;
                        davdac = 1;      // DAC data ACK
                        dacstate = 34;
                     end
                34:  dacstate = 34;
                default: dacstate = 34;
            endcase
        end
    end

endmodule
```

The Pmod DA1 DAC is connected to the 6-pin peripheral connector J1 of the Spartan-3E Starter Board. The two Verilog modules operate in parallel and some independently in the top module. A complete Pmod DA2 DAC Xilinx ISE WebPACK project for the Basys Board is in the *Chapter 3 \da2test\bada2test* folder. The file download procedure is described in the Appendix.

The DAC 2-bit command daccmd is 00 binary which selects normal operation. The DAC controller module gendac.v provides a signal and data to the DAC datapath using an FSM with the 2-bit state register gstate and generates an 12-bit (4096 level) linear ramp. The direct measurement of the period of the 4096 samples in the linear ramp voltage output at the Pmod DA2 DAC header A is 3.031 msec, which indicates that the actual data rate $R_{DAC} = 4096 / 3.031 \times 10^{-3} \approx 1.352$ Msamples/sec (the maximum throughput data rate is 1.515 Msamples/sec).

As described in Chapter 2 Verilog Design Automation, the Design Utilization Summary for the top module da2test.v shows the use of 37 slice flip-flops (1%), 53 4-input LUTs (1%), 43 occupied slices (1%) and a total of 686 equivalent gates in the XC3S500E Spartan-3E FPGA synthesis.

The second National Semiconductor DAC121S101 device on the Pmod DA2 hardware module can be processed by a Verilog HDL module that produces coincident signals for the two data signals DACD0 and DACD1. This da2dac2.v module for the Pmod DA2 DAC using two channels is similar to the da2dac.v module in Listing 3.28 and is in the *\Chapter 3\peripherals* folder. The direct measurement of the period of the 4096 samples in the two concurrent linear ramp voltage outputs at the DAC header A and B remains 3.031 msec, which indicates that the actual data rate $R_{DAC} \approx 1.352$ Msamples/sec for two concurrent channels of operation.

The parallel operation of the Verilog HDL FSM results in efficiency for digital signal processing, digital communications and digital control in embedded designs, as described in Chapter 2 Verilog Design Automation. Complete Xilinx ISE WebPACK projects illustrating the concept are described in Chpater 4 Digital Signal Processing, Communications and Control.

Although the Linear Technology LTC2624 four channel DAC of the Spartan-3E Starter Board also has an actual data rate $R_{DAC} \approx 1.352$ Msamples/sec, only one DAC channel at a time can be loaded. Thus not only the configuration of the SPI components which leads to bus contention but also the choice of peripheral components can affect the performance of an embedded design.

## Analog-to-Digital Converter

An analog-to-digital converter (ADC) provides a discrete, sampled binary data output signal from a continuous analog input signal for embedded system design in audio processing, analog and digital baseband and bandpass communication and digital process control. The n-bit binary data output D[n-1:0] for the ADC that inputs a unipolar (only positive amplitude) analog signal is determined by Equation 3.5.

$$D[n-1:0] = G\frac{V_{IN} - V_{REF}}{V_{FS}}2^{n-1} \qquad (3.5)$$

G is the gain of any analog preamplifier that precedes the ADC, $V_{REF}$ is the common reference voltage of the analog preamplifier and ADC, $V_{FS}$ is the full-scale conversion voltage for the ADC, $V_{IN}$ is the polar analog signal input at the input of the analog preamplifier and n is the number of bits of resolution. Since $V_{IN}$ could be less than $V_{REF}$, the n-bit binary data output D[n-1:0] is in two's complement format with the most significant bit (MSB) representing the sign bit and a number whose integer value is between $-2^{n-1}$ and $2^{n-1} - 1$ [Wakerly00].

The analog preamplifier may also support the function of the ADC by providing a differential to single-ended signal interface, significant gain for an analog transducer or an anti-aliasing filter for discrete data sampling. The maximum rate at which the ADC can sample an analog signal and produce n-bit binary data is its throughput rate RADC in samples per second.

## Spartan-3E Starter Board Preamplifier and ADC

The Spartan-3E Starter Board provides a serial peripheral interface (SPI) compatible, two channel analog capture circuit which consists of a Linear Technology LTC6912 programmable gain analog preamplifier and a 14-bit Linear Technology LTC1407A ADC (*www.linear.com*). The two analog signal inputs to the preamplifier and ADC are on the 6-pin header J7. The SPI bus is a full-duplex, synchronous serial data communication channel. The FPGA operates as the SPI bus master and generates the clock signal SPISCK to synchronously transmit the serial data signal SPIMOSI and to receive the serial data signal SPIMISO.

The SPISCK, SPIMOSI and SPIMISO signals are shared with the serial flash PROM, the parallel flash PROM, the platform flash PROM, the programmable analog preamplifier, the digital-to-analog converter (DAC) and the analog-to-digital converter (ADC) SPI peripherals on the Spartan-3E Starter Board and those devices not in use must be disabled. Although the Strataflash PROM is a parallel device, its least significant bit is shared with the SPIMISO signal used for the ADC binary data output here and must also be disabled.

After these SPI devices are disabled, the analog preamplifier SPI data communication protocol begins with the preamplifier chip select signal AMPCS set to logic 0. The rising edge of the SPI clock signal SPISCK is used to transmit a 8-bit data packet, with the most significant bit (MSB) of the channel B 4-bit gain data sent first, as listed in Table 3.8. The gain range and steps for Channel A are identical to Channel B.

**Table 3.8** Spartan-3E Starter Board Linear Technology LTC6912 analog preamplifier 8-bit data packet

| Bits | Contents (MSB...LSB) | | Gain | Input Voltage Range (V) | |
|---|---|---|---|---|---|
| 7-4 | Channel B Gain: | 0000 | 0 | | |
| | | 0001 | –1 | 0.4 | 2.9 |
| | | 0010 | –2 | 1.025 | 2.275 |
| | | 0011 | –5 | 1.4 | 1.9 |
| | | 0100 | –10 | 1.525 | 1.775 |
| | | 0101 | –20 | 1.5875 | 1.7125 |
| | | 0110 | –50 | 1.625 | 1.675 |
| | | 0111 | –100 | 1.6375 | 1.6625 |
| 3-0 | Channel A Gain | {same as Channel B Gain} | | | |

The input voltage range minimum and maximum are derived from the non-linear steps in gain listed in Table 3.8 and Equation 3.5 with $V_{REF} \approx 1.65$ V DC and $V_{FS} = 1.25$ V DC. Note that the MSB of the 4-bit gain data is logic 0 and 0000 produces a gain of 0 and that the other values of gain are negative. The two analog preamplifier input voltages (VINA, VINB) appear on the 6-pin header J7 (adjacent to the Ethernet RJ-45 connector and the 6-pin DAC header J5) with ground (GND) and a 3.3 V DC signal (VCC).

The module s3eprogamp.v for the programmable analog preamplifier of the Spartan-3E Starter Board, which is in the *Chapter 3\peripherals* folder and is given in abbreviated form in Listing 3.29. The file download procedure is described in the Appendix. This module is a datapath utilizing a finite state machines (FSM) with the 5-bit state register ampstate and provides signals to the controller and the device, as described in Chapter 1 Verilog Hardware Description Language.

The module s3eprogamp.v for the programmable analog preamplifier of the Spartan-3E Starter Board, which is in the *Chapter 3\peripherals* folder and is given in abbreviated form in Listing 3.29. The file download procedure is described in the Appendix. This module is a datapath utilizing a finite state machines (FSM) with the 5-bit state register ampstate and provides signals to the controller and the device, as described in Chapter 1 Verilog Hardware Description Language.

**Listing 3.29** Datapath module for the programmable analog preamplifier of the Spartan-3E Starter Board s3eprogamp.v

```
// Spartan-3E starter Board
// Programmable Pre-Amplifier s3eprogamp.v
// c 2008 Embedded Design using Programmable Gate Arrays  Dennis Silage

module s3eprogamp (input bufclk, input ampdav, output reg davamp, input [3:0] ampcmd0,
                   input [3:0] ampcmd1, output reg ampsck, output reg ampspid,
                   output reg csamp, output reg sdamp);

reg [4:0] ampstate;        // state register
reg [2:0] ampclkreg;       // prog amp clock register
reg ampclk;                // prog amp clock

always@(posedge bufclk) // master clock event driven 50 MHz
    begin
        ampclkreg = ampclkreg + 1;            // increment clock register
        if (ampclkreg == 3)
            begin
                ampclkreg = 0;
                ampclk = ~ampclk;  // 8.33 MHz clock, 60+60=120 ns
            end
    end

always@(posedge ampclk)        // local clock event driven 8.33 MHz
    begin
        if (ampdav == 0)             // prog amp data?
            begin
                ampsck = 0;
                sdamp = 0;        // disable prog amp shutdown
                ampstate = 0;
                csamp = 1;
                davamp = 0;       // prog amp NAK
            end
```

```
        if (ampdav == 1 && davamp == 0)
            begin
                case (ampstate)
                    0:  begin
                            csamp = 1;
                            ampsck = 0;
                            ampstate = 1;
                    end
                    1:  begin
                            csamp = 0;
                            ampspid = ampcmd0[3];
                            ampstate = 2;
                        end
                    2:  begin
                            ampsck = 1;
                            ampstate = 3;
                        end
```

{Programmable amplifier channel A command bits 2 through 0 and channel B command bits 3 through 1 are sent similarly in states 3 through 14}

```
                    15: begin
                            ampsck = 0;
                            ampspid = ampcmd1[0];
                            ampstate = 16;
                        end
                    16: begin
                            ampsck = 1;
                            ampstate = 17;
                        end
                    17: begin
                            csamp = 1;      // set prog amp gain
                            ampstate = 18;
                        end
                    18: begin
                            ampsck = 0;
                            davamp = 1;     // prog amp ACK
                            ampstate = 19;
                        end
                    19: ampstate = 19;
                    default: ampstate = 19;
                endcase
            end
    end

endmodule
```

The ADC of the Spartan-3E Starter Board is the 14-bit Linear Technology LTC1407A ADC device which provides two channels that are concurrently sampled. The ADC is configured to operate as an SPI device which is shared with the serial flash PROM, the parallel flash PROM, the platform flash PROM, the programmable analog preamplifier and the digital-to-analog converter (DAC) SPI peripherals on the Spartan-3E Starter Board and the devices not in use must be disabled.

After SPI devices are disabled, the ADC SPI data communication protocol begins with the conversion signal CONAD set to logic 1. The falling edge of the SPI clock signal SPISCK is used to read a 34-bit data packet consisting of a 2-bit don't care sequence, the 14-bit ADC channel A data, a 2-bit don't care sequence, the 14-bit ADC channel B data and a final 2-bit don't care sequence. The most significant bit (MSB) of the 14-bit ADC data is sent first, as listed in Table 3.9.

**Table 3.9** Spartan-3E Starter Board Linear Technology LTC1407A ADC 34-bit data packet

| Bits | Contents (MSB...LSB) |
|---|---|
| 33-32 | X (don't care) |
| 31-18 | ADC channel A data |
| 17-16 | X (don't care) |
| 15-2 | ADC channel B data |
| 1-0 | X (don't care) |

The 14-bit binary data output D[13-1:0] for the ADC of the Spartan-3E Starter Board is determined by Equation 3.5 with $V_{REF} \approx 1.65$ V DC, $V_{FS} = 1.25$ V DC and G is the gain of the analog preamplifier that precedes the ADC, as listed in Table 3.8. Non-zero values of the gain G are negative and the analog input signal is inverted by the programmable amplifier before ADC conversion. The analog input voltage that produce a 14-bit binary data output of 0 is $V_{REF} \approx 1.65$ V regardless of the gain.

The module s3eadc.v is in the *Chapter 3\peripherals* folder and is given in abbreviated form in Listing 3.30. The file download procedure is described in the Appendix. This module is a datapath utilizing a finite state machine (FSM) with the 7-bit state register adcstate and provides signals to the controller and the ADC device, as described in Chapter 1 Verilog Hardware Description Language.

**Listing 3.30** Datapath module for the ADC of the Spartan-3E Starter Board s3eadc.v

```
// Spartan-3E Starter Board
// Analog-to-Digital Converter s3eadc.v
// c 2008 Embedded Design using Programmable Gate Arrays  Dennis Silage

module s3eadc (input bufclk, input adcdav, output reg davadc, output reg [13:0] adc0data,
        output reg [13:0] adc1data, output reg adcsck, input adcspod, output reg conad);

reg [6:0] adcstate;
reg adcclk;

always@(posedge bufclk) // buffered clock event driven
    begin
        adcclk = ~adcclk;
        if (adcdav == 0)
            begin
                adcstate = 0;
                conad = 0;
                davadc = 0;     // ADC data NAK
            end

        if (adcdav == 1 && davadc == 0 && adcclk == 1)
            begin
                case (adcstate)
                    0:  begin
                            conad = 1;      // ADC convert command
```

```
            adcsck = 0;
            adcstate = 1;
        end
    1:  adcstate = 2;
    2:  begin
            conad = 0;
            adcstate = 3;
        end
    3:  begin
            adcsck = 1;
            adcstate = 4;
        end
    4:  begin
            adcsck = 0;     // 1
            adcstate = 5;
        end
    5:  begin
            adcsck = 1;
            adcstate = 6;
        end
    6:  begin
            adcsck = 0;     // 2
            adcstate = 7;
        end
    7:  begin
            adcsck = 1;
            adcstate = 8;
        end
    8:  begin
            adcsck = 0;     // 3
            adc0data[13] = adcspod;
            adcstate = 9;
        end
    9:  begin
            adcsck = 1;
            adcstate = 10;
        end
```

{ADC channel A data bits 12 through 0, two don't care bits and ADC channel B data bits 13 through 1 are sent similarly in states 8 through 65}

```
    66: begin
            adcsck = 0;     // 32
            adc1data[0] = adcspod;
            adcstate = 67;
        end
    67: begin
            adcsck = 1;
            adcstate = 68;
        end
    68: begin
            adcsck = 0;     // 33
            adcstate = 69;
        end
```

```
                69: begin
                        adcsck = 1;
                        adcstate = 70;
                    end
                70: begin
                        adcsck = 0;       // 34
                        davadc = 1;       // ADC data ACK
                        adcstate = 71;
                    end
                71: adcstate=71;
                default: adcstate=71;
            endcase
        end
    end

endmodule
```

The Linear Technology LTC1407A ADC has a specified maximum SPI bus clock frequency of approximately 50 MHz and the crystal clock oscillator of the Spartan-3E Starter Board is also 50 MHz. However, only a conservative SPI bus clock frequency of 12.5 MHz is found to produce reliable ADC data on the Spartan-3E Starter Board. As a verification of this, another available Spartan-3E Starter Board application for the ADC utilizes an even lower SPI bus clock frequency of 2.08 MHz with the Xilinx PicoBlaze 8-bit soft core embedded processor (*www.xilinx.com*).

The SPI bus clock frequency in the ADC datapath module s3eadc.v is 12.5 MHz because the FSM transitions occur at a frequency of 25 MHz. A toggle register adcclk divides the 50 Mhz crystal clock CCLK by two and generates the FSM clock. The ADC requires a conversion command signal conad for a duration of three FSM transitions (3/2) and a 34-bit data packet to convert two analog input signals to 14-bits of resolution, so the approximate maximum throughput data rate $R_{ADC}$ = 12.5 MHz / (3/2 + 34 bits/packet) = 0.339 Msamples/sec.

The data available signal adcdav from the controller sets the adcstate register to 0 if it is a logic 0 or activates the FSM on the positive edge of the derived clock signal adcclk if it is a logic 1 and if the datapath signal davdac is logic 0. The datapath returns the status signal davadc as a logic 1 to the controller when the process of analog conversion and the output of the data is completed.

The ADC conversion command signal ADCON is controlled by the module output register variable conad. The SPI clock SPISCK is set by the output register variable adcsck and on its positive edge clocks two don't care bits, the 14-bit channel A data adc0data, two don't care bits, the 14-bit ADC channel B data adc1data, and two don't care bits onto the SPI bus, as listed in Table 3.9. The SPI bus serial data input signal SPIMISO is read by the input signal adcspod.

The Spartan-3E Starter Board ADC datapath module s3eadc.v is verified by the Verilog top module s3eadctest.v, which is in the *Chapter 3\ADCtest\s3eadctest* folder and is also given in Listing 3.31. The file download procedure is described in the Appendix.

Initially a timing constraint error occurs here with FPGA hardware synthesis. However, the Xilinx WebPACK project s3eadctest.ise is a controller and datapath construct and, although event driven by a clock, essentially depends only upon the data available signals, as described in Chapter 1 Verilog Hardware Description Language. If the clock transition is late then the state transition merely occurs on the next clock transition with a design penalty of one clock cycle.

The timing constraint error can be ignored by deselecting the *Use Timing Constraints* in the Place & Route Properties window as described in Chapter 2 Verilog Design Automation and shown in Figure 2.58. The place and route process is then rerun and FPGA hardware synthesis is successful.

The application sets the 4-bit programmable amplifier gain with the three slide switches SW0, SW1, and SW2 where the most significant bit (MSB) is logic 0, converts the analog voltages at the inputs VINA and VINA to two 14-bit binary numbers adc0data and adc1data, selects one of the binary numbers with slide switch SW3 and converts it to a four digit binary coded decimal (BCD) number adc1, adc2, adc3, and adc4 and a sign bit adcsign and displays the result on the LCD.

The 14-binary numbers from the ADC output are in 2's complement form [Wakerly00]. The 14-bit register variable value is used to manipulate the 14-bit input binary numbers adc0data or adc1data which cannot be modified [Navabi06]. The sign bit is the MSB (bit 13) and if logic 1 indicates that the binary number is negative. The sign bit output register variable adcsign is then set to logic 1 and the 14-bit number is bit-wise complemented and incremented. The BCD digits are calculated by the successive subtraction algorithm, as described in Chapter 2 Verilog Design Automation.

The top module file includes the controller module genampadc.v. The complete Xilinx ISE WebPACK project is in the in the *Chapter 3\adctest\s3eadctest* folder and uses the UCF adctests3esb.ucf that uncomments the signals CCLK, BTN0, SW0, SW1, SW2, SW3, SPIMISO, SPIMOSI, SPICLK, DACCS, DACCLR, SPISF, AMPCS, AMPSD, ADCON, SFCE, FPGAIB and the LCD signals in the Spartan-3E Starter Board UCF of Listing 3.2. The file download procedure is described in the Appendix. The five Verilog modules operate in parallel and some independently in the top module, as shown in Figure 3.9. However, to reduce the complexity of the configuration of the controllers and datapaths in Figure 3.9, the lcd.v module and its interconnections are not shown.

The device signals for the serial flash PROM SPISF, the digital-to-analog converter (DAC) chip select DACCS and clear DACCLR, the parallel flash PROM SFCE and the platform flash PROM FPGAIB, as given in the Spartan-3E Starter Board UCF in Listing 3.2, are disabled in the top module with continuous assignment statements. The SPI bus clock signal SPISCK is continuously assigned as either the ADC SPI clock adcsck or the programmable amplifier SPI clock ampsck. The controller module genampadc.v insures that these SPI bus clock signals are only active one at a time.

**Listing 3.31** ADC test top module for the Spartan-3E Starter Board s3eadctest.v

```
// Spartan-3E Starter Board
// Analog-to-Digital Converter s3eadctest.v
// c 2008 Embedded Design using Programmable Gate Arrays  Dennis Silage

module s3eadctest (input CCLK, BTN0, SW0, SW1, SW2, SW3, SPIMISO, output SPIMOSI,
                   output SPISCK, DACCS, DACCLR, SPISF, AMPCS, AMPSD, ADCON,
                   output SFCE, FPGAIB, LCDRS, LCDRW, LCDE, output [3:0] LCDDAT);

wire adcdav, davadc, adcsck, adcspod, csamp, conad;
wire CLKFX, CLKOUT, ampdav, davamp, ampsck, ampspid, sdamp;
wire resetlcd, clearlcd, homelcd, datalcd, addrlcd;
wire initlcd, lcdreset, lcdclear, lcdhome, lcddata, lcdaddr;
wire rslcd, rwlcd, elcd;
wire [13:0] adc0data, adc1data;
wire [3:0] ampcmd0, ampcmd1, data, lcdd;
wire [2:0] digitmux;
wire [7:0] lcddatin;

assign LCDDAT[3] = lcdd[3];
assign LCDDAT[2] = lcdd[2];
assign LCDDAT[1] = lcdd[1];
assign LCDDAT[0] = lcdd[0];

assign LCDRS = rslcd;
assign LCDRW = rwlcd;
assign LCDE = elcd;
```

```
assign SPISCK = adcsck | ampsck;    // SPI clock
assign SPIMOSI = ampspid;           // SPI data in
assign adcspod = SPIMISO;           // SPI data out

assign DACCS = 1;                   // disable DAC
assign DACCLR = 1;                  // DAC clear

assign SPISF = 1;                   // disable serial Flash
assign AMPCS = csamp;               // select prog amp
assign AMPSD = sdamp;               // disable prog amp shutdown
assign ADCON = conad;               // ADC convert command
assign SFCE = 1;                    // disable StrataFlash
assign FPGAIB = 1;                  // disable Platform Flash

s3eadc M0 (CCLK, adcdav, davadc, adc0data, adc1data, adcsck, adcspod, conad);
s3eprogamp M1 (CCLK, ampdav, davamp, ampcmd0, ampcmd1, ampsck, ampspid, csamp, sdamp);
adclcd M2 (CCLK, BTN0, resetlcd, clearlcd, homelcd, datalcd, addrlcd, initlcd, lcdreset, lcdclear,
        lcdhome, lcddata, lcdaddr, lcddatin, digitmux, data);
lcd M3 (CCLK, resetlcd, clearlcd, homelcd, datalcd, addrlcd, lcdreset, lcdclear, lcdhome, lcddata,
        lcdaddr, rslcd, rwlcd, elcd, lcdd, lcddatin, initlcd);
genampadc M4 (CCLK, SW0, SW1, SW2, SW3, ampdav, davamp, ampcmd0, ampcmd1, adcdav,
        davadc, adc0data, adc1data, digitmux, data);

endmodule

module genampadc (input genclk, SW0, SW1, SW2, SW3, output reg ampdav, input davamp,
        output reg [3:0] ampcmd0, output reg [3:0] ampcmd1, output reg adcdav,
        input davadc, input [13:0] adc0data, input [13:0] adc1data,
        input [2:0] digitmux, output reg [3:0] data);

reg [2:0] gstate;       // state register
reg [3:0] adc4;         // ADC value BCD digits
reg [3:0] adc3;
reg [3:0] adc2;
reg [3:0] adc1;
reg adcsign;            // ADC sign
reg [13:0] value;       // ADC value

integer i;

always@(digitmux)          // LCD digit mux
    begin
        case (digitmux)
            0:    data = adc4;
            1:    data = adc3;
            2:    data = adc2;
            3:    data = adc1;
            4:    data[0] = adcsign;
            default: data = 0;
        endcase
    end
```

```
always@(posedge genclk)
	begin
		case (gstate)
			0:	begin
					ampdav = 0;
					adcdav = 0;
					if (ampcmd0[2] != SW2 || ampcmd0[1] != SW1
							|| ampcmd0[0] != SW0)
						begin		// set prog amp gain
							ampcmd0[3] = 0;
							ampcmd0[2] = SW2;
							ampcmd0[1] = SW1;
							ampcmd0[0] = SW0;
							ampcmd1[3] = 0;
							ampcmd1[2] = SW2;
							ampcmd1[1] = SW1;
							ampcmd1[0] = SW0;
							gstate = 1;
						end
					else
						gstate = 3;		// no change in gain
				end
			1:	begin
					ampdav = 1;		// prog amp gain
					gstate = 2;
				end
			2:	begin
					if (davamp == 1)
						begin
							ampdav = 0;
							gstate = 3;
						end
				end
			3:	begin
					adcdav = 1;		// ADC conversion
					if (davadc == 1)
						begin
							adcdav = 0;
							gstate = 4;
						end
				end
			4:	begin
					adcsign = 0;
					if (SW3 == 1)		// ADC channel A/B
						begin
							value[13:0] = adc1data[13:0];
							if (adc1data[13] == 1)
								begin
									adcsign = 1;
									value[13:0] = ~value[13:0] + 1;
								end
						end
```

```
                    else
                         begin
                              value[13:0] = adc0data[13:0];
                              if (adc0data[13] == 1)
                                   begin
                                        adcsign = 1;
                                        value[13:0] = ~value[13:0] + 1;
                                   end
                         end

                    adc1 = 0;              // thousands digit
                         for (i = 1; i <= 9; i = i + 1)
                              begin
                                   if (value >= 1000)
                                        begin
                                             adc1 = adc1 + 1;
                                             value = value – 1000;
                                        end
                              end
                    adc2 = 0;              // hundreds digit
                         for (i = 1; i <= 9; i= i + 1)
                              begin
                                   if (value >= 100)
                                        begin
                                             adc2 = adc2 + 1;
                                             value = value – 100;
                                        end
                              end
                    adc3 = 0;              // tens digit
                         for (i = 1; i <= 9; i = i + 1)
                              begin
                                   if (value >= 10)
                                        begin
                                             adc3 = adc3 + 1;
                                             value = value – 10;
                                        end
                              end

                    adc4 = value[3:0];     // units digit
                    gstate = 0;
               end
          default: gstate=0;
     endcase
end

endmodule
```

The adclcd.v module used here is the controller for the lcd.v datapath module, as described in Chapter 1 Verilog Hardware Description Language, and is similar to the genlcd.v module used for keyboard data for the Spartan-3E Starter Board, as given in Listing 3.20. The controller module genampadc.v uses the 3-bit signal digitmux from the adclcd.v module and outputs the ADC digit values adc1, adc2, adc3 and adc4 to the LCD as a signed 4-digit decimal number (+8191 to –8192).

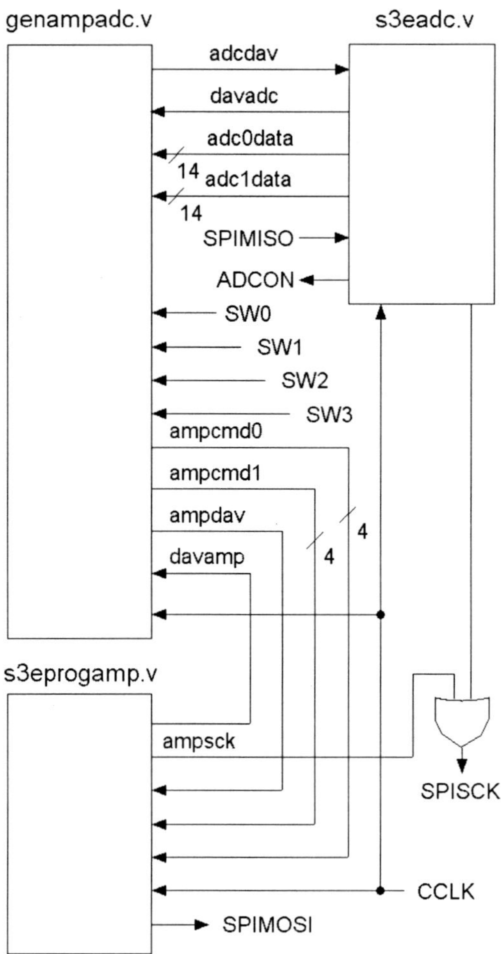


**Figure 3.9** Partial controller and datapath modules for the module s3eadctest.v

As described in Chapter 2 Verilog Design Automation, the Design Utilization Summary for the top module s3eadctest.v shows the use of 266 slice flip-flops (2%), 1488 4-input LUTs (15%), 852 occupied slices (18%) and a total of 13 184 equivalent gates in the XC3S500E Spartan-3E FPGA synthesis.

## Basys Board ADC

The Basys Board does not have an integral analog-to-digital converter (ADC). However, this Spartan-3E FPGA evaluation board has four 6-pin peripheral hardware module connectors and an accessory ADC, the Digilent Pmod AD1, is available (*www.digilentinc.com*), as shown in Figure 3.10. The Spartan-3E Starter Board also has two 6-pin peripheral hardware module connectors and this ADC is used with this evaluation board also for embedded applications.

The Pmod AD1 hardware module features two National Semiconductor (*www.national.com*) ADCS7476 12-bit ADCs which utilize the serial peripheral interface (SPI) bus protocol. The ADCs use an active low ADCCS signal (similar to the DACCS signal for the Linear Technology LTC2624 DAC on the Spartan-3E Starter Board) and an SPI bus clock signal SCLK and output two data signals

ADCD0 and ADCD1. The ADC SPI data communication protocol begins with the chip select signal ADCCS set to logic 0.

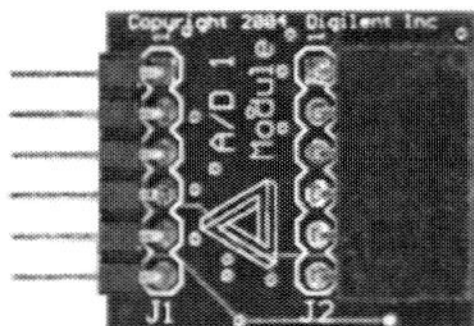

**Figure 3.10** Digilent Pmod AD1 module with two 12-bit ADCs

The Pmod AD1 ADCD7476 ADC uses the falling edge of the SPI clock signal SCLK to transmit a 16-bit data packet, with the first four bits as don't care bits and the most significant bit (MSB) of the data sent first, as listed in Table 3.10. After all 16 bits have been sent, the chip select signal ADCCS is set to logic 1.

**Table 3.10** Pmod AD1 National Semiconductor ADCS7476 ADC 16-bit data packet

| Bits | Contents (MSB...LSB) |
|---|---|
| 15-12 | X (don't care) |
| 11-0 | ADC channel data |

The Pmod AD1 ADC is connected to a 6-pin peripheral hardware module connector. The chip select signal ADCCS is pin 1, the two output data signals ADCD0 and ADCD1 are pins 2 and 3 and the SPI bus clock signal SCLK is pin 4. Pins 5 and 6 are DC ground and power (VDD = +3.3 V DC).

The n-bit binary data output D[n–1:0] for the ADC that inputs a unipolar (only positive amplitude) analog signal is determined by Equation 3.5, where the gain G = 1, $V_{REF} = 0$, $V_{FS} = +3.3$ V DC, $V_{IN}$ is the analog signal input and n = 12 here. The range of the analog input signal is then $0 \leq V_{IN} \leq +3.3$ V and the 12-bit binary data output D[11:0] is in straight signed binary format with an integer value between –2048 and 2047 [Wakerly00].

The module ad1adc.v for the Pmod AD1 ADC, which is in the *Chapter 3\peripherals* and is given in Listing 3.32. The file download procedure is described in the Appendix. This module is a datapath utilizing a finite state machine (FSM) with the 6-bit state register adcstate and provides signals to the controller and the ADC device, as described in Chapter 1 Verilog Hardware Description Language.

The National Semiconductor ASCS7476 ADC has a specified maximum SPI bus clock frequency of approximately 20 MHz and the crystal clock oscillators of the Basys Board and the Spartan-3E Starter Board are 50 MHz. A Digital Clock Manager (DCM) Verilog HDL module, as described in Chapter 2 Verilog Design Automation, could output a 40 MHz clock which is then divided by the transitions of the FSM to produce a 20 MHZ ADC SPI bus clock. However, a conservative SPI bus clock frequency of 12.5 MHz produces reliable ADC data here.

The SPI bus clock frequency in the ADC datapath module s3ead1.v is 12.5 MHz because the FSM transitions occur at a frequency of 25 MHz. A toggle register adcclk divides the 50 Mhz clock bufclk by two and generates the FSM clock. The ADC uses a logic 0 chip select signal adccs for the conversion command. A 16-bit data packet is used to convert two analog input signals to 12-bits of resolution, so the approximate maximum throughput data rate $R_{ADC}$ = 12.5 MHz / 16 bits/packet = 0.781 Msamples/sec.

The data available signal adcdav from the controller sets the adcstate register to 0 if it is a logic 0 or activates the FSM if it and the derived clock signal adcclk is a logic 1 and if the datapath signal davdac is logic 0. The datapath returns the status signal davadc as a logic 1 to the controller when the process of analog conversion and the output of the data is completed. The ADC chip select signal is controlled by the module output register variable adccs. The Pmod AD1 ADC datapath module s3ead1.v is verified by the Verilog top module ad1test.v, which is given in Listing 3.33. The top module ad1test.v is similar to the Verilog top module s3eadctest.v, as given in Listing 3.31, and includes the controller module genad1adc.v.

The complete Xilinx ISE WebPACK project is in the in the *Chapter 3\ad1test\s3ead1test* folder and uses the UCF ad1tests3esb.ucf that uncomments the signals CCLK, BTN0, SW3, JA1, JA2, JA3, JA4 and the LCD signals in the Spartan-3E Starter Board UCF of Listing 3.2. A similar Pmod AD1 ADC datapath module is also verified for the Basys Board and the seven segment display in the *Chapter 3 \ad1test\baad1test* folder but is not given here. The file download procedure is described in the Appendix.

**Listing 3.32** Datapath module for the Pmod AD1 ADC s3ead1.v

```
// Basys Board and Spartan-3E Starter Board
// Analog-to-Digital Converter ad1adc.v
// c 2008 Embedded Design using Programmable Gate Arrays  Dennis Silage

module ad1adc (input bufclk, adcdav, output reg davadc, output reg [11:0] adc0data,
            output reg [11:0] adc1data, output reg adcsck, input adc0d, adc1d, output reg adccs);

reg [6:0] adcstate;
reg adcclk;

always@(posedge bufclk) // buffered clock event driven
    begin
        adcclk = ~adcclk;     // clock divider
            if (adcdav == 0)
                begin
                    adcstate = 0;
                    adcsck = 1;
                    adccs = 1;
                    davadc = 0;     // ADC data NAK
                end

        if (adcdav == 1 && davadc == 0 && adcclk == 1)
            begin
                case (adcstate)
                0:  begin
                        adccs = 0;          // ADC chip select
                        adcsck = 1;
                        adcstate = 1;
                    end
                1:  begin
                        adcsck = 0;         // 1  X
                        adcstate = 2;
                    end
                2:  begin
                        adcsck = 1;
                        adcstate = 3;
                    end
```

```
3:    begin
            adcsck = 0;        // 2  X
            adcstate = 4;
      end
4:    begin
            adcsck = 1;
            adcstate = 5;
      end
5:    begin
            adcsck = 0;        // 3  X
            adcstate = 6;
      end
6:    begin
            adcsck = 1;
            adcstate = 7;
      end
7:    begin
            adcsck = 0;        // 4  x
            adcstate = 8;
      end
8:    begin
            adcsck = 1;
            adcstate = 9;
      end
9:    begin
            adcsck = 0;
            adc0data[11] = adc0d;
            adc1data[11] = adc1d;
            adcstate = 10;
      end
{ADC channels 0 and 1 data bits 10 through 1 are sent similarly in states 10 through 29}
30:   begin
            adcsck = 1;
            adcstate = 31;
      end
31:   begin
            adcsck = 0;
            adc0data[0] = adc0d;
            adc1data[0] = adc1d;
            adcstate = 32;
      end
32:   begin
            adcsck = 1;
            adcstate = 33;
      end
33:   begin
            adccs = 1;
            davadc = 1;
            adcstate = 34;
      end
34:   adcstate = 34;
default: adcstate = 34;
```

```
            endcase
        end
    end

endmodule
```

The slide switch SW3 selects one of the 12-bt straight binary numbers adc0data or adc1data, which is then converted to a four digit binary coded decimal (BCD) number adc1, adc2, adc3, and adc4. The sign bit adcsign in Listing 3.31 is not used here and defaults to logic 0 which allows the design reuse of the adclcd.v module to display the result on the LCD.

The 12-bit register variable value is used to manipulate the 12-bit input binary numbers adc0data or adc1data which cannot be modified. The BCD digits are calculated by the successive subtraction algorithm, as described in Chapter 2 Verilog Design Automation.

The four Verilog modules operate in parallel and some independently in the top module ad1test.v in Listing 3.33. However, to reduce the complexity of the configuration of the controllers and datapaths in Figure 3.11, the lcd.v module and its interconnections are not shown.

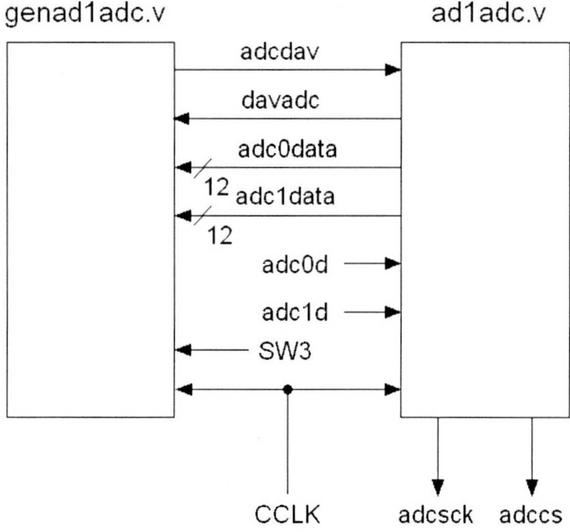


**Figure 3.11** Partial controller and datapath modules for the module ad1test.v

**Listing 3.33** Pmod AD1 ADC test top module for the Spartan-3E Starter Board ad1test.v

```
// Spartan-3E Starter Board
// Analog-to-Digital Converter ad1adctest.v
// c 2008 Embedded Design using Programmable Gate Arrays  Dennis Silage


module ad1test(input CCLK, BTN0, SW3, JA2, JA3, output JA1, JA4, LCDRS, LCDRW,
                    output LCDE, output [3:0] LCDDAT);

wire adcdav, davadc, adcsck, adccs, adc0d, adc1d;
wire resetlcd, clearlcd, homelcd, datalcd, addrlcd;
wire initlcd, lcdreset, lcdclear, lcdhome, lcddata;
wire lcdaddr, rslcd, rwlcd, elcd;
```

```
wire [11:0] adc0data, adc1data;
wire [3:0] data, lcdd;
wire [2:0] digitmux;
wire [7:0] lcddatin;

assign LCDDAT[3] = lcdd[3];
assign LCDDAT[2] = lcdd[2];
assign LCDDAT[1] = lcdd[1];
assign LCDDAT[0] = lcdd[0];

assign LCDRS = rslcd;
assign LCDRW = rwlcd;
assign LCDE = elcd;

assign JA1 = adccs;
assign adc0d = JA2;
assign adc1d = JA3;
assign JA4 = adcsck;

ad1adc M0 (CCLK, adcdav, davadc, adc0data, adc1data, adcsck, adc0d, adc1d, adccs);
adclcd M1 (CCLK, BTN0, resetlcd, clearlcd, homelcd, datalcd, addrlcd, initlcd, lcdreset, lcdclear,
           lcdhome, lcddata, lcdaddr, lcddatin, digitmux, data);
lcd M2 (CCLK, resetlcd, clearlcd, homelcd, datalcd, addrlcd, lcdreset, lcdclear, lcdhome, lcddata,
        lcdaddr, rslcd, rwlcd, elcd, lcdd, lcddatin, initlcd);
genad1adc M3 (CCLK, SW3, adcdav, davadc, adc0data, adc1data, digitmux, data);

endmodule

module genad1adc (input genclk, SW3, output reg adcdav, input davadc, input [11:0] adc0data,
                  input [11:0] adc1data, input [2:0] digitmux, output reg [3:0] data);

reg gstate;            // state register
reg [3:0] adc4;        // ADC value BCD digits
reg [3:0] adc3;
reg [3:0] adc2;
reg [3:0] adc1;
reg [11:0] value;      // ADC value

integer i;

always@(digitmux)          // LCD digit mux
    begin
        case (digitmux)
            0:   data = adc4;
            1:   data = adc3;
            2:   data = adc2;
            3:   data = adc1;
            default: data = 0;
        endcase
    end
```

```
always@(posedge genclk)
    begin
        case (gstate)
            0:  begin
                    adcdav = 1;             // ADC conversion
                    if (davadc == 1)
                        begin
                            adcdav = 0;
                            gstate = 1;
                        end
                end
            1:  begin
                    if (SW3 == 1)           // ADC channel A/B
                        value = adc1data;
                    else
                        value = adc0data;
                    adc1 = 0;               // thousands digit
                        for (i = 1; i <= 9; i  = i + 1)
                            begin
                                if (value >= 1000)
                                    begin
                                        adc1 = adc1+1;
                                        value = value – 1000;
                                    end
                            end
                    adc2 = 0;               // hundreds digit
                        for (i = 1; i <= 9; i = i + 1)
                            begin
                                if (value >= 100)
                                    begin
                                        adc2 = adc2 + 1;
                                        value = value – 100;
                                    end
                            end
                    adc3 = 0;               // tens digit
                        for (i = 1; i <= 9; i = i + 1)
                            begin
                                if (value >= 10)
                                    begin
                                        adc3 = adc3 + 1;
                                        value=value – 10;
                                    end
                            end
                    adc4 = value[3:0];      // units digit
                    gstate = 0;
                end
        endcase
    end

endmodule
```

The device signals for the serial flash PROM SPISF, the digital-to-analog converter (DAC) chip select DACCS and clear DACCLR, the parallel flash PROM SFCE and the platform flash PROM FPGAIB, as given in the Spartan-3E Starter Board UCF in Listing 3.2, are disabled in the top module with continuous assignment statements. The SPI bus clock signal SPISCK is continuously assigned as either the ADC SPI clock adcsck or the programmable amplifier SPI clock ampsck. The controller module genampadc.v insures that these SPI bus clock signals are only active one at a time.

The adclcd.v module used here is the controller for the lcd.v datapath module, as described in Chapter 1 Verilog Hardware Description Language, and is similar to the genlcd.v module used for keyboard data for the Spartan-3E Starter Board, as given in Listing 3.20. The controller module genampadc.v uses the 3-bit signal digitmux from the adclcd.v module and outputs the ADC digit values adc1, adc2, adc3 and adc4 to the LCD as a signed 4-digit decimal number (+8191 to –8192).

As described in Chapter 2 Verilog Design Automation, the Design Utilization Summary for the top module ad1test.v shows the use of 185 slice flip-flops (1%), 1576 4-input LUTs (16%), 876 occupied slices (18%) and a total of 11 527 equivalent gates in the XC3S500E Spartan-3E FPGA synthesis.

## Auxiliary Ports and Peripherals

The Spartan-3E Starter Board includes additional ports and peripherals for embedded system design. These include a clock source port, a Video Graphics Array (VGA) port, an RS-232 serial ports, serial flash memory and hardware expansion ports and connectors. Auxiliary peripherals are also available as hardware modules and are utilized for Xilinx ISE WebPACK projects in Chapter 4 Digital Signal Processing, Communications and Control and Chapter 5 Embedded Processors.

The clock can be supplied from another source and inputted at the SMA connector on the Spartan-3E Starter Board, as shown in Figure 3.2. There is also an 8-pin dual in-line (DIP) socket into which a crystal oscillator device can be placed to derive a clock frequency other than the nominal 50 MHz source. However, the integral Digital Clock Manager (DCM) of the Spartan-3E FPGA can also be used to generate or synthesize other clock frequencies from the 50 MHz source, as given in Listing 3.24.

The hard-wired Video Graphics Array (VGA) port on the Basys Board or the Spartan-3E Starter Board provides a high resolution display in embedded applications. A single DB15 (15-pin) connector can be connected to a standard video cathode ray tube (CRT) or flat-panel liquid crystal display (LCD) monitor using a standard cable.

The Spartan-3E Starter Board has two RS-232 serial data communication ports configured as a data terminal equastion (DTE) and a data communication equipment (DCE) port. An auxiliary RS-232 serial data port is available for the Basys Board (Pmod RS232, *www.digilentinc.com*). A Maxim (*www.maxim-ic.com*) MAX3232 device converts the unipolar FPGA voltage output and input to the bipolar RS232 voltage standard. Hardware flow control for the RS232 serial data communication is not provided. The data carrier detect (DCD), data terminal ready (DTR) and data set ready (DSR) signals are connected together on the DTE and DCE port. The RS-232 serial data communication port is used in an embedded design project in Chapter 4 Digital Signal Processing, Communication and Control.

**Figure 3.12** Digilent Pmod SF serial flash PROM module

The Spartan-3E Starter Board includes a Xilinx 4 megabit (Mb) XCF04S platform flash programmable read-only memory (PROM) and an STMicroelectronics (*www.st.com*) MP25P16 16 Mb serial peripheral interface (SPI) flash PROM. The Basys Board includes a Xilinx 2 Mb XCF02S flash PROM. However, these PROMs are used to configure the volatile interconnections of the Spartan-3 and Spartan-3E FPGA on power-up by downloading the bit file. In addition, the serial flash on the Spartan-3E Starter Board utilizes the SPI bus common to other peripherals, such as the ADC and DAC, which complicates its use. An auxiliary 2 Mb serial flash PROM module is available (Pmod SF, *www.digilientinc.com*), as shown in Figure 3.12.

## Hardware Expansion Port and Connectors

The Spartan-3E Starter Board includes a 100-pin edge connector with 43 FPGA input-output pins and three 6-pin hardware module connector. Auxiliary 6-pin hardware module connector adapters available (*www.digilientinc.com*) for the Basys Board and the Spartan-3E Starter Board include wire screw terminals (Pmod CON1), a dual BNC connector (Pmod CON2), test point header (Pmod TPH) and PS/2 connector (Pmod PS2). These auxiliary adapters are used to conveniently connect test equipment and peripherals during the development of an embedded design and are shown in Figure 3.13.

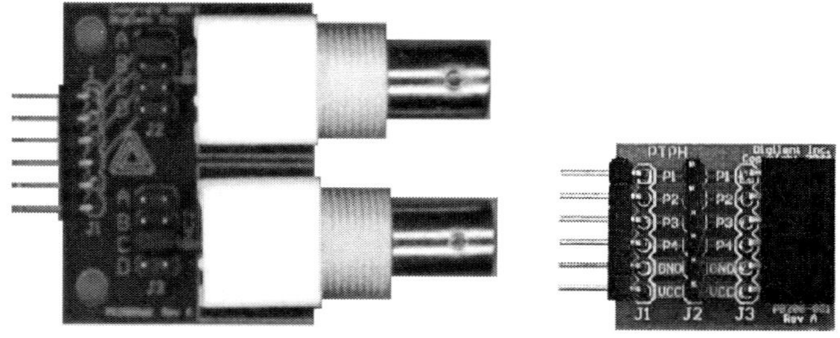


**Figure 3.13** Digilent Pmod BNC and Pmod TPH test point header connector adapters

## Auxiliary Hardware Peripherals

Auxiliary 6-pin hardware module peripherals available from Digilent (*www.digilientinc.com*) for the Basys Board and the Spartan-3E Starter Board include a digital-to-analog converter (Pmod DA1 and Pmod DA2), analog-to-digital converter (Pmod AD1), digital input module (Pmod DIN1), servomotor H-bridge module (Pmod HB5), open-collector driver module (Pmod OC1), RS-232 serial communication module (Pmod RS232), optical sensor interface (Pmod LS1), rotary shaft encoder (Pmod ENC) and high current open-drain driver (Pmod OD1) modules. These auxiliary hardware periperals are used to augment the Basys Board and the Spartan-3E Starter Board in the development of an embedded design and are shown in Figure 3.14 and Figure 3.15.

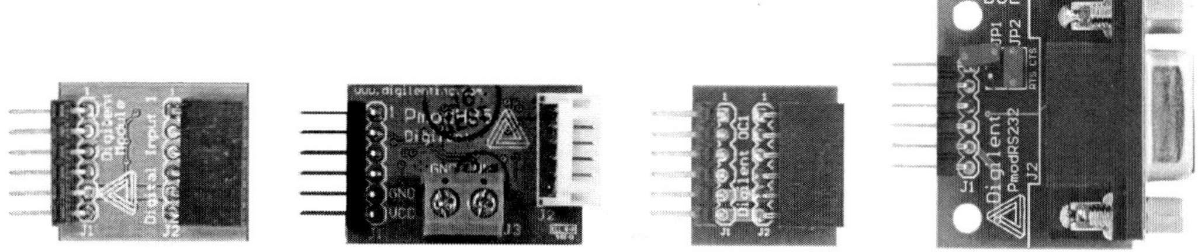


**Figure 3.14** Digilent Pmod DIN1 digital input, Pmod HB5 servomotor H-bridge, Pmod OC1 open-collector driver and Pmod RS232 modules

The Pmod DA1 and Pmod DA2 digital-to-anlaog converters and the Pmod AD1 analog-to-digital converter (ADC) are described in this Chapter. The Pmod HB5 H-bridge module, as shown in Figure 3.14, is used in a DC servomotor digital control embedded design project in Chapter 4 Digital Signal Processing, Communications and Control. The Digilent Pmod HB3 H-bridge module has the same electrical specifications but uses a different type of connector for the servomotor.

The Pmod RS232 module, as shown in Figure 3.14, can provide RS-232 serial data communication for the Basys Board which, unlike the Spartan-3E Starter Board, does not have an integral RS232 standard port. RS232 standard serial data communication is described in Chapter 4 Digital Signal Processing, Communications and Control. Similarly, the Pmod ENC module, as shown in Figure 3.15, is a rotary shaft encoder for an evaluation board like the Basys Board which does not feature this device. The rotary shaft encoder is described in this Chapter.

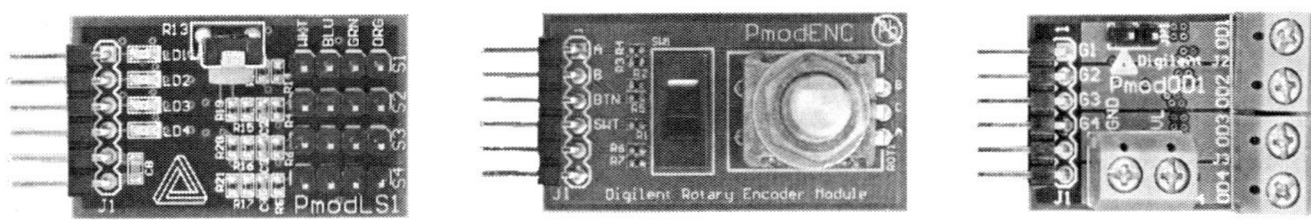


**Figure 3.15** Digilent Pmod LS1 optical sensor interface, Pmod ENC rotary shaft encoder and Pmod OD1 high current open-drain driver modules

The Pmod LS1 optical sensor interface module, as shown in Figure 3.15, is used with an infrared (IR) light emitting diode (LED) and an IR sensitive phototransistor for positional sensing. Finally, the Pmod OD1 high current open-drain module, also shown in Figure 3.15, can switch up to 3 A at 20 V DC for servoactuators and electromechanical relays.

The auxiliary 6-pin hardware module interface consists of four logic signals that are directly connected to the FPGA, +3.3 V DC power and ground. The maximum current available for input or output is approximately 1 mA DC at each of the FPGA pins. Interface integrated circuit (IC) devices must observe the maximum voltage of +3.3 V DC, the maximum current inputted or outputted of 1 mA DC and the resulting overall power dissipation of the FPGA in the intended embedded design application. The data sheet for the Spartan-3E FPGA (DS312, *www.xilinx.com*) provides the completed electrical characteristics of the device.

The four available logic signals of the auxiliary 6-pin hardware peripheral connector limits the extent of the devices that can be used but are suitable for the serial peripheral interface (SPI) bus modules described in this Chapter. The 4 logic signals can be input, output or bidirectional (IOB) signals as determined by the User Constraints File (UCF) for the FPGA evaluation board.

However, the Spartan-3E Starter Board provides a 100-pin, fine pitch FX2 connector with 39 IOB logic signals and 5 dedicated input-only logic signals for clock signals for advanced embedded design projects. The Digilent VDEC1 module (*www.digilientinc.com*) utilizes the Analog Devices (*www.analog.com*) ADV7183B video decoder IC and the Spartan-3E Starter Board FX2 connector to digitize NTSC, PAL and SECAM analog video signals with three 54 MHz clock ADC providing high speed 16-bit parallel data to the FPGA.

## Summary

In this Chapter two evaluation boards that utilize the Xilinx Spartan-3E FPGA are described. The integral components and digital-to-analog and analog-to-digital converters auxiliary hardware modules of the evaluation boards are presented in operation by complete Xilinx ISE WebPACK projects. The projects in this Chapter utilize the controller and datapath and finite state machine construct, as described in Chapter 1 Verilog Hardware Description Language [Ciletti04]. These projects in this Chapter also illustrate the use of the components, ports and external hardware peripherals in embedded design.

Chapter 4 Digital Signal Processing, Communications and Control presents Xilinx ISE WebPACK projects in the Verilog HDL and the tenants of digital filtering, digital modulation in communication systems, digital data transmission and digital control system design. Chapter 5 Embedded Soft Core Processors describes the electronic design automation (EDA) software for the Xilinx PicoBlaze 8-bit soft-core processor and its comparison to the Verilog controller-datapath modules and Xilinx LogiCORE blocks in the Xilinx ISE WebPACK. These two Chapters survey the application of the Verilog HDL, the soft core processor and programmable gate arrays in embedded design.

## References

[Botros06] Botros, Nazeih M., *HDL Programming Fundamentals*. Thomson Delmar, 2006.

[Ciletti99] Cilletti, Michael D., *Modeling, Synthesis and Rapid Prototyping with the Verilog HDL*. Prentice Hall, 1999.

[Ciletti04] Cilletti, Michael D., *Starter's Guide to Verilog 2001*. Prentice Hall, 2004.

[Lee06] Lee, Sunguu, *Advanced Digital Logic Design*. Thomson, 2006.

[Navabi06] Navabi, Zainalabedin, *Verilog Digital System Design*. McGraw-Hill, 1999.

[Wakerly00] Wakerly, John F., *Digital Design Principles and Practice*, Prentice Hall, 2000.

# 4

# Digital Signal Processing, Communications and Control

Electronic design automation (EDA) tools and hardware are available for embedded system design in the Verilog hardware description language (HDL) using a field programmable gate array (FPGA). Structural and behavioral models, finite state machines (FSM) and controller and datapath constructs are presented in Chapter 1 Verilog Hardware Description Language. The Xilinx ISE WebPACK™ EDA is presented in Chapter 2 Verilog Design Automation. The Xilinx Spartan™-3E FPGA evaluation boards and peripheral hardware modules are described in Chapter 3 Programmable Gate Array Hardware with complete Xilinx ISE WebPACK projects.

This Chapter presents the tenants of digital signal processing (DSP), digital communications and digital control with emphasis on embedded design in the Verilog HDL using the FPGA. The Xilinx LogiCORE blocks provide additional functionality for the projects in this Chapter. The LogiCORE FIR Compiler implements the finite impulse response (FIR) digital filter. The LogiCore Sine-Cosine Look-Up Table is used for a dual tone multiple frequency (DTMF) audio signal generator. The LogiCORE DDS Compiler is used for a sinusoidal frequency generator and frequency and phase shift keying modulators in digital communication.

A Verilog HDL construct is used as a binary sequence generator in a linear finite shift register (LFSR) for the generation of pseudo-random numbers and data. Serial data communication is facilitated by the implementation in Verilog HDL of a *soft-core* universal asynchronous receiver transmitter (UART) peripheral as a terminal emulator and a Manchester encoder-decoder. Finally, a closed-loop proportional digital controller in Verilog HDL for a DC servomotor is realized.

The Verilog source modules and project files are located in the Chapter 4 folder as subfolders identified by the name of the appropriate project. The complete contents and the file download procedure are described in the Appendix. The projects in this Chapter illustrate not only the use of the components, ports and external hardware peripherals in applications in DSP, digital communications and digital control, but the versatility of the Xilinx ISE WebPACK EDA and the Verilog HDL for real-time embedded design.

## Sampling and Quantization

Analog sources of information are often derived from transducers which provide continuous electrical voltage signals from physical phenomenon such as light, pressure, temperature, vibration, and acceleration. These analog *baseband* signals are bandlimited to a maximum frequency. Analog baseband signals are continuous in time and amplitude and are *sampled* and *quantized* for digital signal processing (DSP) [Silage06].

Analog signals are first sampled at discrete intervals of time but continuous in amplitude. Quantization then is the *roundoff* of the continuous amplitude sample to a discrete preset value, represented as binary number. The preset values are equally spaced in *uniform quantization* and the total number of binary bits is the resolution. The *ideal* sampling operation can be described as a multiplication of the baseband analog signal $x(t)$ by a periodic series of unit impulse functions $\delta(t - mT_s)$, where $T_s$ is the *sampling interval* and $m$ is an arbitrary index. The ideal sampling process is determined by Equation 4.1.

$$x(mT_s) = \sum_{m} x(t)\, \delta(t - mT_s) \qquad (4.1)$$

The response of the ideal sampling process is described in the spectral domain by the normalized (the load resistor $R_L = 1\ \Omega$) power spectral density (PSD), as determined by Equation 4.2 [Lathi98].

$$\text{PSD} = f_s^2 \sum_k | X(f - k f_s) |^2 \qquad (4.2)$$

The Fourier transform of the baseband analog signal $x(t)$ is $X(f)$ and the *sampling rate* $f_s = 1/T_s$ samples/sec. The normalized PSD of $x(t)$ is $| X(f) |^2$ which is bandlimited to a maximum frequency of $f_{max}$. However, the ideally sampled signal $x(mT_s)$ has a PSD which is not bandlimited but repeats the baseband PSD centered at multiples of the sampling rate $f_s$, as determined by Equation 4.2. The direct implication is that to avoid spectral overlay or *aliasing* the sampling rate $f_s > 2 f_{max}$, where $2 f_{max}$ is the *Nyquist rate* [Haykin01]. To avoid aliasing, an analog filter with a *cutoff frequency* $f_{cutoff} < f_s/2$ is often placed before the ideal sampler.

The output of the ideal sampler $x(mT_s)$ is inputted to an first order sample-and-hold process, whose output $y_{s\text{-}h}(t)$ is determined by Equation 4.3.

$$\begin{aligned} y_{s\text{-}h}(t) &= \sum_m x(mT_s)\, h(t - mT_s) \quad mT_s \le t < (m+1)T_s \\ \text{where} \quad h(t) &= 1 \quad 0 \le t < T_s \\ h(t) &= 0 \quad \text{otherwise} \end{aligned} \qquad (4.3)$$

The sample-and-hold output $y_{s\text{-}h}(t)$ is continuous in time but represents a fixed amplitude signal during the sampling interval $T_s$. This analog signal is then inputted to an analog-to-digital converter (ADC) peripheral to affect the process of quantization. This fixed amplitude signal facilitates the ADC process by maintaining a constant input signal during the finite time required for conversion. The n-bit binary data output D[n-1:0] for the ADC that inputs a unipolar (only positive amplitude) analog signal is determined by Equation 4.4.

$$\text{D[n-1:0]} = \text{G} \frac{\text{V}_{\text{IN}} - \text{V}_{\text{REF}}}{\text{V}_{\text{FS}}} 2^{\text{n-1}} \qquad (4.4)$$

In Equation 4.4 G is the gain of any analog preamplifier that precedes the ADC, $V_{REF}$ is the common reference voltage of the analog preamplifier and ADC, $V_{FS}$ is the *full-scale* conversion voltage for the ADC, $V_{IN}$ is the polar analog signal input at the input of the analog preamplifier and n is the number of bits of resolution. The most significant bit (MSB) of the n-bit binary data output D[n-1:0] represents the sign bit and a number whose integer value is between $-2^{n-1}$ and $2^{n-1} - 1$. The ideal uniform ADC quantizer voltage step size Δ is determined by Equation 4.5.

$$\Delta = \frac{2\,\text{V}_{\text{MAX}}}{\text{L}} \qquad (4.5)$$

$V_{MAX}$ is the equal positive and negative maximum input voltage with respect to $V_{REF}$, $L = 2^n$ is the number of levels in the output of the ideal uniform ADC quantizer and n is the number of bits. The maximum *quantization error* q that can occur in the sampled output of the ideal uniform quantizer is ± Δ/2 V. Assuming that all values of quantization error as a random variable within the range +Δ/2 to −Δ/2 are equally likely and from Equation 4.4, the mean square quantizing error $E_q$ is determined by Equation 4.6 [Haykin01].

$$E_q = \frac{1}{\Delta} \int_{-\Delta/2}^{\Delta/2} q^2 \, dq = \frac{\Delta^2}{12} = \frac{V_{MAX}^2}{3 L^2} \qquad (4.6)$$

$E_q$ is also the normalized power in the resulting quantizing noise. The root mean square (RMS) quantizing noise is $\Delta/\sqrt{12} = \Delta/3.464$. If the normalized power in the analog signal is $S_o$ then, from Equation 4.5 and Equation 4.6, the signal to quantization noise ratio ($SNR_q$) is determined by Equation 4.7.

$$SNR_q = \frac{12\, S_o}{\Delta^2} = 3\, L^2 \frac{S_o}{V_{MAX}^2} \qquad (4.7)$$

$SNR_q$ is a linear function of the normalized power in the signal $S_o$ and a second order function of the number of levels $L = 2^n$ of the ideal uniform quantizer. If $S_o$ and $V_{max}$ remain constant but the number of bits n increases to n+1 (the number of levels L doubles) and from Equation 4.7, $SNR_q$ quadruples or increases by +6.02 dB ($10 \log_{10} 4$). For a sinusoidal input analog signal with a positive and negative maximum input voltage equal to $V_{MAX}$, the normalized power in the signal $S_o = V_{MAX}^2/2$ $V^2$-sec and, from Equation 4.6, the signal to quantization noise ratio $SNR_q = 1.5L^2$.

The two channel ADC of the Spartan-3E Starter Board described in Chapter 3 has 14 bits of resolution or n = 14, $V_{REF} \approx 1.65$ V and $V_{FS} = 1.25$ V. If the gain of the analog preamplifier G = 1, $V_{max}$ = 1.25 V and L = 16384 and $\Delta = 2.5/16384$ V $\approx$ 0.1526 mV. For a sinusoidal input signal with an amplitude of $V_{MAX}$, $SNR_q = 1.5 \times (16384)^2 = 86.05$ dB.

The two channel Digilent Pmod AD1 ADC hardware module described in Chapter 3 has 12 bits of resolution or n = 12, $V_{REF} = 0$ V, $V_{FS} = 2V_{MAX} = 3.3$ V and G = 1. In this instance $\Delta = 3.3/4096$ V $\approx$ 0.8056 mV and for an offset or unipolar sinusoidal input signal of $V_{MAX}$, $SNR_q = 1.5 \times (4096)^2 =$ 74.01 dB. This $SNR_q$ performance is as expected for a 12-bit ADC compared to a 14-bit ADC since 84.05 – 2 × 6.02 dB = 74.01 dB.

The ADC of the Spartan-3E Starter Board has a single pole RC anti-aliasing filter at the analog input with a –3 dB cutoff frequency $f_{-3\,dB} = (2\pi RC)^{-1} \approx 1.54$ MHz. The Pmod AD1 peripheral hardware module ADC has an operational amplifier (*op amp*) two pole, anti-aliasing active filter with $f_{-3\,dB} \approx 500$ kHz at the analog input.

## Discrete Time Sequences

The uniformly sampled and n-bit quantized data output of the ADC is described by a sequence $x_q(mT_s)$ where $m$ is an arbitrary index, $T_s$ is the *sampling interval* and $x_q$ is a discrete binary number ranging from 0 to $2^n - 1$. To facilitate the description of the analysis $x_q(mT_s)$ is often written simply as $x(m)$. An input sequence $x(m)$ is processed by a general DSP system, or *digital filter*, to produce an output sequence $y(m)$, as determined by Equation 4.8 [Ifeachor02].

$$y(m) = \sum_{q}^{Q-1} b_q x(m-q) + \sum_{p=0}^{P-1} a_p y(m-p) \qquad (4.8)$$

Usually the output sequence $y(m)$ is considered to be the sum of the current (indices $q = 0$ and $p = 0$) and a finite number ($Q - q$ and $P$) of the past input and output sequences multiplied by coefficients ($b_q$ and $a_p$). However, the index $q$ for the input sequence $x(m)$ can be a negative ($q = -r$). This implies that output sequence $y(m)$ is also a sum of apparent future input sequences $x(m+r)$ multiplied by coefficients ($b_{-r}$), but this is interpreted as the output sequence is merely delayed by the index $|-r|$. The finite impulse response (FIR) class of digital filter has $a_p = 0$ for all $p$. FIR digital filters are also know as moving average (MA) filters since the output sequence is a weighted average of the input sequence, as determined by Equation 4.9 [Proakis07].

$$y(m) = \sum_{q}^{Q-1} b_q x(m-q) \qquad (4.9)$$

The coefficients $b_q$ of the FIR digital filter are identical to a finite length *impulse response* sequence [Chen01]. The determination of the coefficients an FIR filter can therefore be accomplished in the frequency domain by the *inverse discrete Fourier transform* (IDFT) of the specified frequency response [Mitra06]. In many digital communication and digital image processing applications a linear phase response is important to minimize distortion. The coefficients $b_q$ of the FIR digital filter have a simple relationship that mandates a linear phase sequence response, which is obtained by setting pairs of coefficients equal to each other, as determined by Equation 4.10.

$$b_0 = b_{Q-1} \quad b_1 = b_{Q-2} \quad b_3 = b_{Q-3} \text{ K} \qquad (4.10)$$

The infinite impulse response (IIR) class of digital filter includes the autoregressive (AR) filter, as determined by Equation 4.11 [Mitra06].

$$y(m) = x(m) - \sum_{p=0}^{P-1} a_p y(m-p) \qquad (4.11)$$

The general form of the IIR digital filter is the autoregressive, moving average (ARMA) filter determined by Equation 4.8 [Cavicchi00]. Since the output sequence of the IIR digital filter at least includes a weighted sum of the past output sequences, stability of the output can be problematical.

There is no simple relationship between the coefficients $b_q$ and $a_q$ of the IIR digital filter and the impulse response sequence or that for a linear phase sequence response as there is for the FIR digital filter. However, IIR digital filters require a lesser number of multiplicative coefficients to provide a specified digital filter frequency response than an FIR digital filter. Therefore IIR digital filters are less computationally intense and utilize fewer resources within the FPGA [Mitra06].

## Discrete Frequency Response

DSP requires that the characteristics of the digital filter be assessed in the frequency domain. The discrete time sequence can be readily transformed to the discrete frequency domain [Mitra06]. The discrete time input and output sequences $x(m)$ and $y(m)$ is assumed to have a discrete frequency transforms $X(z)$ and $Y(z)$, where the temporal delay $d$ is replaced by the discrete frequency delay parameter $z^{-d}$, as determined by Equation 4.12. The IIR ARMA digital filter in Equation 4.8 has the discrete frequency transfer or system function $H(z) = Y(z) / X(z)$, as then determined by Equation 4.13.

$$\begin{aligned} Y(z) &= \sum_{q}^{Q} b_q X(z) z^{-q} + \sum_{p=1}^{P} a_p Y(z) z^{-p} \\ Y(z) - \sum_{p=1}^{P} a_p Y(z) z^{-p} &= \sum_{q}^{Q} b_q X(z) z^{-q} \qquad (4.12) \\ Y(z)\left[1 - \sum_{p=1}^{P} a_p z^{-p}\right] &= X(z) \sum_{q}^{Q} b_q z^{-q} \end{aligned}$$

$$H(z)=\frac{Y(z)}{X(z)}=\frac{\sum_{q}^{Q}b_q z^{-q}}{1-\sum_{p=1}^{P}a_p z^{-p}} \qquad (4.13)$$

The discrete frequency system function *H*(*z*) for the FIR or MA digital filter in Equation 4.9 is then determined by Equation 4.14.

$$H(z)=\frac{Y(z)}{X(z)}=\sum_{q}^{Q}b_q z^{-q} \qquad (4.14)$$

Finally, the discrete frequency system function *H*(*z*) for the IIR AR filter in Equation 4.11 is then determined by Equation 4.15.

$$H(z)=\frac{Y(z)}{X(z)}=\frac{1}{1-\sum_{p=1}^{P}a_p z^{-p}} \qquad (4.15)$$

The frequency domain for a discrete frequency transform is the *unit circle* in the complex plane [Proakis07]. The discrete sequence sinusoidal excitation of the discrete system *H*(*z*) exists then as a point *P* on the unit circle in the range $0 \le 2\,\pi f\,T_s \le \pi$, where *f* is the frequency of excitation and $T_s$ is the sampling interval.

As a first example, a simple IIR digital filter rendered as the output sequence *y*(*m*) in terms of the input sequence *x*(*m*) and coefficients from Equation 4.13 will illustrate the discrete frequency response analysis. Here $Q = 0$, $q = -1$, $b_{-1} = 0.25$, $b_0 = 0.5$, $P = 1$, and $a_1 = 0.25$, as determined by Equation 4.16.

$$y(m) = 0.5\;x(m) + 0.25\;x(m+1) + 0.25\;y(m-1) \qquad (4.16)$$

The discrete frequency transfer function *H*(*z*) from Equation 4.12 is determined by Equation 4.17.

$$H(z)=\frac{Y(z)}{X(z)}=\frac{0.5+0.25\,z}{1-0.25\,z^{-1}}$$
$$H(z)=\frac{0.25\,z\,(z+2)}{z-0.25} \qquad (4.17)$$

This simple IIR digital filter has one transfer function *pole* (where the response is infinite) at $z = 0.25$, two transfer function zeros (where the response is zero) at $z = 0$ and $z = -2$ and a multiplicative term of 0.25. The pole (indicated by ×) and the zeros (indicated by ○) of the transfer function are plotted on the complex frequency domain plane, as shown in Figure 4.1. Vectors are drawn from the pole and zeros to an arbitrary point *P* on the unit circle which represents the sinusoidal excitation of the discrete system *H*(*z*) [Mitra06].

The magnitude of the discrete frequency response | *H*(*z*) | is determined by the product of the gain *G* and the length of the vector(s) from the zero(s) divided by the product of the length of the vector(s) from the poles. The angle from the origin of the complex plane to the point *P* on the unit circle where the sinusoidal excitation occurs is in the range $0 \leq 2\,\pi\, f\, T_s \leq \pi$ and the simple geometry allows the length of these vectors and the magnitude of the discrete frequency response to be determined as in Equation 4.18.

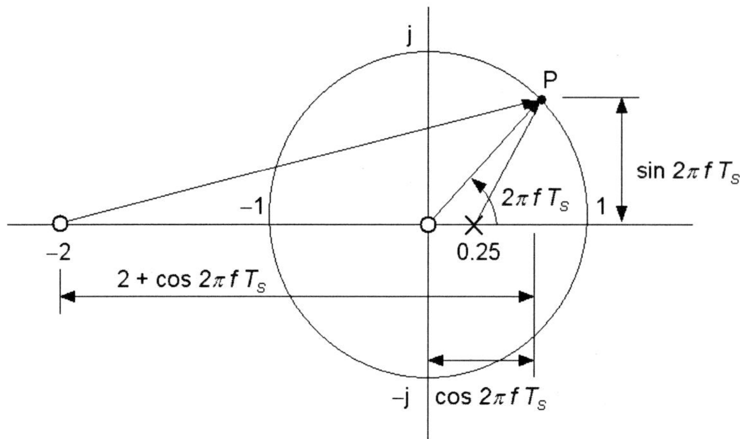


**Figure 4.1** IIR digital filter discrete frequency transfer function plotted on the complex plane

$$
\begin{aligned}
&G = 0.25 \qquad L_{z=0} = 1 \\
&L_{z=-2} = \sqrt{(2+\cos 2\pi f T_s)^2 + \sin^2 2\pi f T_s} \\
&L_{p=0.25} = \sqrt{(\cos 2\pi f T_s - 0.25)^2 + \sin^2 2\pi f T_s} \\
&|H(z)| = 0.25\frac{\sqrt{(2+\cos 2\pi f T_s)^2 + \sin^2 2\pi f T_s}}{\sqrt{(\cos 2\pi f T_s - 0.25)^2 + \sin^2 2\pi f T_s}}
\end{aligned} \tag{4.18}
$$

The magnitude of the discrete frequency response of this IIR digital filter is shown in Figure 4.2 on a *log-log* scale plotted against frequency Hz with $T_s = 4 \times 10^{-6}$ sec or $f_s = 1/\,T_s = 250$ kHz. This IIR digital filter has a *low pass* characteristic with a –3 dB (magnitude = 0.707) cutoff frequency of approximately 44.5 kHz and a low frequency gain of 1. This IIR low pass digital filter is efficient because the coefficients are all negative powers of the binary base 2 ($2^{-n}$) and can be performed with integer scaling (shifting).

The phase angle $\psi$ of the discrete frequency response $H(z)$ is determined by the sum of the angles subtended counter-clockwise with the real axis and the vectors drawn from the zeros of the discrete frequency response $H(z)$ to an arbitrary point $P$ on the unit circle less the angles subtended with the real axis and the vectors drawn from the poles [Cavicchi00]. The phase angle $\psi$ of the IIR low pass digital filter, derived from Figure 4.1 and Equation 4.18, is determined by Equation 4.19.

$$\psi = 2\pi f T_s + \tan^{-1}\left(\frac{\sin 2\pi f T_s}{2+\cos 2\pi f T_s}\right) - \tan^{-1}\left(\frac{\sin 2\pi f T_s}{\cos 2\pi f T_s - 0.25}\right) \tag{4.19}$$

The ambiguous calculation of the arctangent ($\tan^{-1}$) must also be considered in the analytical determination of the phase angle. A requirement for no distortion in the *baseband* region is that the phase angle should be a linear function of frequency there. The phase angle $\psi$ is only approximately linear in frequency $f$ due to the first term of Equation 4.18 ($2\pi f T_s$) for this simple IIR low pass digital filter.

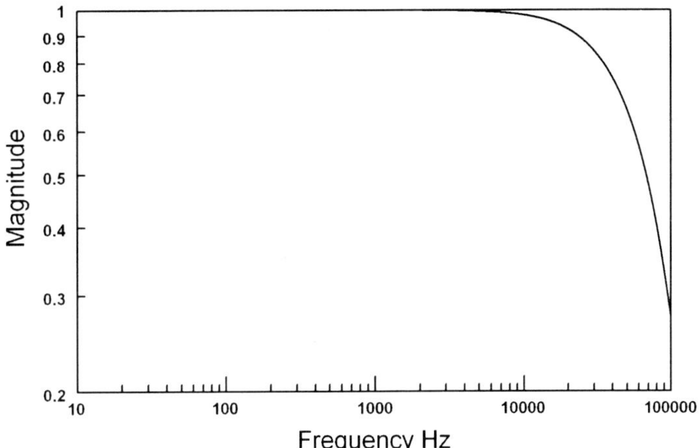


**Figure 4.2** IIR low pass digital filter discrete frequency response magnitude for $T_s = 4 \times 10^{-6}$ sec ($f_s = 250$ kHz)

The IIR low pass digital filter from Equation 4.15 can be rendered as a *building block* in a *direct* form, as shown in Figure 4.3 [Ifeachor02]. The building block uses delay elements ($z^{-1}$) from storage registers, the gain elements from multipliers, dividers or shifters and an adder from an accumulator. Although the direct form usually requires more storage registers than the *canonic* form, it is less susceptible to internal overflow error and may demonstrate an improved noise performance [Mitra06].

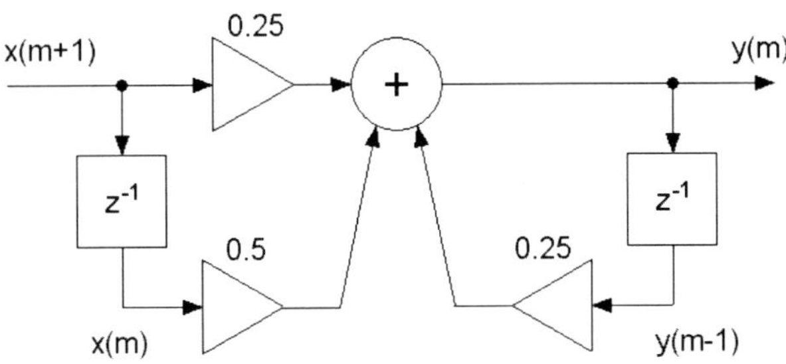


**Figure 4.3** IIR low pass digital filter direct form

As a second example, a simple FIR digital filter rendered as the output sequence $y(m)$ in terms of the input sequence $x(m)$ and coefficients from Equation 4.14 will illustrate the discrete frequency response analysis. Here $Q = 1$, $q = -1$, $b_{-1} = -0.5$, $b_0 = 2$ and $b_1 = -1.5$ as determined by Equation 4.20.

$$y(m) = 2\,x(m) - 1.5\,x(m-1) - 0.5\,x(m+1) \qquad (4.20)$$

The discrete frequency transfer function $H(z)$ from Equation 4.13 is determined by Equation 4.21.

$$H(z) = \frac{Y(z)}{X(z)} = 2 - 1.5\,z^{-1} - 0.5\,z = \frac{2\,z - 1.5 - 0.5\,z^2}{z}$$
$$H(z) = \frac{0.5\,(z-3)\,(1-z)}{z} \tag{4.21}$$

This simple FIR digital filter has one transfer function *pole* (where the response is infinite) at $z = 0$, two transfer function zeros (where the response is zero) at $z = 3$ and $z = 1$ and a multiplicative term of 0.5. The pole (indicated by ×) and the zeros (indicated by ○) of the transfer function are plotted on the complex frequency domain plane, as shown in Figure 4.4. Vectors are drawn from the pole and zeros to an arbitrary point $P$ on the unit circle which represents the sinusoidal excitation of the discrete system $H(z)$.

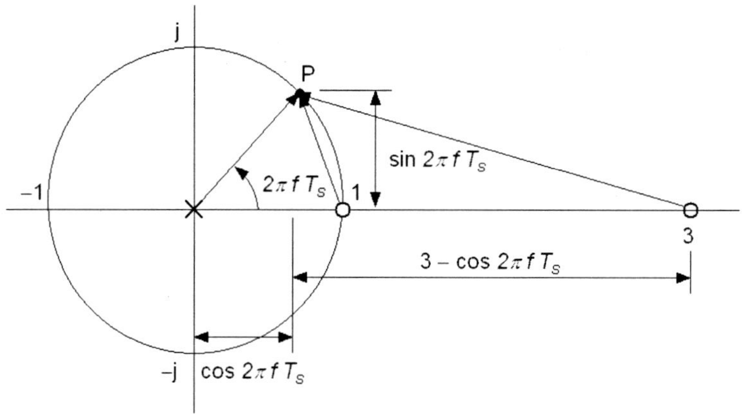


**Figure 4.4** FIR digital filter discrete frequency transfer function plotted on the complex plane

As for the IIR digital filter analyzed previously, the magnitude of the discrete frequency response | $H(z)$ | is determined by the product of the gain $G$ and the length of the vector(s) from the zero(s) divided by the product of the length of the vector(s) from the poles. The simple geometry allows the length of these vectors and the magnitude of the discrete frequency response to be determined as in Equation 4.22.

$$\begin{aligned}
&G = 0.5 \qquad L_{p=0} = 1 \\
&L_{z=3} = \sqrt{(3 - \cos 2\pi f\,T_s)^2 + \sin^2 2\pi f\,T_s} \\
&L_{z=1} = \sqrt{(1 - \cos 2\pi f\,T_s)^2 + \sin^2 2\pi f\,T_s} \\
&|\,H(z)\,| = 0.5\sqrt{(3 - \cos 2\pi f\,T_s)^2 + \sin^2 2\pi f\,T_s} \;\times \\
&\qquad\qquad \sqrt{(1 - \cos 2\pi f\,T_s)^2 + \sin^2 2\pi f\,T_s}
\end{aligned} \tag{4.22}$$

The magnitude of the discrete frequency response of this FIR digital filter is shown in Figure 4.5 on a *log-log* scale plotted against frequency Hz with $T_s = 4 \times 10^{-6}$ sec or $f_s = 1/T_s = 250$ kHz. This FIR digital filter has a *derivating* (*d* / *dt*) characteristic because of its approximately linear magnitude response with frequency [Proakis07]. As for the IIR digital filter analyzed previously, this FIR derivating digital filter is efficient because the coefficients and multiplicative gain term are positive and negative powers of the binary base 2 ($2^n$ or $2^{-n}$) and can be performed with integer scaling or are formed with a low order addition.

The phase angle $\psi$ of the FIR derivating digital filter then derived from Figure 4.1 and is determined by Equation 4.23.

$$\psi = \pi - 2\pi f T_s - \tan^{-1}\left(\frac{\sin 2\pi f T_s}{3 - \cos 2\pi f T_s}\right) + \tan^{-1}\left(\frac{\sin 2\pi f T_s}{1 - \cos 2\pi f T_s}\right) \quad (4.23)$$

The phase angle of a derivating filter should be a constant $\pi / 2$ but the phase angle $\psi$ of the simple FIR derivating digital filter does not strictly meet this requirement.

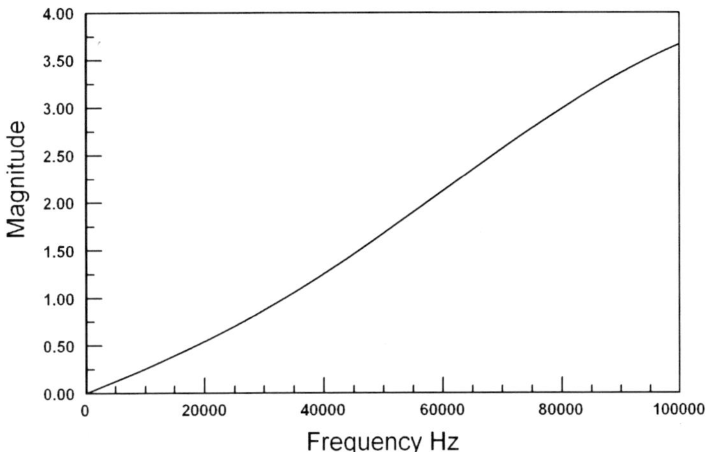


**Figure 4.5** FIR derivating digital filter discrete frequency response magnitude for $T_s = 4 \times 10^{-6}$ sec ($f_s = 250$ kHz)

The FIR derivating digital filter from Equation 4.20 can be rendered as a building block in a direct or *transversal* form, as shown in Figure 4.6 [Ifeachor02]. The transversal FIR digital filter represents a *tapped delay line* and is particularly simple to implement. [Proakis07]. The building block uses delay elements ($z^{-1}$) from storage registers, the gain elements from multipliers, dividers or shifters and an adder/subtractor from an accumulator.

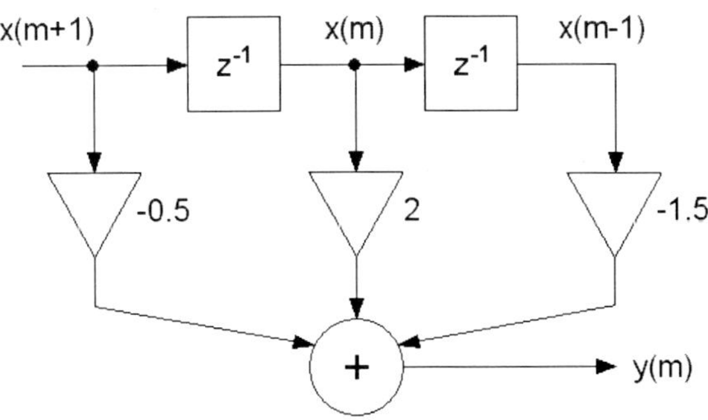


**Figure 4.6** FIR derivating digital filter direct or transversal form

IIR and FIR digital filters generally do not have coefficients that are relatively small integer numbers. Often the coefficients are expressed as floating-point numbers during the digital filter design process which must be approximately scaled for integer processing [Chen01]. DSP using floating-point numbers for the coefficients of a digital filter and the data obviates the concern to some degree for *overflow* but may execute relatively slowly.

The determination of the frequency response magnitude and phase angle is tedious if the number of terms in the discrete frequency system function $H(z)$ is large. The difficulty in factoring the resulting expression to determine the location of the poles and zeros is daunting. However, a direct substitution of the equivalence $z = \exp(2\pi f T_s)$ provides a complex value for the discrete frequency system function $H(z)$, as determined here by Equation 4.24 for the FIR digital filter in Equation 4.14 with $q = 0$.

$$
\begin{aligned}
H(z) &= \sum_{q=0}^{Q} b_q z^{-q} = b_0 + b_1 z^{-1} + b_2 z^{-2} + \ldots + b_Q z^{-Q} \\
H(e^{j2\pi f T_s}) &= b_0 + b_1 e^{-j2\pi f T_s} + b_2 e^{-j4\pi f T_s} + \ldots + b_Q e^{j2Q\pi f T_s}
\end{aligned}
\qquad (4.24)
$$

Substitution of Euler's Identity transforms the sum of complex exponentials in Equation 4.24 to a discernable form in Equation 4.25 where $A$ and $B$ represent the real and imaginary parts of the expansion of $H(\exp(2\pi f T_s)$ evaluated at a frequency $f$. The computation of the frequency response magnitude and phase angle from Equation 4.25 can be implemented easily in a computer language such as C [Ifeachor02].

MATLAB™ (*www.mathworks.com*) provides a direct function *zplane* which computes and plots the poles and zeros of the discrete frequency system function $H(z)$ directly from the $a_p$ and $b_q$ coefficients. MATLAB also provides a function *freqz* that utilizes a fast Fourier transform (FFT) technique to compute the frequency response magnitude and phase angle [Mitra06].

$$
\begin{aligned}
&\text{Euler's Identity} \quad e^{-j2\pi f T_s} = \cos 2\pi f T_s - j \sin 2\pi f T \\
&H(e^{j2\pi f T_s}) = b_0 + b_1[\cos 2\pi f T_s - j\sin 2\pi f T_s] + \\
&\qquad b_2[\cos 4\pi f T_s - j\sin 4\pi f T_s] + \ldots + \\
&\qquad b_Q[\cos 2Q\pi f T_s - j\sin 2Q\pi f T_s] \\
&H(e^{j2\pi f T_s}) = [b_0 + b_1 \cos 2\pi f T_s + b_2 \cos 4\pi f T_s + \ldots + \\
&\qquad b_Q \cos 2Q\pi f T_s] - j[b_1 \sin 2\pi f T_s + \\
&\qquad b_2 \sin 4\pi f T_s + \ldots + b_Q \sin 2Q\pi f T_s] \\
&H(e^{j2\pi f T_s}) = A - jB = \sqrt{A^2 + B^2} \angle \tan^{-1}[-B/A]
\end{aligned}
\qquad (4.25)
$$

## Analog Output

The output sequence $y(m)$ is converted to a *step-wise* but continuous analog output signal $y(t)$ by a digital-to-analog converter (DAC) peripheral. The analog output voltage resolution (or *step size*) $\Delta V$ for a DAC that inputs an unsigned n-bit binary values is determined by Equation 4.26.

$$
\Delta V = \frac{V_{REF}}{2^n} \qquad (4.26)
$$

$V_{REF}$ is the reference voltage for the DAC and n is the number of bits of resolution. The number of bits of resolution of the discrete time input sequence from the ADC, the resolution in bits of the DSP digital filter or the number of bits of resolution for the DAC need not all be equal.

The analog output signal $y(t)$ from the DAC is a sample-and-hold process, as determined by Equation 4.3. The normalized PSD of the analog output signal from the DAC then is determined by Equation 4.27 [Haykin01].

$$\text{PSD} = f_s^2 \sum_k | Y(f - k f_s) |^2 T_s^2 \, \text{sinc}^2(2\pi f T_s) = \sum_k | Y(f - k f_s) |^2 \, \text{sinc}^2(2\pi f T_s) \qquad (4.27)$$

The normalized PSD of the DAC analog output signal $y(t)$ is $| \; Y(f) \; |^2$. The PSD is not theoretically bandlimited but repeats the baseband PSD centered at multiples of the sampling rate $f_s$, as determined by Equation 4.26. However, the PSD is practically bandlimited because of the $\text{sinc}^2$ (sinc x = sin x/x) term which decreases in magnitude as the frequency increases. An analog low pass filter is often placed after the DAC analog output to *smooth* the discrete time step-wise response.

The four channel DAC of the Spartan-3E Starter Board, as described in Chapter 3 Programmable Gate Array Hardware, has 12 bits of resolution or n = 12, $V_{REF}$ = 3.3 V DC for DAC A and DAC B with $\Delta V = 3.3/4096 \approx 0.806$ mV or $V_{REF}$ = 2.5 V DC for DAC C and D with $\Delta V = 2.5/4096 \approx 0.61$ mV.

The four channel Digilent Pmod DA1 DAC hardware module described in Chapter 3 has n = 8 and $V_{REF}$ = 3.3 V DC with $\Delta V = 3.3/256 \approx 12.89$ mV. The two channel Digilent Pmod DA2 DAC hardware module, as described in Chapter 3, has n = 12 and $V_{REF}$ = 3.3 V DC with $\Delta V = 3.3/4096 \approx 0.806$ mV. None of these DACs provide an integral analog output filter to smooth the discrete time to analog step-wise response.

## Digital Signal Processing Embedded System

The digital signal processing (DSP) embedded hardware system consists of an ADC, field programmable gate array (FPGA) and DAC, as shown in Figure 4.7. The ADC provides n-bit data to the FPGA processor and receives an a-bit data packet for command and control. The DAC receives both m-bit data and a d-bit data packet for command and control from the FPGA processor. A crystal oscillator provides a clock signal to the FPGA for synchronization and timing of the data transfers and to establish the sampling rate $f_s$ of the DSP system. The DSP embedded hardware system can optionally have input/output (I/O) devices such as light emitting diodes (LED), liquid crystal display (LCD), switches and data communication ports, as described in Chapter 3 Programmable Gate Array Hardware.

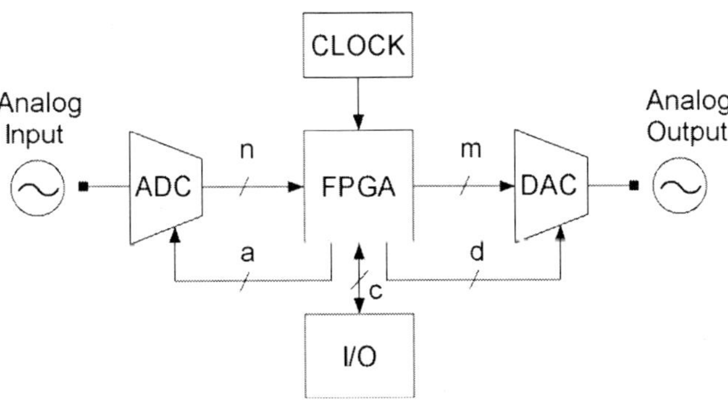


**Figure 4.7** DSP embedded hardware system

The DSP system executes on the FPGA hardware using Verilog HDL structural and behavioral synthesis modules that are developed using software electronic design automation (EDA) tools, as described in Chapter 2 Verilog Design Automation. The Verilog modules are configured as finite state machines (FSM) and the controller and datapath constructs, as described in Chapter 1 Verilog Hardware Description Language. These Verilog HDL modules and simple applications were introduced in Chapter 3 Programmable gate Array Hardware but are extended here to applications in DSP.

A DSP system is initially implemented as a *straight-through* processor that inputs and outputs an analog signal without any manipulation to assess the maximum data throughput rate. The two DSP systems to be benchmarked are the integral ADC and DAC of the Spartan-3E Starter Board and the Digilent Pmod AD1 and Pmod DA2 hardware modules both processed by the FPGA, as described in Chapter 3 Programmable Gate Array Hardware.

The first of the straight-through DSP systems is implemented with the Verilog *top* module s3eadcdac.v in Listing 4.1, which is in the *Chapter 4\adcdac\s3eadcdac* folder. The Xilinx ISE WebPACK project uses the User Constraints File (UCF) adcdacs3esb.ucf which uncomments the signals CCLK, SW0, SPIMISO, SPIMOSI, SPISCK, DACCS, DACCLR, SPISF, AMPCS, AMPSD, ADCON, SFCE, FPGAIB, and JC1 (J4-1) in the Spartan-3E Starter Board UCF of Listing 3.2. The five Verilog modules operate in parallel and some independently in the top module. The file download procedure is described in the Appendix.

**Listing 4.1** DSP ADC DAC system top module for the Spartan-3E Starter Board s3eadcdac.v

```
// Spartan-3E Starter Board
// Digital-to-Analog Converter s3eadcdac.v
// c 2008 Embedded Design using Programmable Gate Arrays  Dennis Silage

module s3eadcdac (input CCLK, SW0, SPIMISO, output SPIMOSI, SPISCK, DACCS, DACCLR,
                  output SPISF, AMPCS, AMPSD, ADCON, SFCE, FPGAIB, JC1);

wire adcdav, davadc, adcsck, adcspod, ampdav, davamp, ampsck, ampspid, sdamp;
wire dacdav, davdac, dacsck, dacspid, clrdac, csdac, csamp, conad;
wire [11:0] dacdata;
wire [13:0] adc0data, adc1data;
wire [3:0] ampcmd0, ampcmd1, daccmd, dacaddr;
wire CLKOUT, CLKFX;

assign SPISCK = adcsck | ampsck | dacsck;      // SPI clock
assign SPIMOSI = ampspid | dacspid;            // SPI data in
assign adcspod = SPIMISO;                      // SPI data out

assign DACCS = csdac;          // disable DAC
assign DACCLR = clrdac;        // DAC clear
assign AMPCS = csamp;          // select prog amp
assign AMPSD = sdamp;          // disable prog amp shutdown
assign ADCON = conad;          // ADC convert command

assign SPISF = 1;              // disable serial Flash
assign SFCE = 1;               // disable StrataFlash
assign FPGAIB = 1;             // disable Platform Flash

assign JC1 = conad;            // monitor sampling rate
```

```
s3eadc M0 (CLKOUT, adcdav, davadc, adc0data, adc1data, adcsck, adcspod, conad);
s3eprogamp M1 (CLKOUT, ampdav, davamp, ampcmd0, ampcmd1, ampsck, ampspid, csamp,
                    sdamp);
s3edac M2 (CLKFX, dacdav, davdac, dacdata, dacaddr, daccmd, dacsck, dacspid, csdac, clrdac);
genadcdac M3 (CLKOUT, SW0, ampdav, davamp, ampcmd0, ampcmd1, adcdav, davadc, adc0data,
                    adc1data, dacdav, davdac, dacdata, dacaddr, daccmd);
dacs3edcm M4 (.CLKIN_IN(CCLK), .RST_IN(0), .CLKFX_OUT(CLKFX),
                        .CLKIN_IBUFG_OUT(CLKINIBO), .CLK0_OUT(CLKOUT),
                        .LOCKED_OUT(LOCK));

endmodule

module genadcdac (input genclk, SW0, output reg ampdav = 0, input davamp,
                    output reg [3:0] ampcmd0 = 1, output reg [3:0] ampcmd1 = 1,
                    output reg adcdav = 0, input davadc, input [13:0] adc0data,
                    input [13:0] adc1data, output reg dacdav =  0, input davdac,
                    output reg [11:0] dacdata, output reg [3:0] dacaddr = 0,
                    output reg [3:0] daccmd = 3);

reg [1:0] gstate = 0;          // state register
reg [13:0] value;

always@(posedge genclk)
    begin
        case (gstate)
            0:  begin
                    ampdav = 1;             // programmable amplifier
                    if (davamp == 1)
                        gstate = 1;
                end
            1:  begin
                    adcdav = 1;             // ADC conversion
                    if (davadc == 1)
                        begin
                            adcdav = 0;
                            if (SW0 == 0)  // select one ADC channel
                                value = adc0data;
                            else
                                value = adc1data;
                            value[13] = ~value[13];          // sign bit
                            dacdata[11:0] = ~value[13:2];   // inversion
                            gstate = 2;
                        end
                end
            2:  begin
                    dacdav = 1;             // DAC conversion
                    if (davdac == 1)
                        begin
                            dacdav = 0;
                            gstate = 1;
                        end
                end
```

```
            default: gstate = 1;
        endcase
    end

endmodule
```

The Verilog top module s3eadcdac.v utilizes the se3adc.v and se3progamp.v modules for the ADC and the se3dac.v and dacs3edcm.v modules for the DAC of the Spartan-3E Starter Board, as described in Chapter 3 Programmable Gate Array Hardware. The 6-pin peripheral hardware module connector JC1 (J4-1) is used to monitor the ADC conversion command signal conad and the sampling rate $f_s \approx 282$ ksamples/sec here.

This maximum sampling rate is far below the cutoff frequency of the anti-aliasing filter of the Spartan-3E Starter Board ADC (1.54 MHz). To avoid aliasing, an analog filter with a cutoff frequency $f_{cutoff} < f_s/2$ or $f_{cutoff} < 141$ kHz would be appropriate here.

Although data from the two ADC channels are available simultaneously, the DAC can only accept one channel of data at this sampling rate. In addition, the Spartan-3E Starter Board DAC shares access to the serial peripheral interface (SPI) bus with the ADC which leads to SPI *bus contention* and an even lower sampling rate.

The Spartan-3E Starter Board ADC outputs 14-bit two's complement binary data and the DAC requires 12-bit straight binary data, which are incompatible representations. The conversion is accomplished by complementing the most significant bit (MSB) of the ADC output (bit 13) as the sign bit [Wakerly00]. The twelve MSBs of the ADC output (bit 13 through bit 2) are outputted to the DAC. Since the programmable amplifier of the Spartan-3E Starter Board inverts the analog input signal, the resulting straight binary data is complemented for comparison of the analog input and outputs signals on an oscilloscope.

The module genadcdac.v utilizes a simple three state FSM controller and for the DSP system. The ADC 14-bit signed 2's complement data is converted to a 12-bit straight binary data for the DAC. Slide switch SW0 selects one of the two available ADC channels. The Digital Clock Manager (DCM) frequency synthesizer module dacs3edcm.v inputs the 50 MHz crystal clock oscillator and outputs a 50 MHz × 5/3 = 83.333 MHz clock for the DAC, as described in Chapter 3 Programmable Gate Array Hardware.

As described in Chapter 2 Verilog Design Automation, the Design Utilization Summary for the module s3eadcdac.v shows the use of 202 slice flip-flops (2%), 265 4-input LUTs (2%), 177 occupied slices (3%) and a total of 10 250 equivalent gates in the XC3S500E Spartan-3E FPGA synthesis.

Figure 4.8 shows the DAC output of the straight-through DSP system s3eadcdac.v for a sinusoidal input frequency of 5 kHz (top, 1 V/div and 100 μsec/div) and 50 kHz (bottom, 1 V/div and 10 μsec/div). Since the sampling rate $f_s \approx 282$ kHz here, the step-wise voltage output of the Spartan-3E Starter Board, DAC is quite noticeable at an input frequency of 50 kHz. The DAC provides no smoothing of the output.

The second straight-through DSP system utilizes the Pmod AD1 ADC and DA2 DAC and is implemented with the Verilog top module s3ead1da2.v in Listing 4.2, which is in the *Chapter 4 \adcdac\s3ead1da2* folder. A similar straight-through DSP system for the Basys Board is in the *Chapter 4\adcdac\baad1da2* folder but not described here.

The Xilinx ISE WebPACK project uses the UCF ad1da2s3esb.ucf which uncomments the signals CCLK, SW0, JA1, JA2, JA3, JA4, JB1, JB2, JB3, JB4 and JC1 (J4-1) in the Spartan-3E Starter Board UCF of Listing 3.2. The three Verilog modules operate in parallel and some independently in the top module. The file download procedure is described in the Appendix.

The Verilog top module s3ead1da2.v utilizes the ad1adc.v module for the Pmod AD1 ADC and the da2dac.v module for the Pmod DA2 DAC, as described in Chapter 3 Programmable Gate Array Hardware. The Pmod AD1 hardware module is connected to J1 (as JA) on the Spartan-3E starter Board. The Pmod DA2 hardware module is connected to J2 (as JB).

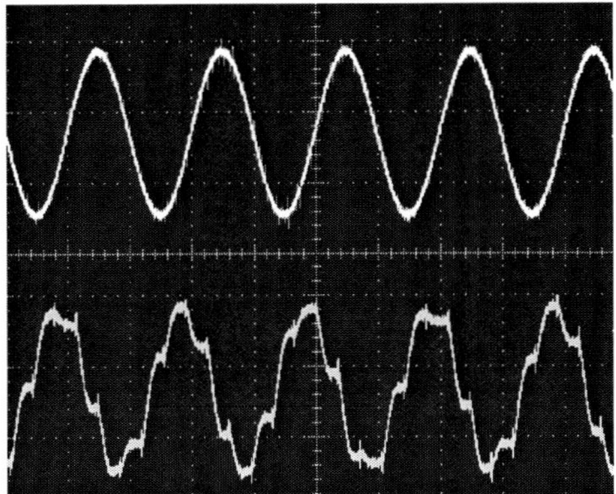

**Figure 4.8** DAC output of the DSP system s3eadcdac.v for a sinusoidal input at 1 V/div and 100 μsec/div (top) and 1 V/div and 10 μsec/div (bottom)

The 6-pin peripheral hardware module connector JC1 (J4-1) is used to monitor the ADC conversion command signal adccs and the sampling rate is $f_s \approx$ 714 ksamples/sec. This maximum sampling rate is above the cutoff frequency of the anti-aliasing filter of the Digilent Pmod AD1 ADC (500 kHz). To avoid aliasing, an analog filter with a cutoff frequency $f_{cutoff} < f_s/2$ or $f_{cutoff}$ < 357 kHz would be appropriate here.

Although data from the two ADC channels are available simultaneously, the DAC can only accept one channel of data at this sampling rate. However, unlike the ADC and DAC of the Spartan-3E Starter Board, the Pmod AD1 and Pmod DA2 do not share a serial data communication channel and can be processed in parallel. The output data of the ADC and the input data of the DAC are both 12-bit and utilize a straight binary data representation and no data conversion is thus required.

The module genad1da2.v does not use an FSM controller for the DSP system. The Pmod AD1 ADC and Pmod DA2 DAC status signals davadc and davdac from the datapath modules ad1adc.v and da2dac.v are processed in parallel and both provide a logic 1 signal to indicate that processing has completed. The DAC is outputting the previous analog signal sampled by the ADC while the ADC is obtaining the current analog signal sample. Slide switch SW0 selects one of the two available ADC channels.

This second straight-through DSP system, utilizing the Pmod AD1 ADC and DA2 DAC, illustrates *wavefront* processing and results in a sampling rate nearly twice that of the straight-through DSP system using the Spartan-3E Starter Board ADC and DAC. This primarily occurs because there is no SPI bus contention for the Pmod AD1 ADC and DA2 DAC peripherals on the separate 6-pin ports JA and JB of the Spartan-3E Starter Board or the Basys Board.

**Listing 4.2** DSP Pmod AD1 Pmod DA2 system top module s3ead1da2.v

```
// Spartan-3E Starter Board
// Digital-to-Analog Converter s3ead1da2.v
// c 2008 Embedded Design using Programmable Gate Arrays  Dennis Silage

module s3ead1da2 (input CCLK, SW0, JA1, JA3, output JA1, JA4, JB1, JB2, JB3, JB4, JC1);

wire adcdav, davadc, adcsck, adc0d, adc1d, adccs;
wire dacdav, davdac, dacsck, dacout, dacsync;
wire [11:0] dacdata;
wire [1:0] dacmd;
wire [11:0] adc0data, adc1data;
```

```
assign JA1 = adccs;         // ADC AD1 on J1
assign adc0d = JA2;
assign adc1d = JA3;
assign JA4 = adcsck;
assign JB1 = dacsync;       // DAC DA1 on J2
assign JB2 = dacout;
assign JB3 = 0;
assign JB4 = dacsck;

assign JC1 = adccs;         // monitor sampling rate

ad1adc M0 (CCLK, adcdav, davadc, adc0data, adc1data, adcsck, adc0d, adc1d, adccs);
da2dac M1 (CCLK, dacdav, davdac, dacout, dacsck, dacsync, daccmd, dacdata);
genad1da2 M2 (CCLK, SW0, adcdav, davadc, adc0data, adc1data, dacdav, davdac, dacdata, daccmd);

endmodule

module genad1da2 (input genclk, SW0, output reg adcdav = 0, input davadc, input [11:0] adc0data,
                  input [11:0] adc1data, output reg dacdav = 0, input davdac,
                  output reg [11:0] dacdata, output reg [1:0] daccmd = 0);

always@(posedge genclk)
    begin
        adcdav = 1;             // ADC conversion
        dacdav = 1;             // DAC conversion, parallel processing

        if (davadc == 1 && davdac == 1)     // ADC and DAC status
            begin
                adcdav = 0;
                davdac = 0;
                if (SW0 == 0)   // select one ADC channel
                    dacdata = adc0data;
                else
                    dacdata = adc1data;
            end
    end

endmodule
```

As described in Chapter 2 Verilog Design Automation, the Design Utilization Summary for the module s3ead1da2.v shows the use of 114 slice flip-flops (1%), 142 4-input LUTs (2%), 104 occupied slices (2%) and a total of 1789 equivalent gates in the XC3S500E Spartan-3E FPGA synthesis. A comparison of this Design Utilization Summary with that of the module s3eadcdac.v shows that this second straight-through DSP systems is more efficient in FPGA resources since it does not utilize either the programmable amplifier module se3progamp.v or the DCM module se3dacdcm.v.

## IIR Digital Filter

The simple IIR low pass digital filter in Equation 4.16 is implemented as a DSP system, as shown in Figure 4.7. The Verilog top module s3eadclpfdac.v is given in Listing 4.3, which is in the *Chapter 4\adcdac\s3eadclpfdac* folder. The Xilinx ISE WebPACK project uses the UCF adclpfdacs3esb.ucf which uncomments the signals CCLK, SW0, SPIMISO, SPIMOSI, SPISCK,

DACCS, DACCLR, SPISF, AMPCS, AMPSD, ADCON, SFCE, FPGAIB, and JC1 (J4-1) in the Spartan-3E Starter Board UCF of Listing 3.2. The six Verilog modules operate in parallel and some independently in the top module. The file download procedure is described in the Appendix.

The Verilog top module s3eadclpfdac.v utilizes the se3adc.v and se3progamp.v modules for the ADC and the se3dac.v and dacs3edcm.v modules for the DAC of the Spartan-3E Starter Board, as described in Chapter 3 Programmable Gate Array Hardware. The clock module clock.v is used to set the sampling rate $f_s$ = 250 kHz with a clock scale of 100, as given in Listing 3.3.

**Listing 4.3** Simple IIR low pass digital filter DSP system top module s3eadclpfdac.v

```
// Spartan-3E Starter Board
// Digital-to-Analog Converter s3eadclpfdac.v
// c 2008 Embedded Design using Programmable Gate Arrays  Dennis Silage

module s3eadclpfdac (input CCLK, SW0, SW1, SPIMISO, output SPIMOSI, SPISCK, DACCS,
                     output DACCLR, SPISF, AMPCS, AMPSD, ADCON, SFCE, FPGAIB,
                     output JC1);

wire adcdav, davadc, adcsck, adcspod, csamp, conad;
wire ampdav, davamp, ampsck, ampspid, sdamp;
wire dacdav, davdac, dacsck, dacspid, clrdac, csdac;
wire [11:0] dacdata;
wire [13:0] adc0data, adc1data;
wire [3:0] ampcmd0, ampcmd1, daccmd, dacaddr;
wire CLKOUT, CLKFX, smpclk;

assign SPISCK = adcsck | ampsck | dacsck;     // SPI clock
assign SPIMOSI = ampspid | dacspid;           // SPI data in
assign adcspod = SPIMISO;                     // SPI data out
assign DACCS = csdac;           // disable DAC
assign DACCLR = clrdac;         // DAC clear
assign AMPCS = csamp;           // select prog amp
assign AMPSD = sdamp;           // disable prog amp shutdown
assign ADCON = conad;           // ADC convert command

assign SPISF = 1;               // disable serial Flash
assign SFCE = 1;                // disable StrataFlash
assign FPGAIB = 1;              // disable Platform Flash
assign JC1 = conad;             // monitor sampling rate

s3eadc M0 (CLKOUT, adcdav, davadc, adc0data, adc1data, adcsck, adcspod, conad);
s3eprogamp M1 (CLKOUT, ampdav, davamp, ampcmd0, ampcmd1, ampsck, ampspid, csamp,
               sdamp);
clock M2 (CLKOUT, 100, smpclk);
s3edac M3 (CLKFX, dacdav, davdac, dacdata, dacaddr, daccmd, dacsck, dacspid, csdac, clrdac);
genadclpfdac M4 (CLKOUT, SW0, SW1, smpclk, ampdav, davamp, ampcmd0, ampcmd1,adcdav,
               davadc, adc0data, adc1data, dacdav, davdac, dacdata, dacaddr, daccmd);
dacs3edcm M5 (.CLKIN_IN(CCLK), .RST_IN(0), .CLKFX_OUT(CLKFX),
                     .CLKIN_IBUFG_OUT(CLKINIBO), .CLK0_OUT(CLKOUT),
                     .LOCKED_OUT(LOCK));

endmodule
```

```
module genadclpfdac (input genclk, SW0, SW1, smpclk, output reg ampdav = 0, input davamp,
                    output reg [3:0] ampcmd0 = 1, output reg [3:0] ampcmd1 = 1,
                    output reg adcdav = 0, input davadc, input [13:0] adc0data, input [13:0] adc1data,
                    output reg dacdav = 0, input davdac, output reg [11:0] dacdata,
                    output reg [3:0] dacaddr = 0, output reg [3:0] daccmd = 3);

reg [1:0] gstate = 0;  // state register
reg [13:0] value;
reg [13:0] xa;         // x(m)
reg [13:0] xb;         // x(m+1)
reg [13:0] ya;         // y(m)
reg [13:0] yc;         // y(m-1)

always@(posedge genclk)
    begin
        case (gstate)
            0:  begin
                    ampdav = 1;           // programmable ampliifer
                    if (davamp == 1)
                        gstate = 1;
                end
            1:  begin
                    if (smpclk == 1)      // sample clock
                        gstate = 2;
                    else
                        gstate = 1;
                end
            2:  begin
                    adcdav = 1;           // ADC conversion
                        if (davadc == 1)
                            begin
                                adcdav = 0;
                                if (SW0 == 0)  // select one ADC channel
                                    value = adc0data;
                                else
                                    value = adc1data;
                                value[13] = ~value[13];    // sign bit
                                if (SW1 == 0)              // bypass filtering
                                    begin
                                        xa = xb;
                                        xb = ~value;       // inversion
                                        yc = ya;
                                        ya = (xa/2) + (xb/4) + (yc/4);
                                        dacdata[11:0] = ya[13:2];
                                    end
                                else
                                    dacdata[11:0] = ~value[13:2];
                                gstate = 3;
                            end
                        else
                            gstate = 2;
                end
```

```
            3:   begin
                      dacdav = 1;              // DAC conversion
                      if (davdac == 1)
                           begin
                                dacdav = 0;
                                gstate = 1;
                           end
                 end
        endcase
    end

endmodule
```

The maximum sampling rate for the straight-through or ADC-DAC DSP system was $f_{ADC\text{-}DAC} \approx 282$ ksamples/sec. The in-sequence computation time available for the IIR low pass digital filter here is 0.46 μsec, since the sampling period $T_s = 1/f_s = 4$ μsec and $T_{ADC\text{-}DAC} \approx 3.54$ μsec, and is sufficient. If the available time is found to be not sufficient, then the capability of the Verilog HDL to evoke parallel sequence controller and datapath modules for data acquisition, computation and data output can be used.

However, the SPI bus contention of the Spartan-3E Starter Board ADC and DAC peripheral obviates the full degree of parallelism for data acquisition and data output here. The 6-pin peripheral hardware module connector JC1 (J4-1) is used to monitor the ADC conversion command signal conad. If the in-sequence computation available is not sufficient then the period of the ADC conversion command signal conad would not be at the sampling period $T_s$.

The module genadclpfdac.v utilizes a four state FSM controller for the IIR low pass digital filter DSP system. Logic 1 on the sample clock signal smpclk sets the state transition to begin the DSP process. As from Equation 4.16, the IIR filter utilizes four 14-bit registers which represent $x(m+1)$ as xb, $x(m)$ as xa, $y(m)$ as ya and $y(m-1)$ as yc. The registers must be *pushed-down* to store the immediate values before being updated. The register xb is stored as the register xa before being updated with the current ADC output signal as the input register value. Similarly, the immediate output register ya is stored as the register yc before the IIR filter calculation is performed.

The conversion of the 14-bit two's complement binary data from the ADC of the Spartan-3E Starter Board to a straight binary representation is accomplished by complementing the MSB before the IIR digital filter calculation since only arithmetic summation is required. The inversion of the binary data accommodates the programmable amplifier of the Spartan-3E Starter Board which inverts the analog input signal. The IIR low pass digital filter here avoids arithmetic register overflow because the 14-bit registers xb, xa and yc are scaled before addition. The most significant bits of the 14-bit output register ya is set equal to the 12-bit register dacdata for output to the DAC.

Slide switch SW0 selects one of the two available ADC channels and SW1 is used to bypass the IIR low pass digital filter to provide the straight-through response. The frequency response can then be assessed on an oscilloscope by comparing the input sinusoidal excitation as the straight-through response to that of the output of the IIR low pass digital filter.

As described in Chapter 2 Verilog Design Automation, the Design Utilization Summary for the module s3eadclpfdac.v shows the use of 259 slice flip-flops (2%), 321 4-input LUTs (3%), 235 occupied slices (3%) and a total of 11 507 equivalent gates in the XC3S500E Spartan-3E FPGA synthesis.

## FIR Digital Filter

The simple FIR derivating digital filter in Equation 4.20 is implemented as a DSP system, as shown in Figure 4.7. The Verilog top module s3eadcddtdac.v is given in Listing 4.4, which is in the *Chapter 4\adcdac\s3eadcddtdac* folder. The Xilinx ISE WebPACK project uses the UCF adcddtdacs3esb.ucf which uncomments the signals CCLK, SW0, SPIMISO, SPIMOSI, SPISCK, DACCS, DACCLR, SPISF, AMPCS, AMPSD, ADCON, SFCE, FPGAIB, and JC1 (J4-1) in the Spartan-3E Starter Board UCF of Listing 3.2. The six Verilog modules operate in parallel and some independently in the top module. The file download procedure is described in the Appendix.

The Verilog top module s3eadcdtdac.v is similar to the s3eadcdlpfdac.v module in Listing 4.3. However, the module genadcddtdac.v is different and utilizes a four state FSM controller for the FIR derivating digital filter DSP system, as given in Listing 4.4. The clock module clock.v is used to set the sampling rate $f_s$ = 250 kHz with a clock scale of 100, as given in Listing 3.3.

The 6-pin peripheral hardware module connector JC1 (J4-1) is used to monitor the ADC conversion command signal conad. The in-sequence computation time available for the FIR derivating digital filter is 0.46 μsec, since the sampling period $T_s$ = 4 μsec and the sampling period for the straight-through or ADC-DAC DSP system $T_{ADC\text{-}DAC} \approx$ 3.54 μsec.

**Listing 4.4** Simple FIR derivating digital filter DSP system s3eadcddtdac.v

```
// Spartan-3E Starter Board
// Digital-to-Analog Converter s3eadcddtdac.v
// c 2008 Embedded Design using Programmable Gate Arrays  Dennis Silage

module s3eadcddtdac (input CCLK, SW0, SW1, SPIMISO, output SPIMOSI, SPISCK, DACCS,
                     output DACCLR, SPISF, AMPCS, AMPSD, ADCON, SFCE, FPGAIB,
                     output JC1);

wire adcdav, davadc, adcsck, adcspod, csamp, conad;
wire ampdav, davamp, ampsck, ampspid, sdamp;
wire dacdav, davdac, dacsck, dacspid, clrdac, csdac;
wire [11:0] dacdata;
wire [13:0] adc0data, adc1data;
wire [3:0] ampcmd0, ampcmd1, daccmd, dacaddr;
wire CLKOUT, CLKFX;

assign SPISCK = adcsck | ampsck | dacsck;       // SPI clock
assign SPIMOSI = ampspid | dacspid;             // SPI data in
assign adcspod = SPIMISO;                       // SPI data out

assign DACCS = csdac;            // disable DAC
assign DACCLR = clrdac;          // DAC clear

assign AMPCS = csamp;            // select prog amp
assign AMPSD = sdamp;            // disable prog amp shutdown
assign ADCON = conad;            // ADC convert command

assign SPISF = 1;                // disable serial Flash
assign SFCE = 1;                 // disable StrataFlash
assign FPGAIB = 1;               // disable Platform Flash

assign JC1 = conad;              // monitor sampling rate

s3eadc M0 (CLKOUT, adcdav, davadc, adc0data, adc1data, adcsck, adcspod, conad);
s3eprogamp M1 (CLKOUT, ampdav, davamp, ampcmd0, ampcmd1, ampsck, ampspid, csamp,
               sdamp);
clock M2 (CLKOUT, 100, smpclk);
```

```
s3edac M3 (CLKFX, dacdav, davdac, dacdata, dacaddr, daccmd, dacsck, dacspid, csdac, clrdac);
genadcddtdac M4 (CLKOUT, SW0, SW1, smpclk, ampdav, davamp, ampcmd0, ampcmd1,adcdav,
                davadc, adc0data, adc1data, dacdav, davdac, dacdata, dacaddr, daccmd);
dacs3edcm M5 (.CLKIN_IN(CCLK), .RST_IN(0), .CLKFX_OUT(CLKFX),
                .CLKIN_IBUFG_OUT(CLKINIBO), .CLK0_OUT(CLKOUT),
                .LOCKED_OUT(LOCK));

endmodule

module genadcddtdac (input genclk, SW0, SW1, smpclk, output reg ampdav = 0, input davamp,
                output reg [3:0] ampcmd0 = 1, output reg [3:0] ampcmd1 = 1,
                output reg adcdav = 0, input davadc, input [13:0] adc0data,
                input [13:0] adc1data, output reg dacdav = 0, input davdac,
                output reg [11:0] dacdata, output reg [3:0] dacaddr = 0,
                output reg [3:0] daccmd = 3);

reg [1:0] gstate = 0;       // state register
reg [13:0] value;
reg signed [15:0] xa;       // x(m)
reg signed [15:0] xb;       // x(m+1)
reg signed [15:0] xc;       // x(m-1)
reg signed [15:0] ya;       // y(m)

always@(posedge genclk)
    begin
        case (gstate)
            0:  begin
                    ampdav = 1;         // programmable amplifier
                    if (davamp == 1)
                        gstate = 1;
                end
            1:  begin
                    if (smpclk == 1)    // sample clock
                        gstate = 2;
                    else
                        gstate = 1;
                end
            2:  begin
                    adcdav = 1;         // ADC conversion
                        if (davadc == 1)
                            begin
                                adcdav = 0;
                                if (SW0 == 0)  // select ADC channel
                                    value = adc0data;
                                else
                                    value = adc1data;
                                if (SW1 == 0)  // bypass filter
                                    begin
                                        xc = xa;
                                        xa = xb;
                                        xb[13:0] = value;
                                        xb[15] = value[13];  // sign extension
```

```
                                        xb[14] = value[13];
                                        ya = (2 * xa) – (xb / 2) – ((xc + xc + xc) / 2);
                                        dacdata[11:0] = ya[13:2];
                                        dacdata = 2048 – dacdata; // ADC inversion
                                    end                         // and DAC offset
                                else
                                    begin
                                        value[13] = ~value[13];
                                        dacdata[11:0] = ~value[13:2];
                                    end
                                gstate = 3;
                            end
                        else
                            gstate = 2;
                end
            3:  begin
                    dacdav = 1;         // DAC conversion
                    if (davdac == 1)
                        begin
                            dacdav = 0;
                            gstate = 1;
                        end
                end
            default: gstate = 1;
        endcase
    end

endmodule
```

From Equation 4.20, the FIR filter utilizes three 14-bit registers which represent $x(m+1)$ as xb, $x(m)$ as xa and $x(m–1)$ as xc. The registers must also be *pushed-down* to store the immediate values before being updated. The register xb is stored as the register xa before being updated with the current ADC output signal as the input register value.

The 14-bit two's complement binary data from the ADC of the Spartan-3E Starter Board is converted to a signed 16-bit representation to avoid overflow by extending the sign bit (bit 13) to bit 14 and bit 15 of the *signed* register xb for the input data [Mano07]. The FIR digital filter calculation is performed using signed registers for xb, xa and xc since arithmetic summation and subtraction are both required.

The most significant bits of the 16-bit signed output register ya is subtracted from the midpoint (2048) of the 12-bit register dacdata for output to the DAC. This inversion of the binary data accommodates the programmable amplifier of the Spartan-3E Starter Board which inverts the analog input signal.

Slide switch SW0 selects one of the two available ADC channels and SW1 is used to bypass the FIR derivating digital filter to provide the *straight-through* response. The derivating response can then be assessed on an oscilloscope by comparing an input sinusoidal excitation as the straight-through response to that of the output of the FIR digital filter.

Figure 4.9 shows the derivating response (top) to a 1.1 V peak-to-peak, 500 Hz sinusoid (bottom). Although noisy, averaging shows that the sinusoidal response has the approximate phase relationship of $\pi / 2 = 90°$ to the excitation that is expected for a derivative.

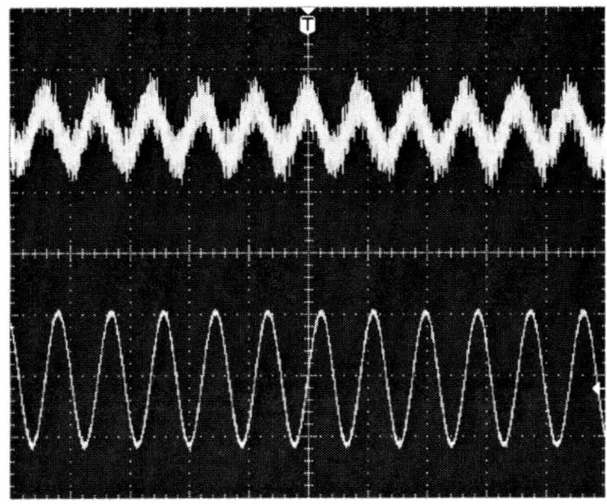

**Figure 4.9** DSP system s3eadcddtdac.v for the derivative output at 50 mV/div (top) with a 500 Hz sinusoidal input at 0.5 V/div (bottom) and 2 msec/div

Figure 4.10 shows the derivating response (top) to a 1.2 V peak-to-peak, 5 kHz sinusoid (bottom). Although the response is still sinusoidal, the phase relationship of the derivating output of 90° (50 μsec at 5 KHz) is delayed by approximately 20 μsec. This is due in part to the non-ideal phase response of the derivating filter, as described in Equation 4.23. A small amount of this delay though is in the processing inherent in the FIR digital filter. The output data $y(m)$ is delayed by one sample interval $T_s = 4$ μsec from the input data $x(m+1)$, as determined by Equation 4.20.

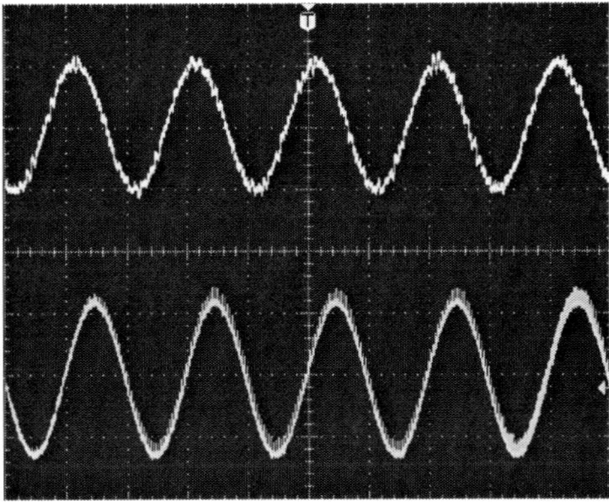

**Figure 4.10** DSP system s3eadcddtdac.v for the derivative output at 200 mV/div (top) with a 5 kHz sinusoidal input at 0.5 V/div (bottom) and 100 μsec/div

As described in Chapter 2 Verilog Design Automation, the Design Utilization Summary for the module s3eadcddtdac.v shows the use of 263 slice flip-flops (2%), 346 4-input LUTs (3%), 261 occupied slices (3%) and a total of 11 984 equivalent gates in the XC3S500E Spartan-3E FPGA synthesis.

## FIR Compiler LogiCORE Block

The Xilinx CORE Generator provides a FIR Compiler LogiCORE block, as in Chapter 2 Verilog Design Automation. The Xilinx FIR Compiler is a common interface for the design of FPGA resource efficient FIR digital filters with either multiply-accumulate (MAC) or distributed arithmetic (DA) architectures [Ifeachor02]. The Xilinx FIR Compiler provides sufficient arithmetic precision to avoid overflow and is described in data sheet DS534 (*www.xilinx.com*).

FIR digital filters can be implemented as single-rate, interpolated or multi-rate [Mitra06]. The single-rate FIR digital filter is conceptualizes as a tapped-delay line, as shown in Figure 4.11 and where the index $q = 0$ as determined by Equation 4.9 [Proakis07]. Although correct in essence, the actual implementation using an FPGA is different. If the MAC architecture is selected in the FIR Compiler, one or more MAC functional units are used to provide the $Q$ sum-of-product calculations for the specified data throughput. If the DA architecture is selected only look-up tables (LUT), shift registers and an accumulator are used.

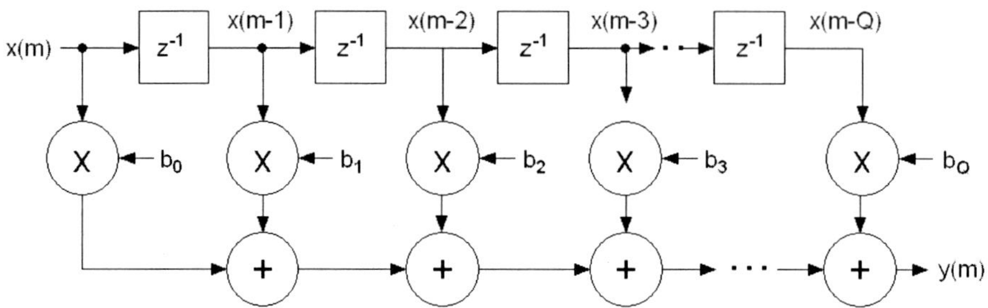


**Figure 4.11** Tapped-delay line, single rate FIR digital filter

The Xilinx LogiCORE block FIR Compiler provides from 2 to 1024 *taps* (equivalent to $Q$+1 in Figure 4.11) and integer signed or unsigned coefficients that range from 1-bit to 32-bit precision. The signed or unsigned integer data widths can be as large as 32-bit and are accommodate by accumulator widths up to 74-bits. The FIR Compiler LogiCORE block is evoked from within a Xilinx ISE WebPACK project, as described in Chapter 2 Verilog Design Automation.

The first design window of the LogiCORE FIR Compiler is shown in Figure 4.12. The component name (fir1) and coefficient file (fir1.coe) is specified, a single-rate FIR digital filter type is chosen, one data channel is used (although up to 256 channels are possible), the clock frequency is 50 MHz and the sample frequency (or sampling rate) is 0.25 MHz (250 kHz). The clock and sample frequency parameters specified here are those utilized with the Spartan-3E Starter Board and the integral analog-to-digital converter (ADC) and digital-to-analog converter (DAC).

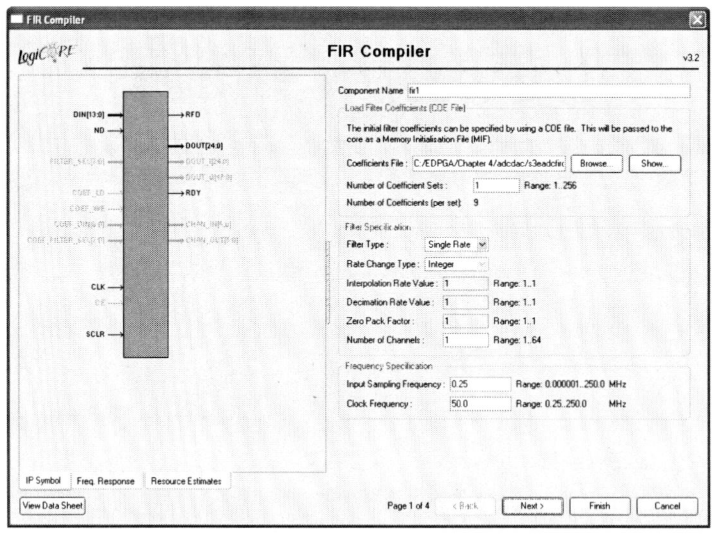


**Figure 4.12** First design window of the LogiCORE FIR Compiler

The second design window of the LogiCORE FIR Compiler specifies the multiplier-accumulator (MAC) filter architecture, the coefficient options, the input and output data type and width and the frequency response of the digital filter, as shown in Figure 4.13. The input data is a signed 14-bit integer from the ADC of the Spartan-3E Starter Board. The FIR Compiler has the capability to reload coefficients on a separate data bus and change the digital filter function with an external command. The FIR Compiler can also utilize different coefficient sets for each data channel. However, these advanced capabilities are not used here.

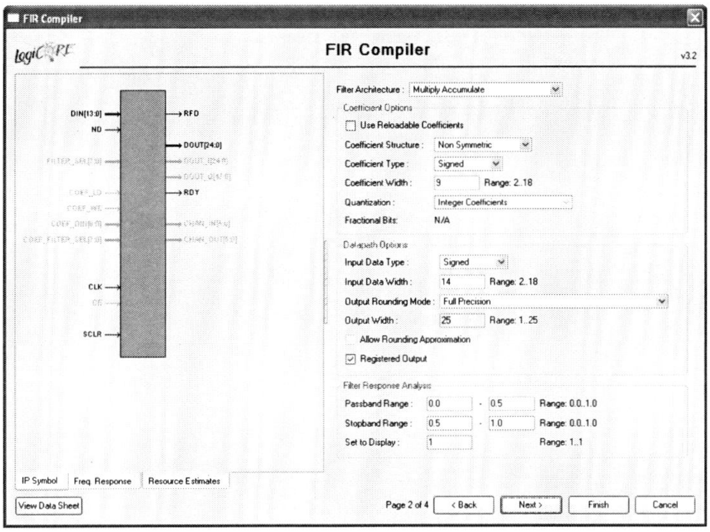

**Figure 4.13** Second design window of the LogiCORE FIR Compiler

The nine non-symmetrical coefficients here are signed integers specified by an ASCII text data file with the file extension *.coe*. The first line of the file is the declaration of the coefficient radix, which can be 2 (binary), 10 (decimal) or 16 (hexadecimal), terminated by a semicolon ( ; ). The integer FIR digital filter coefficients are then separated by a comma ( , ) and terminated with a semicolon, as given by Listing 4.5. This arbitrary set of non-symmetrical, signed integer coefficients is essentially the sampled impulse response of the digital filter [Mitra06]. These coefficients are presented in the Xilinx FIR Compiler data sheet DS534 as Figure 75.

**Listing 4.5** FIR Compiler coefficent file fir1.coe

```
radix = 10;
coefdata = 255, 200, –180, 80, 220, 180, 100, –48, 40;
```

The minimum bit precision of the non-symmetrical coefficients is 9 bits, since the maximum positive integer is 255 and the maximum negative integer is greater than –256. The *Show...* button displays the coefficients selected. If the number, type, width or structure of the coefficients does not agree with the text data file, the text boxes of the second design window of the LogiCORE FIR Compiler are displayed in *red*. The FIR Compiler determines that the digital filter full precision data output dout is 25 bits wide to avoid overflow and is selected to be a register, as shown in Figure 4.13.

The third design window of the LogiCORE FIR Compiler specifies the control and implementation options of the digital filter, as shown in Figure 4.14. The FIR Compiler is a controller and datapath construct, as described in Chapter 1 Verilog Hardware Description Language. The synchronous clear input signal sclr resets the internal state machine of the FIR digital filter with the clock input clk. The new data input signal nd loads the data sample input into the FIR digital filter core.

The ready for input data status signal rfd is asserted as a logic 1 to load new data. Finally, the data ready status signal rdy indicates that the FIR digital filter data is available. The storage buffer for the data and coefficients of the FIR digital filter can utilize block or distributed FPGA random access memory (RAM), as described in Chapter 2 Verilog Design Automation.

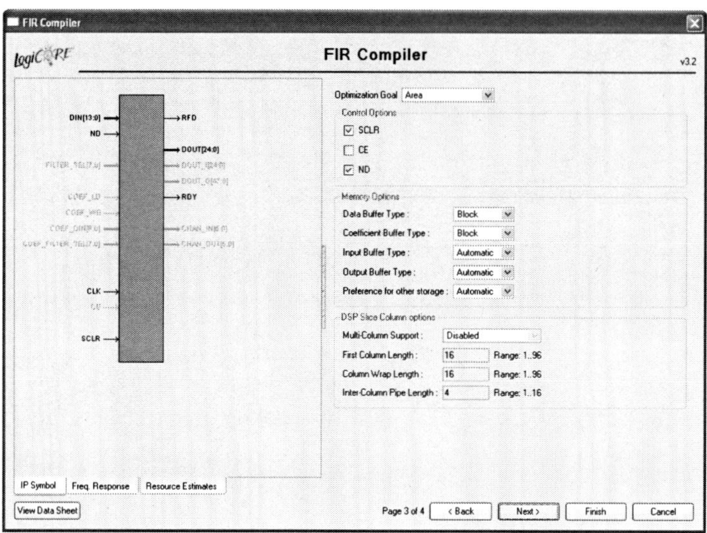


**Figure 4.14** Third design window of the LogiCORE FIR Compiler

Finally, the fourth design window of the LogiCORE FIR Compiler is the summary screen for the parameters of the digital filter, as shown in Figure 4.15. A *tab menu* can display the magnitude of the discrete frequency response of the digital filter. The frequency axis is normalized and 1.0 represents a frequency of $f_s$ /2 Hz, where $f_s$ is the sampling rate. Here $f_s$ = 250 kHz and the FIR digital filter displays a –20 dB *notch* at a frequency of approximately 0.22 × $f_s$ /2 kHz = 27.5 kHz.

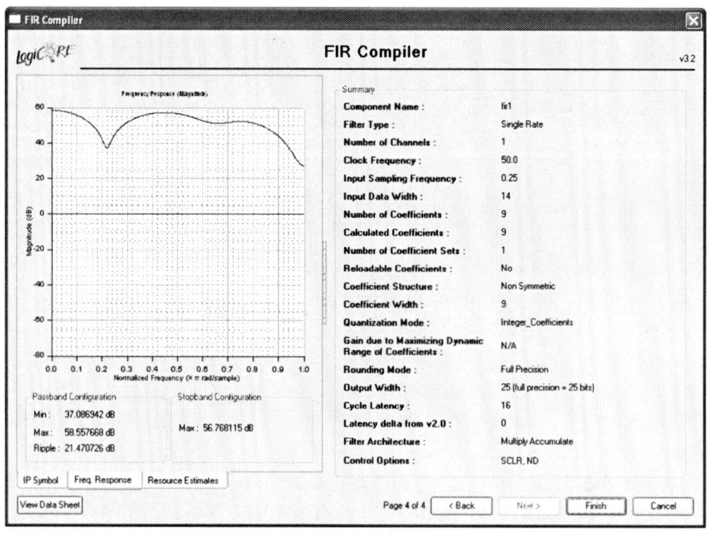


**Figure 4.15** Fourth design window of the LogiCORE FIR Compiler

Another tab menu in Figure 4.15 provides a resource estimate for the implementation of the digital filter. The generation of the FIR Compiler LogiCORE block provides Verilog components, including the module instantiation template file fir1.veo, as given in Listing 4.6. The template file for the FIR Compiler LogiCORE block is instantiated into the parent design by adding the module to the project, as described in Chapter 2 Verilog Design Automation.

**Listing 4.6** FIR Compiler LogiCORE instantiation template file fir1.veo

```
// The following must be inserted into your Verilog file for this
// core to be instantiated. Change the instance name and port connections
// (in parentheses) to your own signal names.

//----------- Begin Cut here for INSTANTIATION Template ---// INST_TAG
fir1 YourInstanceName (
	.sclr(sclr),
	.clk(clk),
	.nd(nd),
	.din(din),
	.rfd(rfd),
	.rdy(rdy),
	.dout(dout));

// INST_TAG_END ------ End INSTANTIATION Template ---------
```

*YourInstanceName*, a dummy name from the instantiation template file fir1.veo in Listing 4.6, is changed to M5 in the top module in Listing 4.7 and the port connections by name, as described in Chapter 1 Verilog Hardware Description Language, are changed to reflect the actual connections in the Xilinx ISE WebPACK project.

## Digital Signal Processing System

The FIR Compiler digital filter is implemented as a digital signal processing (DSP) system, as shown in Figure 4.7. The Verilog top module s3eadcfirdac.v is given in Listing 4.7, which is also located in the *Chapter 4 \adcdac\s3eadcfirdac* folder. The Xilinx ISE WebPACK project uses the UCF adcfirdacs3esb.ucf which uncomments the signals CCLK, SW0, SW1, SW2, SW3, SPIMISO, SPIMOSI, SPISCK, DACCS, DACCLR, SPISF, AMPCS, AMPSD, ADCON, SFCE, FPGAIB, JC1, JC2, JC3 and JC4 in the Spartan-3E Starter Board UCF of Listing 3.2. The seven Verilog modules operate in parallel and some independently in the top module. The file download procedure is described in the Appendix.

The Verilog top module s3eadcfirdac.v is similar to the se3adclpfdac.v in Listing 4.3. The 6-pin peripheral hardware module connector JC1 (J4-1) is used to monitor the ADC conversion command signal conad. The estimated cycle latency of the FIR filter from the FIR Compiler LogiCORE block summary, as shown in Figure 4.15, is $16 \times T_{clock} = 0.32$ µsec, where $T_{clock} = 20$ nsec since $f_{clock} = 50$ MHz.

The actual latency of the digital filter calculation can be determined by measuring the time interval between the assertion of the new data input signal nd at the hardware module connector JC4 and the appearance of the data ready output signal rdy on the hardware module connector JC3. The latency is measured to be 0.39 µsec, which is less than the available in-sequence computation time of 0.46 µsec, since the sampling period $T_s = 4$ µsec and the sampling period for the straight-through or ADC-DAC DSP system s3eadcfac.v in Listing 4.1 is $T_{ADC\text{-}DAC} \approx 3.54$ µsec.

The controller module genadcfirdac.v utilizes a five state FSM controller for the digital filter DSP system. Logic 1 on the FIR filter ready for data signal rfd sets the sampling rate of 250 kHz and is the FIR Compiler digital filter datapath signal. The controller module responses with the FIR filter new input data available signal nd. The data ready status signal rdy indicates that the 25-bit FIR digital filter data firdout is available.

The conversion of the 25-bit two's complement binary digital filter data output firdout to the 12-bit straight binary data input dacdata of the DAC of the Spartan-3E Starter Board is accomplished by first selecting a 12-bit range. Slide switches SW2 and SW3 are inputted to the register swreg which selects the bit ranges to be outputted to the DAC. Next, the MSB of the selected 12-bit register value is then complemented. Finally, the entire register value is inverted and set equal to the DAC data input dacdata. The inversion of the binary data accommodates the programmable amplifier of the Spartan-3E Starter Board which inverts the analog input signal.

Slide switch SW0 selects one of the two available ADC channels and SW1 is used to bypass the FIR digital filter to provide the straight-through response. The frequency response can then be assessed on an oscilloscope by comparing the input sinusoidal excitation as the straight-through response to that of the output of the FIR digital filter.

As described in Chapter 2 Verilog Design Automation, the Design Utilization Summary for the module s3eadcfirdac.v shows the use of 331 slice flip-flops (3%), 416 4-input LUTs (4%), 298 occupied slices (6%), 1 18 × 18 multiplier, 1 block RAM and a total of 79 376 equivalent gates in the XC3S500E Spartan-3E FPGA synthesis.

**Listing 4.7** FIR Compiler LogiCORE block DSP system s3eadcfirdac.v

```
// Spartan-3E Starter Board
// Digital-to-Analog Converter s3eadcfirdac.v
// c 2008 Embedded Design using Programmable Gate Arrays  Dennis Silage

module s3eadcfirdac (input CCLK, SW0, SW1, SW2, SW3, SPIMISO, output SPIMOSI, SPISCK,
                     output DACCS, DACCLR, SPISF, AMPCS, AMPSD, ADCON, SFCE,
                     output FPGAIB, JC1, JC2, JC3, JC4);

wire adcdav, davadc, adcsck, adcspod, csamp, conad, ampdav, davamp;
wire ampsck, ampspid, sdamp, dacdav, davdac, dacsck, dacspid, clrdac, csdac;
wire sclr, rfd, rdy, nd, CLKOUT, CLKFX;
wire [11:0] dacdata;
wire [13:0] adc0data, adc1data, firdin;
wire [24:0] firdout;
wire [3:0] ampcmd0, ampcmd1, daccmd, dacaddr;

assign SPISCK = adcsck | ampsck | dacsck;      // SPI clock
assign SPIMOSI = ampspid | dacspid;            // SPI data in
assign adcspod = SPIMISO;                      // SPI data out

assign DACCS = csdac;          // disable DAC
assign DACCLR = clrdac;        // DAC clear

assign AMPCS = csamp;          // select prog amp
assign AMPSD = sdamp;          // disable prog amp shutdown
assign ADCON = conad;          // ADC convert command

assign SPISF = 1;              // disable serial Flash
assign SFCE = 1;               // disable StrataFlash
assign FPGAIB = 1;             // disable Platform Flash
```

```
assign JC1 = conad;           // monitor sampling rate
assign JC2 = rfd;             // FIR ready for data
assign JC3 = rdy;             // FIR data ready
assign JC4 = nd;              // FIR new data available

s3eadc M0 (CLKOUT, adcdav, davadc, adc0data, adc1data, adcsck, adcspod, conad);
s3eprogamp M1 (CLKOUT, ampdav, davamp, ampcmd0, ampcmd1, ampsck, ampspid, csamp,
                sdamp);
s3edac M2 (CLKFX, dacdav, davdac, dacdata, dacaddr, daccmd, dacsck, dacspid, csdac, clrdac);
genadcfirdac M3 (CLKOUT, SW0, SW1, ampdav, davamp, ampcmd0, ampcmd1,adcdav, davadc,
                adc0data, adc1data, dacdav, davdac, dacdata, dacaddr, daccmd, sclr, rfd, rdy, nd,
                firdout, firdin, SW2, SW3);
dacs3edcm M4 (.CLKIN_IN(CCLK), .RST_IN(0), .CLKFX_OUT(CLKFX),
                .CLKIN_IBUFG_OUT(CLKINIBO), .CLK0_OUT(CLKOUT),
                .LOCKED_OUT(LOCK));
fir1 M5 (.sclr(sclr), .clk(CLKOUT), .nd(nd), .din(firdin), .rfd(rfd), .rdy(rdy), .dout(firdout));

endmodule

module genadcfirdac (input genclk, SW0, SW1, output reg ampdav = 0, input davamp,
                output reg [3:0] ampcmd0 = 1, output reg [3:0] ampcmd1 = 1,
                output reg adcdav = 0, input davadc, input [13:0] adc0data, input [13:0] adc1data,
                output reg dacdav = 0, input davdac, output reg [11:0] dacdata,
                output reg [3:0] dacaddr = 0, output reg [3:0] daccmd = 3, output reg sclr = 0,
                input rfd, rdy, output reg nd = 0, input [26:0] firdout, output reg [13:0] firdin,
                input SW2, SW3);

reg [2:0] gstate = 0;         // state register
reg [13:0] value;             // temporary value
reg [1:0] swreg;              // switch register

always@(posedge genclk)
    begin
        case (gstate)
            0:  begin
                    ampdav = 1;             // set programmable amplifier
                    if (davamp == 1)
                        gstate = 1;
                end
            1:  begin
                    adcdav = 1;             // ADC conversion
                        if (davadc == 1)
                            begin
                                adcdav = 0;
                                if (SW1 == 1)
                                    begin
                                        if (SW0 == 0)
                                            value = adc0data;
                                        else
                                            value = adc1data;
                                        value[13] = ~value[13];
                                        dacdata[11:0] = ~value[13:2];
```

```
                                        end

                                    else
                                        begin
                                            if (SW0 == 0)
                                                firdin = adc0data;
                                            else
                                                firdin = adc1data;
                                        end
                                    gstate = 2;
                                end
                    end
                2:  begin
                        if (rfd == 1)           // FIR ready for data
                            begin
                                nd = 1;         // FIR input data
                                if (SW1 == 1)
                                    gstate = 4;
                                else
                                    gstate = 3;
                            end
                    end
                3:  begin
                        if (rdy == 1)           // FIR output data
                            begin
                                nd = 0;
                                swreg[1] = SW3;
                                swreg[0] = SW2;
                                case (swreg)
                                    0:  value[11:0] = firdout[24:13];
                                    1:  value[11:0] = firdout[21:10];
                                    2:  value[11:0] = firdout[18:7];
                                    4:  value[11:0] = firdout[15:4];
                                endcase
                                value[11] = ~value[11];
                                dacdata[11:0] = ~value[11:0];
                                gstate = 4;
                            end
                    end
                4:  begin
                        dacdav = 1;             // DAC conversion
                        if (davdac == 1)
                            begin
                                dacdav = 0;
                                gstate = 1;
                            end
                    end
                default: gstate = 1;
            endcase
        end

endmodule
```

## Wavefront Digital Signal Processing System

The FIR Compiler digital filter can also be implemented as a *wavefront* digital signal processing (DSP) system in which the processing tasks of the ADC, FIR filter and DAC essentially occur in parallel, as shown in Figure 4.7. The Pmod AD1 ADC and Pmod DA2 DAC are used because the SPI bus contention of the Spartan-3E Starter Board ADC and DAC precludes a parallel implementation, as described in Chapter 3 Programmable Gate Array Hardware.

The Verilog top module s3ead1firda2.v is given in Listing 4.8, which is also located in the *Chapter 4\adcdac\s3ead1firda2* folder. The Xilinx ISE WebPACK project uses the UCF ad1firda2s3esb.ucf which uncomments the signals CCLK, SW0, SW1, SW2, SW3, JA1, JA2, JA3, JA4, JB1, JB2, JB3, JB4, JC1 JC2, JC3 and JC4 in the Spartan-3E Starter Board UCF of Listing 3.2. The four Verilog modules operate in parallel and some independently in the top module. The file download procedure is described in the Appendix.

**Listing 4.8** FIR Compiler LogiCORE block wavefront DSP system s3ead1firda2.v

```
// Spartan-3E Starter Board
// Digital-to-Analog Converter s3ead1da2.v
// c 2008 Embedded Design using Programmable Gate Arrays  Dennis Silage

module s3ead1firda2 (input CCLK, SW0, SW1, SW2, SW3, JA2, JA3, output JA1, JA4, JB1, JB2,
                     output JB3, JB4, JC1, JC2, JC3, JC4);

wire adcdav, davadc, adcsck, adc0d, adc1d, adccs;
wire dacdav, davdac, dacsck, dacout, dacsync;
wire sclr, rfd, rdy, nd;
wire [11:0] dacdata;
wire [1:0] daccmd;
wire [11:0] adc0data, adc1data;
wire [11:0] firdin;
wire [22:0] firdout;

assign JA1 = adccs;          // ADC AD1 on J1
assign adc0d = JA2;
assign adc1d = JA3;
assign JA4 = adcsck;

assign JB1 = dacsync;        // DAC DA2 on J2
assign JB2 = dacout;
assign JB3 = 0;
assign JB4 = dacsck;

assign JC1 = adccs;          // monitor sampling rate
assign JC2 = rfd;            // FIR ready for data
assign JC3 = rdy;            // FIR data ready
assign JC4 = nd;             // FIR new data available

ad1adc M0 (CCLK, adcdav, davadc, adc0data, adc1data, adcsck, adc0d, adc1d, adccs);
da2dac M1 (CCLK, dacdav, davdac, dacout, dacsck, dacsync, daccmd, dacdata);
genad1firda2 M2 (CCLK, SW0, SW1, adcdav, davadc, adc0data, adc1data, dacdav, davdac, dacdata,
                 daccmd, sclr, rfd, rdy, nd, firdout, firdin, SW2, SW3);
```

```
fir1 M3 (.sclr(sclr), .clk(CCLK), .nd(nd), .din(firdin), .rfd(rfd), .rdy(rdy), .dout(firdout));

endmodule

module genad1firda2 (input genclk, SW0, SW1, output reg adcdav = 0, input davadc,
                    input [11:0] adc0data, input [11:0] adc1data, output reg dacdav = 0, input davdac,
                    output reg [11:0] dacdata, output reg [1:0] daccmd = 0, output reg sclr = 0,
                    input rfd, rdy, output reg nd, input [22:0] firdout, output reg [11:0] firdin,
                    input SW2, SW3);

reg [2:0] gstate=0;        // state register
reg [11:0] value;          // temporary value
reg [1:0] swreg;           // switch register
reg [11:0] firsscnt;       // FIR startup sequence count
reg firss;                 // FIR startup sequence flag

always@(posedge genclk)
    begin
        case (gstate)
            0:  begin
                    gstate = 1;
                    firdin =0 ;
                    firsscnt = 0;
                    firss = 0;
                end
            1:  begin
                    firsscnt = firsscnt + 1;
                    if (firsscnt == 0)
                        firss = 1;
                    dacdav = 1;     // DAC conversion
                    adcdav = 1;     // ADC conversion
                    if (rfd == 1)   // FIR filter ready
                        begin
                            nd = 1;     //load FIR data
                            gstate = 2;
                        end
                    else
                        gstate = 1;
                end
            2:  begin
                    if (davdac == 1)        // DAC complete
                        begin
                            if (davadc == 1)        // ADC complete
                                begin
                                    dacdav = 0;
                                    adcdav = 0;
                                    gstate = 3;
                                end
                        end
                end
```

```
                3:  begin
                        if (SW1 == 1)                   // DAC output ADC
                            begin
                                if (SW0 == 0)
                                    value = adc0data;
                                else
                                    value = adc1data;
                                dacdata = value;
                                gstate = 1;
                            end
                        else                            // FIR data input
                            begin
                                if (firss == 1)
                                    begin
                                        if (SW0 == 0)
                                            firdin = adc0data;
                                        else
                                            firdin = adc1data;
                                        gstate = 4;
                                    end
                                else
                                    gstate = 1;
                            end
                    end
                4:  begin
                        nd = 0;
                        if (rdy == 1)               // DAC output FIR data
                            begin
                                swreg[1] = SW3;
                                swreg[0] = SW2;
                                case (swreg)
                                    0:  value[11:0] = firdout[22:11];
                                    1:  value[11:0] = firdout[18:7];
                                    2:  value[11:0] = firdout[15:6];
                                    4:  value[11:0] = firdout[11:0];
                                endcase
                                value[11] = ~value[11];
                                dacdata = value;
                                gstate = 1;
                            end
                        else
                            gstate = 4;
                    end
            endcase
        end

endmodule
```

The Verilog top module s3ead1firda2.v is similar to the se3adcfirdac.v for the Spartan-3E Starter Board ADC and DAC in Listing 4.7. The 6-pin peripheral hardware module connector JC (J4) is used to monitor the ADC conversion command signal adccs as JC1 (J4-1), the FIR Compiler digital filter ready for input data signal rfd as JC2, the data ready output signal rdy as JC3 and the new data input signal nd as JC4. The sampling rate $f_s$ = 700 kHz here which is selected to be near the maximum possible for this wavefront Pmod AD1 and Pmod DA2 DSP system of 714 kHz.

The estimated cycle latency of the FIR filter from the FIR Compiler LogiCORE block, as shown in Figure 4.15, is 21 × $T_{clock}$ = 0.42 μsec, where $T_{clock}$ = 20 nsec since $f_{clock}$ = 50 MHz. The estimated cycle latency here is greater with $f_s$ = 700 kHz and 12-bit input data than that previously estimated with $f_s$ = 250 kHz and 14-bit input data at 0.32 μsec.

The actual latency of the digital filter calculation can be determined from the time interval between the assertion of the new data input signal nd on the hardware module connector JC4 and the appearance of the data ready output signal rdy on JC3 and is measured to be 0.46 μsec.

The sampling period here $T_s = 1/f_s \approx$ 1.43 μsec is nearly the sampling period for the straight-through AD1-DA2 DSP system s3ead1da2.v in Listing 4.2 with $T_{AD1\text{-}DA2} \approx$ 1.40 μsec. The sampling rate for the wavefront Pmod ADC and Pmod DA2 DSP system is limited by the execution of the longest task. The data conversion for the Pmod DA2 DAC requires 34 × $T_{clock}$ = 0.68 μsec and the FIR filter calculation utilizes 0.46 μsec but the data conversion for the Pmod AD1 ADC requires 34 × $T_{clock}$ / 2 = 1.36 μsec, where $T_{clock}$ = 20 nsec since $f_{clock}$ = 50 MHz.

The controller module genad1firda2.v utilizes a five state FSM controller for the digital filter DSP system. Logic 1 on the FIR filter ready for data signal rfd sets the sampling signal of 700 kHz and is the FIR Compiler digital filter datapath signal. The sampling rate was specified in the first design window of the FIR Compiler, as shown in Figure 4.12.The controller module responds with the FIR filter new data input signal nd. The data ready status signal rdy indicates that the 23-bit FIR digital filter data firdout is available.

The FIR digital filter using the Verilog top module s3ead1firda2.v displays a *startup sequence instability* at the 700 kHz sampling rate with 12-bit input data that was not evident with the top module s3eadcfirdac.v at the 250 kHz sampling rate with 14-bit input data. The instability of the FIR filter is obviated by a short startup sequence of input data set to 0 determined by the length of the count register firsscnt and cleared with the flag register firss.

The conversion of the 23-bit two's complement binary digital filter data output firdout to the 12-bit straight binary data input dacdata of the Pmod DA2 DAC is accomplished by first selecting a 12-bit range. Slide switches SW2 and SW3 are inputted to the register swreg which selects the bit ranges to be outputted to the DAC. The MSB of the selected 12-bit register value is then complemented and set equal to the DAC data input dacdata.

Slide switch SW0 selects one of the two available ADC channels and SW1 is used to bypass the FIR digital filter to provide the straight-through response. The frequency response can then be assessed on an oscilloscope by comparing the input sinusoidal excitation as the straight-through response to that of the output of the FIR digital filter. Since the coefficient file remains the same but $f_s$ = 700 kHz, the FIR digital filter now displays a –20 dB *notch* at a frequency of approximately 0.22 × $f_s$ / 2 kHz = 77 kHz.

As described in Chapter 2 Verilog Design Automation, the Design Utilization Summary for the module s3eadcfirdac.v shows the use of 324 slice flip-flops (3%), 317 4-input LUTs (3%), 285 occupied slices (6%), 1 18 × 18 multiplier, 1 block RAM and a total of 72 456 equivalent gates in the XC3S500E Spartan-3E FPGA synthesis.

## Implementations

The FIR Compiler coefficient file facilitates the convenient restructuring of the digital filter without the necessity of modifying a Verilog HDL structure, as is required for the IIR digital filter in Listing 4.3 and the FIR digital filter in Listing 4.4. For example, a new coefficient file generates new frequency response parameters and reconfigures the FIR filter. The FIR Compiler also provides the half-band, Hilbert transform, interpolator and decimator implementations of the FIR digital filter [Chen01].

The half-band FIR digital filter provides a low-pass filter response centered at $f_s$ / 4 Hz, where $f_s$ is the sampling rate, with equal *passband* $\delta_p$ and *stopband* $\delta_s$ amplitude *ripples*, as shown in Figure 4.16. Approximately half of the half-band digital filter coefficients are *interleaved* zeros which can be exploited to provide an efficient implementation [Ifeachor02]. The FIR Compiler 11-tap half-band coefficient file hb1.coe is given in Listing 4.8. These coefficients are presented in the FIR Compiler data sheet DS534 as Figure 79 (*www.xilinx.com*).

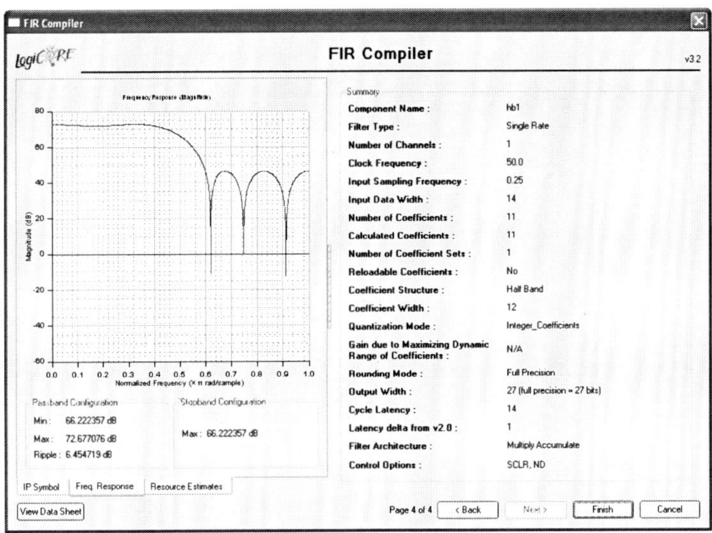


**Figure 4.16** LogiCORE FIR Compiler half-band filter frequency response and summary

**Listing 4.8** FIR Compiler 11-tap half-band coefficient file hb1.coe

```
radix = 10;
coefdata = 220, 0, –375, 0, 1283, 2047, 1283, 0, –375, 0, 220;
```

The FIR Compiler half-band digital filter is implemented as a DSP system, as shown in Figure 4.7. The Verilog top module s3eadchbdac.v is similar to Listing 4.7 and is in the *Chapter 4 \adcdac\s3eadchbdac* folder. The Xilinx ISE WebPACK project is also located in the *Chapter 4 \adcdac\s3eadchbdac* folder.

The FIR Compiler half-band digital filter has 27-bit two's complement binary digital filter data output. The minimum bit precision of the symmetrical coefficients is 12 bits, since the maximum positive integer is 2047 and the maximum negative integer is greater than –2048. The estimated cycle latency of the FIR filter is 14 × $T_{clock}$ = 0.28 μsec and the total latency is measured to be 0.35 μsec, which is less than the available in-sequence computation time of 0.46 μsec.

As described in Chapter 2 Verilog Design Automation, the Design Utilization Summary for the module s3eadchbdac.v shows the use of 421 slice flip-flops (4%), 442 4-input LUTs (4%), 375 occupied slices (8%), 1 18 × 18 multiplier, 1 block RAM and a total of 147 488 equivalent gates in the XC3S500E Spartan-3E FPGA synthesis.

The Hilbert transform FIR digital filter is utilized in digital communication systems [Lathi98]. An ideal Hilbert transform provides a 90° phase shift for positive frequencies and a –90° phase shift for (so-called) negative frequencies, which converts a real signal into its in-phase and quadrature components [Silage06]. The coefficients of the Hilbert transform FIR digital filter have odd symmetry and interleaved zeros.

The interpolated FIR digital filter replaces the discrete frequency unit delay parameter $z^{-1}$ with $k–1$ units of delay $z^{-k+1}$, where $k$ is the *zero-packing factor* [Chen01]. Interpolated FIR digital filters provide efficient implementation of both narrow-band and wide-band discrete frequency response. Finally, the decimator FIR digital filter is a parallel construct in which the discrete signal input $x(m)$ is provided in sequence to $M$ sub-filters and the output is obtained as the sum with an output sampling rate of $f_s$ / $M$ Hz [Mitra06]. The decimator FIR digital filter is well suited to the parallel architecture of the FPGA.

The requisite number of MACs required for the FIR digital filter is a function of the number of implicit multiplications in the tapped delay line representation, as shown in Figure 4.11, divided by the number of clock cycles available to process each input sample. The input data and coefficient storage of a single MAC can be reloaded and the MAC reused for efficiency, if sufficient processing time is available. A conceptual block diagram of the MAC architecture is shown in Figure 4.17.

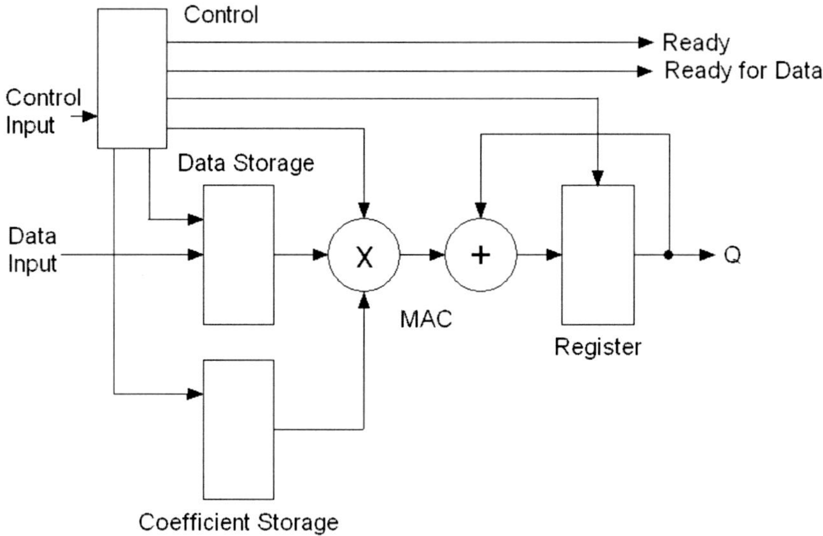


**Figure 4.17** Conceptual block diagram of the MAC architecture

The FIR Complier also provides the distributed arithmetic (DA) bit serial architecture which requires look-up tables (LUT), adders, subtractors and shift registers and can be efficiently mapped to fine-grained architecture of the FPGA. With the MAC architecture the maximum sample rate of the FIR digital filter is related to the number of coefficients. However, with the DA architecture the maximum sample rate is related to the bit precision of the input data [Ifeachor02].

For $N$-bit precision $N$ clock cycles are required to process an output sample for a non-symmetrical coefficient FIR digital filter. As the number of coefficients increases the DA FIR digital filter utilizes more FPGA resources to maintain the sample rate.

The discrete finite impulse response of many digital filters displays a degree of symmetry in the coefficients which can be exploited to provide FPGA resource efficient realizations. The tapped-delay line, single rate FIR digital filter, as shown in Figure 4.11, can be reconfigured when coefficient symmetry occurs [Chen01].

The tapped-delay line FIR digital filter requires approximately $N$ multiplications and $N$–1 additions. The FIR digital filter architecture with an odd number of symmetrical coefficients requires only approximately $N$ / 2 multiplications and $N$ additions, as shown in Figure 4.18 with $N = 7$.

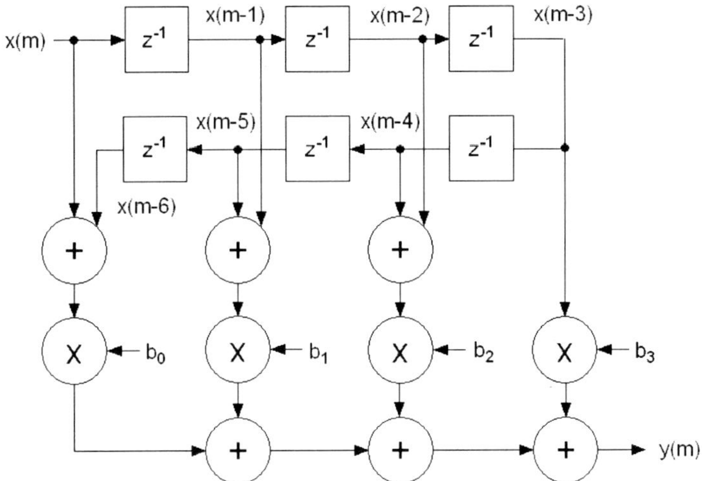


**Figure 4.18** FIR digital filter architecture with an odd number of symmetrical coefficients ($N = 7$)

The FIR digital filter architecture with negative coefficient symmetry is similar to Figure 4.18 but with subtractors in the intermediate layer. The FIR digital filter architecture with an even number of symmetrical coefficients also requires only approximately $N / 2$ multiplications and $N$ additions, as shown in Figure 4.19 with $N = 8$.

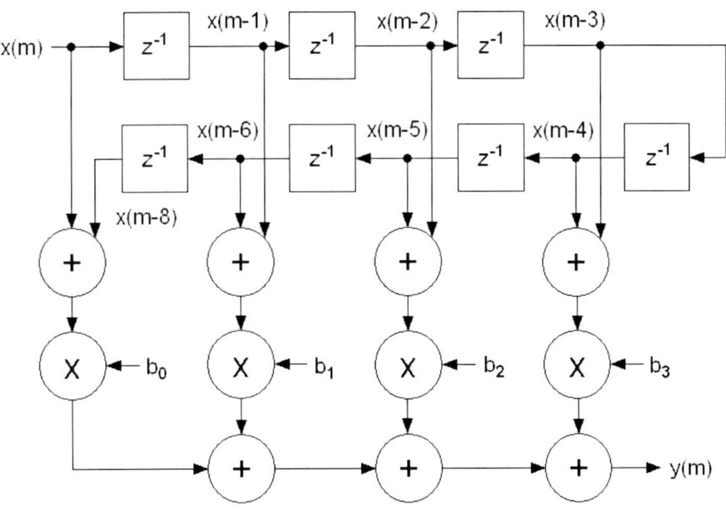


**Figure 4.19** FIR digital filter architecture with an even number of symmetrical coefficients ($N = 8$)

## Sine-Cosine Look-Up Table LogiCORE Block

The Xilinx CORE Generator Sine-Cosine Look-Up Table LogiCORE block accepts an unsigned input angle *theta* and produces a two's complement output of sine(*theta*) and cosine(*theta*). The input *theta* is specified as *theta_width* of 3 to 10 bits for the distributed ROM and 3 to 16 bits for the block ROM of the Spartan-3E FPGA, as described in Chapter 3 Programmable Gate Array Hardware. The sine and cosine *output_width* can be from 4 to 32 bits and either a full or a quarter of a sinusoid is store in the block ROM. The quarter sinusoid can be used to generate a full sinusoid using additional internal logic and the selection is determined by the LogiCORE block for the most efficient implementation. The actual radian angle θ is determined by Equation 4.28.

$$\theta = \text{theta}\,\frac{2\,\pi}{2^{\text{theta\_width}}} \qquad (4.28)$$

The two's complement sine and cosine output are fractional fixed-point values and in the range for the non-symmetrical form determined by Equation 4.29. For the non-symmetrical form with an arbitrary output width of 4 bits the range is – 1 to + 0.875. The symmetrical form of the sine and cosine output uses one bit to provide an effective range –1 to +1 but with a concomitant one bit reduction in resolution.

$$\frac{-2^{\text{output\_width}-1}}{2^{\text{output\_width}-1}} \rightarrow \frac{2^{\text{output\_width}-1}-1}{2^{\text{output\_width}-1}} \qquad (4.29)$$

The Sine-Cosine Look-Up Table LogiCORE block is described in a data sheet DS275 (*www.xilinx*.com) and implemented with three LogiCORE design windows, as in Chapter 2 Verilog Design Automation. The first design window of the LogiCORE Sine-Cosine Look-Up Table specifies the name of the component, the *output_width* as 10 bits, *theta_width* as 7 bits, sine only output using the block ROM of the Spartan-3E FPGA, as shown in Figure 4.21. The second design window specifies non-registered theta input and symmetrical, registered sine output, as shown in Figure 4.22. Finally, third design window disables the clock enable, asynchronous and synchronous clear signals but enables *handshaking* with the new data nd input signal and the ready for data rfd and ready rdy output signals, as shown in Figure 4.23.

## DTMF Generator

The Sine-Cosine Look-Up Table LogiCORE block is used to generate a dual-tone multiple frequency (DTMF) audio signal. DTMF is the standard method for instructing a telephone switching system of the telephone number to be dialed or to issue commands to switching systems or related telephony equipment. The common DTMF keypad is a 4 by 4 matrix where each row represents a low frequency and each column represents a high frequency, as shown in Figure 4.20. Depressing a single key causes a low and a high frequency audio tone to be generated.

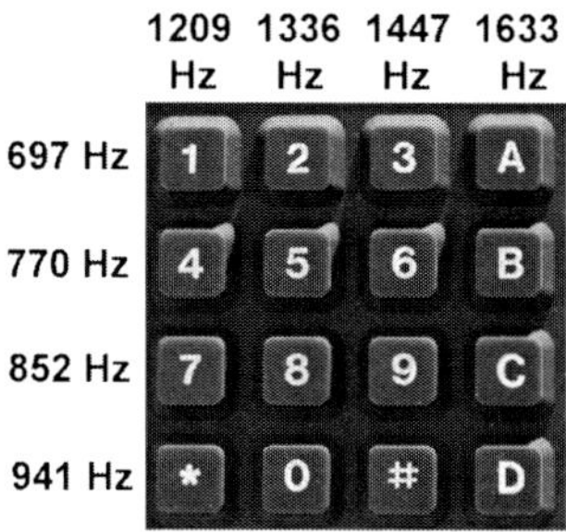


**Figure 4.20** DTMF keypad and dual tone frequencies

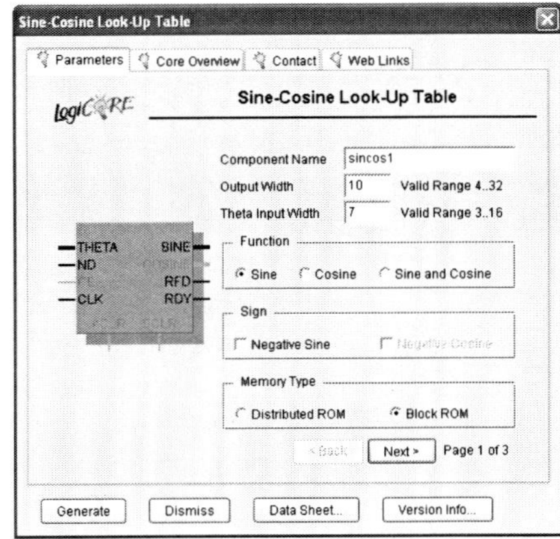


**Figure 4.21** First design window of the LogiCORE Sine-Cosine Lookup Table

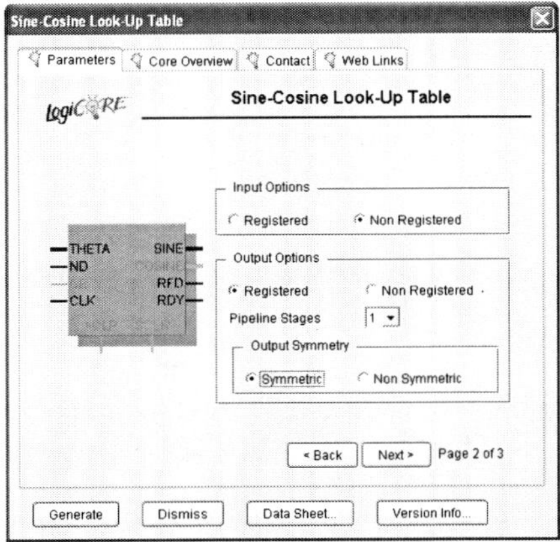


**Figure 4.22** Second design window of the LogiCORE Sine-Cosine Lookup Table

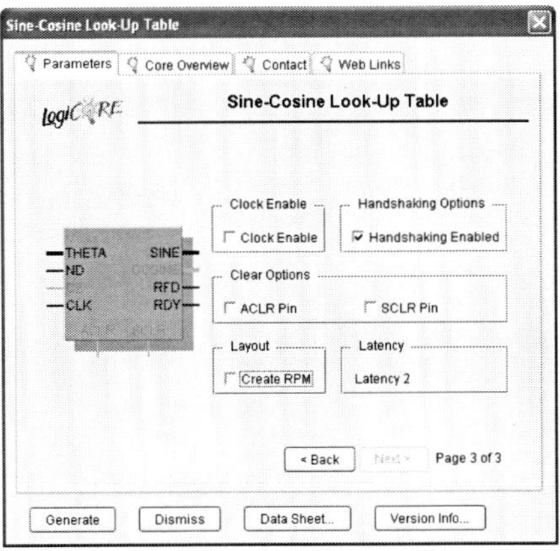


**Figure 4.23** Third design window of the LogiCORE Sine-Cosine Lookup Table

The tone frequencies are selected such that harmonics and intermodulation products do not cause an unreliable decoding of the DTMF signal. No frequency is a multiple of another and the difference between and the sum of any two frequencies does not equal any of the individual frequencies. The ITU-T Recommendation Q.23 for the DTMF signal specifies that the frequencies may not vary more than ± 1.5%. The signal amplitude ratio between the high and low frequencies must be within 3 decibels (0.707). The minimum duration of the DTMF signal must be at least 70 msec.

The generation of the Sine-Cosine Lookup Table LogiCORE block provides Verilog components, including the module instantiation template file sincos1.veo, as given in Listing 4.9. The template file for the Sine-Cosine Lookup Table LogiCORE block is instantiated into the parent design by adding the module to the project, as described in Chapter 2 Verilog Design Automation.

**Listing 4.9** Sine-Cosine Lookup Table LogiCORE instantiation template file sincos1.veo

```
// The following must be inserted into your Verilog file for this
// core to be instantiated. Change the instance name and port connections
// (in parentheses) to your own signal names.

//----------- Begin Cut here for INSTANTIATION Template ---// INST_TAG
sincos1 YourInstanceName (
	.THETA(THETA),	// Bus [6 : 0]
	.CLK(CLK),
	.ND(ND),
	.RFD(RFD),
	.RDY(RDY),
	.SINE(SINE));		// Bus [9 : 0]

// INST_TAG_END ------ End INSTANTIATION Template ---------
```

*YourInstanceName*, a dummy name from the instantiation template file sincos1.veo in Listing 4.9, is changed to M5 and M6 in the top module in Listing 4.10 and the port connections by name, as described in Chapter 1 Verilog Hardware Description Language, are changed to reflect the actual connections in the Xilinx ISE WebPACK project.

The DTMF generator Verilog top module s3esincosdtmf.v is given in Listing 4.10, which is in the *Chapter 4\sincoslut\s3esincosdtmf* folder. The Xilinx ISE WebPACK project uses the UCF sincosdtmfs3esb.ucf which uncomments the signals CCLK, SW0, SW1, SW2, SW3, BTN0, BTN1, BTN2, SPIMISO, SPIMOSI, SPISCK, DACCS, DACCLR, SPISF, AMPCS, AMPSD, ADCON, SFCE and FPGAIB in the Spartan-3E Starter Board UCF of Listing 3.2. The seven Verilog modules operate in parallel and some independently in the top module. The file download procedure is described in the Appendix.

**Listing 4.10** Sine-Cosine Look-Up Table LogiCORE block DTMF generator s3esincosdtmf.v

```
// Spartan-3E Starter Board
// Digital-to-Analog Converter s3esincosdtmf.v
// c 2008 Embedded Design using Programmable Gate Arrays  Dennis Silage

module s3esincosdtmf (input CCLK, SW0, SW1, SW2, SW3, BTN0, BTN1, BTN2, output SPIMOSI,
				output SPISCK, DACCS, DACCLR, SPISF, AMPCS, ADCON, SFCE,
				output FPGAIB);

wire dacdav, davdac, dacsck, dacspid, clrdac, csdac, clka, clkb;
wire [11:0] dacdata;
wire [31:0] clkdiva, clkdivb;
wire [3:0] daccmd, dacaddr;
wire [6:0] theta1, theta2;
```

```
wire [9:0] sine1, sine2;
wire nd1, rfd1, rdy1, nd2, rfd2, rdy2;
wire [1 :0] lowreg, highreg ;
wire [2 :0] dtmfctrl ;

assign SPISCK = dacsck;         // SPI clock
assign SPIMOSI = dacspid;       // SPI data in

assign DACCS = csdac;           // disable DAC
assign DACCLR = clrdac;         // DAC clear

assign AMPCS = 1;               // disable prog amp
assign ADCON = 0;               // disable ADC

assign SPISF = 1;               // disable serial Flash
assign SFCE = 1;                // disable StrataFlash
assign FPGAIB = 1;              // disable Platform Flash

assign dtmfctrl = {BTN2, BTN1, BTN0};
assign lowreg = {SW1, SW0};
assign highreg = {SW3, SW2};

clock M0 (CLKOUT, clkdiva, clka);
clock M1 (CLKOUT, clkdivb, clkb);
s3edac M2 (CLKFX, dacdav, davdac, dacdata, dacaddr, daccmd, dacsck, dacspid, csdac, clrdac);
gendtmf M3 (CLKOUT, lowreg, highreg, dtmfctrl, clkdiva, clka, clkdivb, clkb, theta1, nd1, rfd1, rdy1,
        sine1, theta2, nd2, rfd2, rdy2, sine2, dacdav, davdac, dacdata, dacaddr, daccmd);
dacs3edcm M4 (.CLKIN_IN(CCLK), .RST_IN(0), .CLKFX_OUT(CLKFX),
        .CLKIN_IBUFG_OUT(CLKINIBO), .CLK0_OUT(CLKOUT),
        .LOCKED_OUT(LOCK));
sincos1 M5 (.THETA(theta1), .CLK(CLKOUT), .ND(nd1), .RFD(rfd1), .RDY(rdy1),
        .SINE(sine1));
sincos1 M6 (.THETA(theta2), .CLK(CLKOUT), .ND(nd2), .RFD(rfd2), .RDY(rdy2),
        .SINE(sine2));

endmodule

module gendtmf (input genclk, input [1:0] lowreg, input [1:0] highreg, input [2:0] dtmfctrl,
        output reg [31:0] clkdiva, input clka, output reg [31:0] clkdivb, input clkb,
        output reg [6:0] theta1=0, output reg nd1=0, input rfd1, rdy1,
        input [9:0] sine1, output reg [6:0] theta2=0, output reg nd2=0,
        input rfd2, rdy2, input [9:0] sine2, output reg dacdav=0, input davdac,
        output reg [11:0] dacdata, output reg [3:0] dacaddr=0,
        output reg [3:0] daccmd=3);

reg [2:0] gstate1 = 0;      // state registers
reg [2:0] gstate2 = 0;
reg gstate3 = 0;
reg [11:0] lowf;            // low frequency tone
reg [11:0] highf;           // high frequency tone
```

```
always@(posedge genclk)
      begin
            case (lowreg)
                  0:    clkdiva = 280;  // 697 Hz
                  1:    clkdiva = 254;  // 770 Hz
                  2:    clkdiva = 229;  // 852 Hz
                  3:    clkdiva = 208;  // 941 Hz
            endcase
            case (highreg)
                  0:    clkdivb = 162;  // 1209 Hz
                  1:    clkdivb = 146;  // 1336 Hz
                  2:    clkdivb = 133;  // 1471 Hz
                  3:    clkdivb = 120;  // 1633 Hz
            endcase

            case (gstate1)                  // generate low frequencies
                  0:    begin
                              if (clka == 1)
                                    gstate1 = 1;
                        end
                  1:    begin
                              if (dtmfctrl[2] == 0)
                                    theta1 = 0;
                              else
                                    theta1 = theta1 + 1;
                              gstate1 = 2;
                        end
                  2:    begin
                              if (rfd1 == 1)
                                    gstate1 = 3;
                        end
                  3:    begin
                              nd1 = 1;
                                    gstate1 = 4;
                        end
                  4:    begin
                              if (rdy1 == 1 && clka == 0)
                                    begin
                                          nd1 = 0;
                                          gstate1 = 0;
                                    end
                         end
            endcase

            case (gstate2)                  // generate high frequencies
                  0:    begin
                              if (clkb == 1)
                                    gstate2 = 1;
                        end
                  1:    begin
                              if (dtmfctrl[2] == 0)
                                    THETA2 = 0;
```

```
                    else
                          THETA2 = THETA2 + 1;
                    gstate2 = 2;
               end
          2:   begin
                    if (rfd2 == 1)
                          gstate2 = 3;
               end
          3:   begin
                    nd2 = 1;
                    gstate2 = 4;
               end
          4:   begin
                    if (rdy2 == 1 && clkb == 0)
                          begin
                                nd2 = 0;
                                gstate2 = 0;
                          end
               end
     endcase

     case (gstate3)                      // output DTMF
          0:   begin
                    lowf[9:0] = sine1[9:0];     // output low frequencies
                    lowf[9] = ~lowf[9];
                    lowf[11:10] = 0;
                    highf[9:0] = sine2[9:0];    // output high frequencies
                    highf[9] = ~highf[9];
                    highf[11:10] = 0;
                    if (dtmfctrl[0] == 1)       // DAC output only low frequencies
                          dacdata = lowf;
                    if (dtmfctrl[1] == 1)       // DAC output only high frequencies
                          dacdata = highf;
                    if ((dtmfctrl[0] || dtmfctrl[1]) == 0)    // DAC output DTMF
                          dacdata = lowf + highf;
                    gstate3 = 1;
               end
          1:   begin
                    dacdav = 1;                 // DAC conversion
                          if (davdac == 1)
                                begin
                                      dacdav = 0;
                                      gstate3 = 0;
                                end
               end
          default: gstate3 = 0;
     endcase
end

endmodule
```

The low and high frequency tones are generated by two Sine-Cosine Look-Up Table LogiCORE blocks. Two clock modules provide the rate at which the 128 (theta_width of 7 bits) sine table values are retrieved. For the highest tone of 1633 Hz, there would be 128 × 1633 = 209 024 sine table data retrievals per second. The clock scale factor for the clock module is determined by Equation 3.1 and would be 25 000 000 / 209 024 = 119.60.

Since the tone frequency need only be accurate to ± 1.5% the clock scale factor for 1633 Hz is rounded to 120. If 256 (theta_width of 8 bits) or more sine table entries are used, rounding of the clock scale factor can produce an inaccurate tone. Additionally, the data throughput rate of the Spartan-3E Starter Board DAC $R_{DAC} \approx 1.517$ Msamples/sec, as described in Chapter 3 Programmable Gate Array Hardware, and would be exceeded for the highest tone of 1633 Hz with 1024 (10-bit) sine table entries.

The datapath module gendtmf.v utilizes three FSM controllers for the DTMF system. Separate clock signals clka and clkb are provided by the two modules clock.v. The requisite timing for the low and high frequency tones of the DTMF signal is set with the clock scale factor clkdiva and clkdivb. The DTMF system here uses two 2-bit signals from the four slide switches to select one of the 16 possible sets of dual tones. The two sets of two control switches are concatenated and continuously assigned to the 2-bit control inputs lowreg and highreg.

Two five state FSM controllers, event driven with the clock signals clka and clkb and the control signals rfd1, rdy1, rfd2 and rdy2, increments the two inputs theta1 and theta2 to the two Sine-Cosine Look-Up Table LogiCORE blocks if pushbutton BTN2 is logic 1. If BTN2 is logic 0 theta1 and theta2 are set to 0 which effectively cancels the DTMF tone output. The three control pushbuttons are concatenated and continuously assigned to the 3-bit control input dtmfctrl.

The 10-bit two's complement sine outputs sine1 and sine2 are converted to 12-bit straight binary data for the DAC by complementing bit 9 and setting the two most significant bits (MSB) to 0. Pushbuttons BTN0 and BTN1 are used to output only one of the two tones. The 10-bit low and high frequency tones are *padded* to 12 bits then added together to form the DTMF signal. The 12-bit Spartan-3E Starter Board DAC is used to output the DTMF audio tones. The DAC output data rate is its maximum rate of 1.47 Msamples/sec and greater than the 209 024 sine table data retrievals per second for the highest frequency in the DTMF audio signal at 1633 Hz.

Figure 4.24 shows the DAC output of the DTMF system s3esincosdtmf.v for the dual tone of 770 and 1209 Hz which corresponds to keypad 4. In an embedded design the pushbutton and slide switch inputs of the DTMF datapath module gendtmf.v are replaced with control signal inputs.

As described in Chapter 2 Verilog Design Automation, the Design Utilization Summary for the module s3esincosdtmf.v shows the use of 188 slice flip-flops (2%), 215 4-input LUTs (2%), 201 occupied slices (4%), 2 block RAMs (10%) and a total of 141 953 equivalent gates in the XC3S500E Spartan-3E FPGA synthesis.

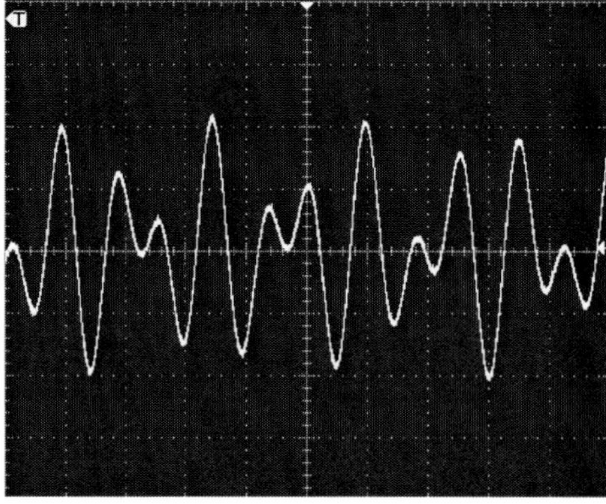

**Figure 4.24** DTMF generator s3esincosdtmf.v output at 200 mV/div and 1 msec/div for the dual tone 770 Hz and 1209 Hz (keypad 4)

## Direct Digital Synthesis Compiler LogiCORE Block

The Xilinx CORE Generator Direct Digital Synthesis (DDS) LogiCORE block or numerically controlled oscillator (NCO) is an essential building block in a digital communication system. DDS forms the basis for digital frequency synthesizers, up and down converters, coherent demodulators and modulators [Haykin01]. The DDS LogiCORE block is described in data sheet DS558 (*www.xilinx.com*) and accepts a phase increment $\Delta\theta$ and clock signal of frequency $f_{clock}$ to a phase accumulator of $B_\theta$ bits and produces a discrete sinusoidal signal with an output frequency $f_{out}$ determined by Equation 4.30.

$$f_{out} = \frac{f_{clock}\,\Delta\theta}{2^{B_\theta}} \text{ Hz} \qquad (4.30)$$

Equation 4.31 gives the phase increment $\Delta\theta$ required to generate an output frequency $f_{out}$ Hz.

$$\Delta\theta = \frac{f_{out}\,2^{B_\theta}}{f_{clock}} \qquad (4.31)$$

The spectral purity of the output sinusoidal is affected by the number of bits in the phase accumulator and the amplitude quantization of the samples from the look-up table. The clock frequency and the bit width of the phase accumulator determine the frequency resolution of the DDS. As a result, a large number of bits are usually allocated to the phase accumulator to satisfy these requirements. If the required frequency resolution is $\Delta f$ then the number of bits $B_\theta$ of the phase accumulator is determined by Equation 4.32.

$$B_\theta = \log_2\left\lceil \frac{f_{clock}}{\Delta f} \right\rceil \qquad (4.32)$$

The number of bits $B_\theta$ from Equation 4.32 is *rounded-up*. If $f_{clock}$ = 50 MHz and $\Delta f$ = 0.5 Hz then $B_\theta$ = 27 bits. However, a phase accumulator of 27 bits requires an excessive amount of entries in the sine-cosine look-up table and a truncated version or quantized version of the phase angle is used. Although this quantization allows a reasonable memory requirement, the resulting *jitter* in the time base produces an undesired phase modulation.

The DDS Compiler LogiCORE block is evoked from within a Xilinx ISE WebPACK project with six design windows, as described in Chapter 2 Verilog Design Automation. The first design window of the LogiCORE DDS Compiler is shown in Figure 4.25. The component name (dds1) is specified, a sinusoidal output is chosen, the DDS clock rate $f_{clock}$ is 50 MHz, the spurious free dynamic range (SFDR) is 66 dB (decibel) and the frequency resolution $\Delta f$ is 0.5 Hz. The clock frequency here is that provided directly by the clock oscillator of the Spartan-3E Starter Board without the use of the Spartan-3E FPGA Digital Clock Manager (DCM).

The specification of the SFDR determines the number of bits in the discrete sinusoidal output signal of the DDS. Since an analog sinusoidal output is derived from this discrete signal by using the 12-bit DAC of the Spartan-3E Starter Board, 11 bits of output provides a spurious free dynamic range of 66 dB, as shown in Figure 4.25.

The second and third design windows of the LogiCORE DDS Compiler, not shown here, specifies the output frequency $f_{out}$ as 25 kHz, allows a programmable phase increment $\Delta\theta$ and sets no phase offset angle. The phase increment is programmable in the second design window which allows the output frequency to be set or provides frequency shift keying (FSK) in digital modulation [Silage06].

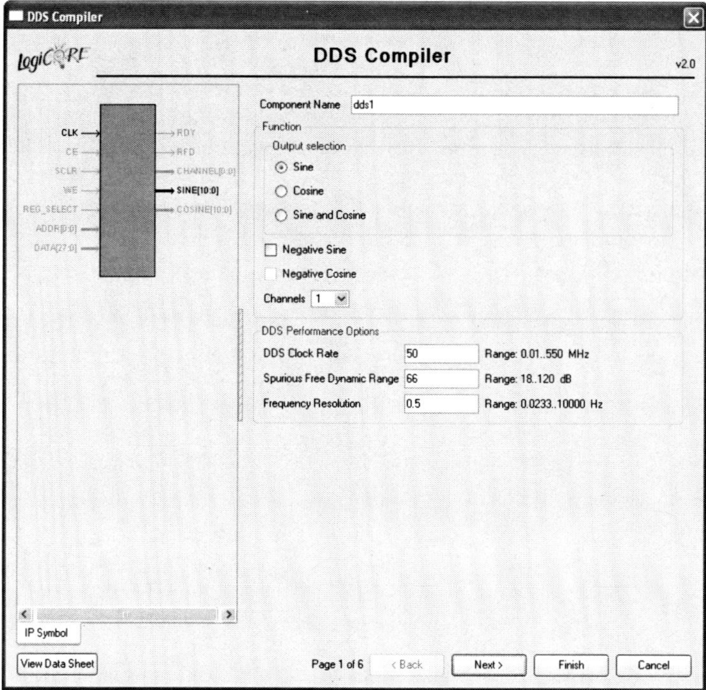


**Figure 4.25** First design window of the LogiCORE DDS Compiler

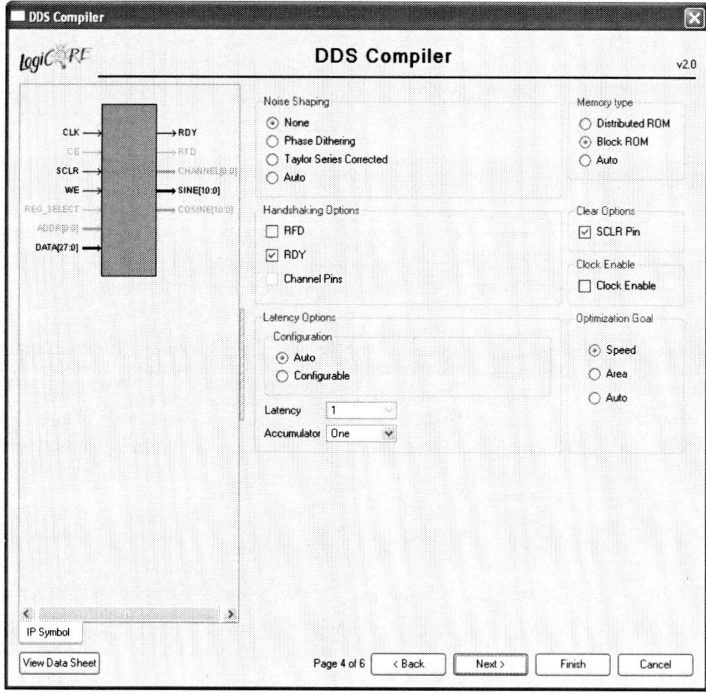


**Figure 4.26** Fourth design window of the LogiCORE DDS Compiler

The phase offset can be programmable in the third design window of the LogiCORE DDS Compiler. A programmable phase offset provides phase shift keying (PSK) in digital modulation or adjusts the phase of a reference sinusoid in the matched filter or correlator to that of the received signal in bandpass digital demodulation [Sklar01].

The fourth design window, as shown in Figure 4.26, specifies no noise shaping, block RAM for the sine-cosine look-up table, the data ready rdy and synchronous clear sclr interface signals and optimization for speed of execution. The write enable we signal enables an updated phase increment $\Delta\theta$ to be written to the DDS. The LogiCORE DDS Compiler does not use the ready for data rfd signal and set its to logic 1, although it is included for compatibility with other LogiCORE blocks. The design here does not enable the output of the logic 1 rfd signal.

The generation of the DDS Compiler LogiCORE block provides Verilog components, including the module instantiation template file dds1.veo, as given in Listing 4.11. The template file for the DDS Compiler LogiCORE block is instantiated into the parent design by adding the module to the Xilinx ISE WebPACK project, as described in Chapter 2 Verilog Design Automation.

**Listing 4.11** DDS Compiler LogiCORE instantiation template file dds1.veo

```
// The following must be inserted into your Verilog file for this
// core to be instantiated. Change the instance name and port connections
// (in parentheses) to your own signal names.

//----------- Begin Cut here for INSTANTIATION Template ---// INST_TAG
dds1 YourInstanceName (
     .clk(clk),
     .sclr(sclr),
     .we(we),
     .data(data),      // Bus [26 : 0]
     .rdy(rdy),
     .sine(sine));     // Bus [10 : 0]

// INST_TAG_END ------ End INSTANTIATION Template ---------
```

*YourInstanceName*, a dummy name from the instantiation template file dds1.veo in Listing 4.11, is changed to M1 in the top module in Listing 4.12 and the port connections by name, as described in Chapter 1 Verilog Hardware Description Language, are changed to reflect the actual connections in the Xilinx ISE WebPACK project.

The DDS Compiler uses a quantized version of the phase angle producing jitter and an undesired phase modulation. The LogiCORE block here does not use either the phase dithered or Taylor series corrected DDS, referred to as noise shaping, to ameliorate the jitter. The phase dithered DDS provides an additional 12 dB of SFDR even with two fewer bits in the quantized phase angle. Additional logic implements a randomizing or dithering sequence to improve performance. The Taylor series corrected DDS use FPGA multiplier resources to decrease the undesired phase modulation.

The fifth and sixth design windows of the LogiCORE DDS Compiler provides a summary with the latency of the design and the value of the phase increment $\Delta\theta$ required for the output frequency $f_{out}$ specified, as determined by Equation 4.31, and are not shown here. For an output frequency $f_{out}$ = 25 kHz with $B_\theta$ = 27 bits and $f_{clock}$ = 50 MHz the phase increment $\Delta\theta$ = 10624h (67 108). The estimated latency for the DDS task is $1 \times T_{clock}$ = 20 nsec.

The LogiCORE DDS Compiler is inherently more efficient and versatile that the Sine-Cosine Look-Up LogiCORE block since the discrete sinusoidal data is accessed continuously without providing argument angles. The synthesized discrete sinusoid from the LogiCORE DDS Compiler can be used to generate an analog sinusoidal signal by outputting the samples to a DAC.

A carrier sinusoid from the LogiCORE DDS Compiler can be frequency and phase modulated from a digital information source [Slar01]. Thus the implementation of baseband frequency generation and bandpass digital modulation as a transmitter is straightforward and the output can be conveniently viewed on an oscillographic display. However, the development of a baseband or bandpass receiver requires complex bit and carrier synchronization in the demodulation process [Silage06] and is not attempted here.

## Frequency Generator

A synthesized frequency generator is implemented with the LogiCORE DSS Compiler in the Verilog top module s3eddsfreqgen.v in Listing 4.12, which is in the *Chapter 4\dds\s3eddsfreqgen* folder. The Xilinx ISE WebPACK project uses the UCF ddsfreqgens3esb.ucf which uncomments the signals CCLK, ROTA, ROTB, ROTCTR, BTN0, SPIMOSI, SPISCK, DACCS, DACCLR, SPISF, AMPCS, AMPSD, ADCON, SFCE, FPGAIB, and JC1 (J4-1) in the Spartan-3E Starter Board UCF of Listing 3.2. The five Verilog modules operate in parallel and some independently in the top module. The file download procedure is described in the Appendix.

The DDS Compiler LogiCORE block initially generates a 11-bit discrete 25 kHz sinusoid and outputs an analog signal using the Spartan-3E Starter Board DAC. The maximum data throughput rate of the DAC $R_{DAC}$ is 1.517 Msamples/sec (660 nsec period), as described in Chapter 3 Programmable Gate Array Hardware. The data throughput rate of the synthesized frequency generator $R_{freqgen}$ is measured at the 6-pin peripheral hardware module connector JC as 1.429 Msamples/sec (700 nsec period) and reduced by the latency of the DDS and the datapath module genddsfreq.v.

**Listing 4.12** Synthesized frequency generator top module for the Spartan-3E Starter Board s3eddsfreqgen.v

```
// Spartan-3E Starter Board
// Digital-to-Analog Converter s3eddsfreqgen.v
// c 2008 Embedded Design using Programmable Gate Arrays  Dennis Silage

module s3eddsfreqgen (input CCLK, ROTA, ROTB, ROTCTR, BTN0, output SPIMOSI, SPISCK,
                      output DACCS, DACCLR, SPISF, AMPCS, ADCON, SFCE, FPGAIB,
                      output JC1);

wire dacdav, davdac, dacsck, dacspid, clrdac, csdac, rdy, sclr, we, rotAreg, rotBreg, rotCTRreg;
wire [11:0] dacdata;
wire [3:0] daccmd, dacaddr;
wire [2:0] dtmfctrl;
wire [1:0] lowreg, highreg;
wire [26:0] data;
wire [10:0] sine;

assign SPISCK = dacsck;         // SPI clock
assign SPIMOSI = dacspid;       // SPI data in

assign DACCS = csdac;           // disable DAC
assign DACCLR = clrdac;         // DAC clear
assign AMPCS = 1;               // disable prog amp
assign ADCON = 0;               // disable ADC

assign SPISF = 1;               // disable serial Flash
assign SFCE = 1;                // disable StrataFlash
```

```
assign FPGAIB = 1;            // disable Platform Flash

assign JC1 = dacdav;          // monitor DAC rate

s3edac M0 (CLKFX, dacdav, davdac, dacdata, dacaddr, daccmd, dacsck, dacspid, csdac, clrdac);
dds1 M1 (.clk(CLKOUT),.sclr(sclr),.we(we),.data(data),.rdy(rdy),.sine(sine));
dacs3edcm M2 (.CLKIN_IN(CCLK), .RST_IN(0), .CLKFX_OUT(CLKFX),
                        .CLKIN_IBUFG_OUT(CLKINIBO), .CLK0_OUT(CLKOUT),
                        .LOCKED_OUT(LOCK));
rotary M3 (CLKOUT, ROTA, ROTB, ROTCTR, rotAreg, rotBreg, rotCTRreg);
genddsfreq M4 (CLKOUT, rotAreg, rotBreg, rotCTRreg, BTN0, sclr, we, data, rdy, sine,
                    dacdav, davdac, dacdata, dacaddr, daccmd);

endmodule

module genddsfreq (input genclk, rotAreg, rotBreg, rotCTRreg, BTN0, output reg sclr = 0,
                    output reg we = 0, output reg [26:0] data = 27'h10624, input rdy,
                    input [10:0] sine, output reg dacdav = 0, input davdac,
                    output reg [11:0] dacdata = 0, output reg [3:0] dacaddr = 0,
                    output reg [3:0] daccmd = 3);

reg [2:0] gstate = 0;  // state register

always@(posedge rotAreg)      // debounced rotary switch
    begin
        if (rotBreg == 0 && rotCTRreg == 0)
            begin
                if (BTN0 == 0)
                    data = data + 1;        // fine increment
                else
                    data = data + 260;      // coarse increment
            end
        else
            begin
                if (BTN0 == 0)
                    data = data – 1;        // fine decrement
                else
                    data = data – 260;      // coarse decrement
            end
        if (rotCTRreg == 1)
            data = 27'h10624;               // reset frequency to 25 kHz
    end

always@(posedge genclk)
    begin
        case (gstate)
            0:  begin
                    we = 1;                 // write DDS data
                    gstate = 1;
                end
            1:  gstate = 2;
```

```
            2:   begin
                     we = 0;
                     if (rdy == 1)
                         begin
                             dacdata[10:0] = sine[10:0];
                             dacdata[10] = ~dacdata[10];
                             gstate = 3;
                         end
                 end
            3:   begin
                     dacdav = 1;            // DAC conversion
                         if (davdac == 1)
                             begin
                                 dacdav = 0;
                                 gstate = 0;
                             end
                 end
            default: gstate = 0;
        endcase
    end

endmodule
```

The datapath module genddsfreq.v utilizes one FSM controller for the synthesized frequency generator. The phase increment $\Delta\theta$ is incremented or decremented from its initial value of 10624h (67 108), which provides an 11-bit discrete 25 kHz sinusoidal, by the debounced rotary shaft encoder described in Chapter 3 Programmable Gate Array Hardware. The observation of the rotation of the shaft encoder is an event driven task independent of the FSM.

While depressing push button 0 the rotary shaft encoder provides coarse incrementation or decrementation of the phase increment $\Delta\theta$ by ± 260 resulting in an approximate ± 100 Hz change in the analog sinusoidal output frequency. The output frequency can be changed in the first design window of the LogiCORE DDS Compiler and the resultant change in the phase increment $\Delta\theta$ can be read in the sixth design window without finishing the design to ascertain the value needed. The 11-bit two's complement sinusoidal output sine is converted to 12-bit straight binary data for the DAC of the Spartan-3E Starter Board by complementing bit 10 and setting the most significant bits (MSB) to 0.

As described in Chapter 2 Verilog Design Automation, the Design Utilization Summary for the module s3eddsfreqgen.v shows the use of 238 slice flip-flops (2%), 158 4-input LUTs (1%), 155 occupied slices (3%), 1 block RAM (5%) and a total of 75 895 equivalent gates in the XC3S500E Spartan-3E FPGA synthesis.

## Frequency Shift Keying Modulator

Frequency shift keying (FSK) is a digital bandpass modulation technique that encodes information as the frequency of a sinusoidal carrier. Binary FSK shifts the carrier frequency to one of two discrete frequencies during the bit time $T_b$ for the representation of binary logic signals for the transmission of information. The modulated sinusoidal carrier signal has an amplitude of $A$ V, a frequency of $f_c$ Hz, and a 0° reference phase angle, as given by the *analytical expression* in Equation 4.33 [Silage06].

$$s_j(t) = A\sin(2\pi(f_c + k_f m_j(t))\,t) \quad (i-1)T_b \le t \le iT_b \quad j = 0, 1 \qquad (4.33)$$

The information signal or data source is $m_j(t)$ ($j$ = 0, 1) and for binary FSK $m_j(t)$ = ± 1 V for one bit time $T_b$. The factor $k_f$, whose units are Hz/V, is the *frequency deviation factor* (or the modulation gain) and the *frequency deviation* $\Delta F$ is given by Equation 4.34.

$$\Delta F = k_f m_j(t) \qquad (4.34)$$

Since $m_j(t)$ = ± 1 V the magnitude of the frequency deviation $\Delta F$ is equal on either side of the carrier frequency $f_c$.

An FSK modulator is implemented with the LogiCORE DSS Compiler and programmable phase increments $\Delta\theta$ in the Verilog top module s3eddsfsk.v in Listing 4.13, which is in the *Chapter 4 \dds\s3eddsfsk* folder. The six Verilog modules operate in parallel and some independently in the top module. The file download procedure is described in the Appendix.

The FSK modulator uses the Bell type 103 standard for an originating modem with a data rate $r_b$ = 300 b/sec, $f_c$ = 1170 Hz and $\Delta F$ = 100 Hz. The LogiCORE DSS Compiler has $f_{clock}$ = 50 MHz, $B_\theta$ = 26 bits, $\Delta f$ = 1 Hz, SFDR = 66 dB and phase increments $\Delta\theta$ of 6A8h for 1270 Hz and 59Ch for 1070 Hz. A clock module clock.v divides the 50 MHz crystal oscillator and provides the 300 b/sec data rate to a linear feedback shift register (LFSR) module lfsr.v [Sklar01]. The output of the LFSR is a *pseudorandom* binary bit stream as test data to be inputted to the datapath module genddsfsk.v.

The 11-bit two's complement sinusoidal output sine is converted to 12-bit straight binary data for the DAC of the Spartan-3E Starter Board by complementing bit 10 and setting the most significant bits (MSB) to 0. The output of the DAC is the Bell type 103 standard FSK modulator signal and is shown in Figure 4.27. A binary 1 is *semaphored* with 1270 Hz and binary 0 with 1070 Hz for a period of 3.33 msec reprogramming the phase offset data to the DDS LogiCORE block at each bit time $T_b$.

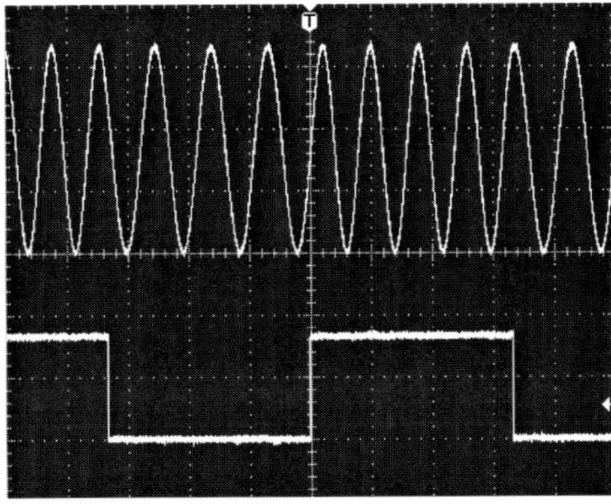

**Figure 4.27** FSK modulator output with 1270 Hz for binary 1 and 1070 Hz for binary 0 at 500 mV/div (top) and 300 b/sec data at 2 V/div (bottom) and 1 msec/div

As described in Chapter 2 Verilog Design Automation, the Design Utilization Summary for the module s3eddsfsk.v shows the use of 200 slice flip-flops (2%), 119 4-input LUTs (1%), 156 occupied slices (3%), 1 block RAM (5%) and a total of 75 646 equivalent gates in the XC3S500E Spartan-3E FPGA synthesis.

**Listing 4.13** FSK modulator top module for the Spartan-3E Starter Board s3eddsfsk.v

```
// Spartan-3E Starter Board
// Digital-to-Analog Converter s3eddsfsk.v
// c 2008 Embedded Design using Programmable Gate Arrays  Dennis Silage
```

```
module s3eddsfsk (input CCLK, output SPIMOSI, SPISCK, DACCS, DACCLR, SPISF, AMPCS,
                    output ADCON, SFCE, FPGAIB, JC1, JC2);

wire dacdav, davdac, dacsck, dacspid, clrdac, csdac, rdy, sclr, we, lfsrclk, lfsrout;
wire [11:0] dacdata;
wire [3:0] daccmd, dacaddr;
wire [2:0] dtmfctrl;
wire [1:0] lowreg, highreg;
wire [25:0] data;
wire [10:0] sine;

assign SPISCK = dacsck;         // SPI clock
assign SPIMOSI = dacspid;       // SPI data in

assign DACCS = csdac;           // disable DAC
assign DACCLR = clrdac;         // DAC clear

assign AMPCS = 1;               // disable prog amp
assign ADCON = 0;               // disable ADC

assign SPISF = 1;               // disable serial Flash
assign SFCE = 1;                // disable StrataFlash
assign FPGAIB = 1;              // disable Platform Flash

assign JC1 = dacdav;            // monitor DAC rate
assign JC2 = lfsrout;           // monitor LFSR data

s3edac M0 (CLKFX, dacdav, davdac, dacdata, dacaddr, daccmd, dacsck, dacspid, csdac, clrdac);
dds2 M1 (.clk(CLKOUT),.sclr(sclr),.we(we),.data(data),.rdy(rdy),.sine(sine));
dacs3edcm M2 (.CLKIN_IN(CCLK), .RST_IN(0), .CLKFX_OUT(CLKFX),
                    .CLKIN_IBUFG_OUT(CLKINIBO), .CLK0_OUT(CLKOUT),
                    .LOCKED_OUT(LOCK));
clock M3 (CLKOUT, 83333, lfsrclk);      // 300 Hz
lfsr M4 (lfsrclk, lfsrout);
genddsfsk M5 (CLKOUT, lfsrout, sclr, we, data, rdy, sine, dacdav, davdac, dacdata, dacaddr,
              daccmd);

endmodule

module genddsfsk (input genclk, lfsrout, output reg sclr = 0, output reg we = 0,
                    output reg [25:0] data=26'h6A8, input rdy, input [10:0] sine,
                    output reg dacdav=0, input davdac, output reg [11:0] dacdata=0,
                    output reg [3:0] dacaddr=0, output reg [3:0] daccmd=3);

reg [2:0] gstate=0;   // state register

always@(posedge genclk)
    begin
        case (gstate)
            0:  begin
                    if (lfsrout == 1)
                        data = 26'h6A8;         // mark 1270 Hz
```

```
                else
                    data = 26'h59C;         // space 1070 Hz
                we = 1;                     // write DDS data
                gstate = 1;
            end
        1:  gstate = 2;
        2:  begin
                we = 0;
                if (rdy == 1)
                    begin
                        dacdata[10:0] = sine[10:0];
                        dacdata[10] = ~dacdata[10];
                        gstate = 3;
                    end
            end
        3:  begin
                dacdav = 1;                 // DAC conversion
                if (davdac == 1)
                    begin
                        dacdav = 0;
                        gstate = 0;
                    end
            end
        default: gstate = 0;
    endcase
end

endmodule
```

## Phase Shift Keying Modulator

Phase shift keying (PSK) is a digital bandpass modulation technique that encodes information as the phase of a sinusoidal carrier. Binary phase shift keying (BPSK) shifts the phase angle of the carrier frequency to one of two discrete phases during the bit time $T_b$ for the representation of binary logic signals for the transmission of information. The modulated sinusoidal carrier signal has an amplitude of $A$ V, a frequency of $f_c$ Hz, as given by the *analytical expression* in Equation 4.35 [Silage06].

$$s_j(t) = A\sin(2\pi f_c t + k_p m_j(t)) \quad (i-1)T_b \le t \le iT_b \quad j = 0, 1 \qquad (4.35)$$

The information signal or data source is $m_j(t)$ ($j$ = 0, 1) and for PSK $m_j(t)$ = 0 V and 1 V for one bit time $T_b$. The factor $k_p$, whose units are 2π (radians)/V, is the *phase deviation factor* (or the modulation gain) and the *phase deviation* $\Delta\varphi$ is given by Equation 4.36.

$$\Delta\varphi = k_p m_j(t) \qquad (4.36)$$

A PSK modulator is implemented with the LogiCORE DSS Compiler and programmable phase deviations $\Delta\varphi$ or phase offsets in the Verilog top module s3eddspsk.v, which is in the *Chapter 4 \dds\s3eddspsk* folder. The PSK modulator top module s3eddspsk.v is similar to the FSK modulator top module s3eddsfsk.v in Listing 4.13. The six Verilog modules operate in parallel and some independently in the top module. The file download procedure is described in the Appendix.

The generation of the DDS LogiCore block here specifies a fixed phase increment $\Delta\theta$ of 53Eh for an output sinusoidal carrier frequency of 1 kHz but a programmable phase offset in the third design window of the LogiCORE DDS Compiler, as shown in Figure 4.28. The 26-bit phase offset is an unsigned integer with 0 representing 0° and 2000000h as 180°. Other phase offsets are 1000000h representing 90° and 3000000h as 270°.

The phase offset, in the range ± 1 which is multiplied by $2\pi$ radians, can be changed in the third design window of the LogiCORE DDS Compiler and the resultant phase offset can be read in the sixth design window without finishing the design to ascertain the value needed. The phase offset input of 0.5 in Figure 4.28 represents $0.5 \times 2\pi = \pi$ (180°).

The binary PSK (BPSK) modulator uses a data rate $r_b$ = 250 b/sec, $f_c$ = 1 kHz and $\Delta\varphi$ = 0, 180°. The LogiCORE DSS Compiler has $f_{clock}$ = 50 MHz, $B_\theta$ = 26 bits, $\Delta f$ = 1 Hz, SFDR = 66 dB, a phase increment $\Delta\theta$ of 53Eh for 1 kHz and an initial phase offset of 0°. A clock module clock.v divides the 50 MHz crystal oscillator and provides the 250 b/sec data rate to a linear feedback shift register (LFSR) module lfsr.v [Sklar01]. The output of the LFSR is a pseudorandom binary bit stream as test data to be inputted to the datapath module genddspsk.v and is described in this Chapter.

The 11-bit two's complement sinusoidal output sine is converted to 12-bit straight binary data for the DAC of the Spartan-3E Starter Board by complementing bit 10 and setting the most significant bits (MSB) to 0. The output of the DAC is the BPSK modulator signal and is shown in Figure 4.29. A binary 1 is semaphored with 1 kHz at 180° and binary 0 with 1 kHz at 0° for a period of 0.4 msec by reprogramming the phase offset data to the DDS LogiCORE block at each bit time $T_b = 1/r_b$.

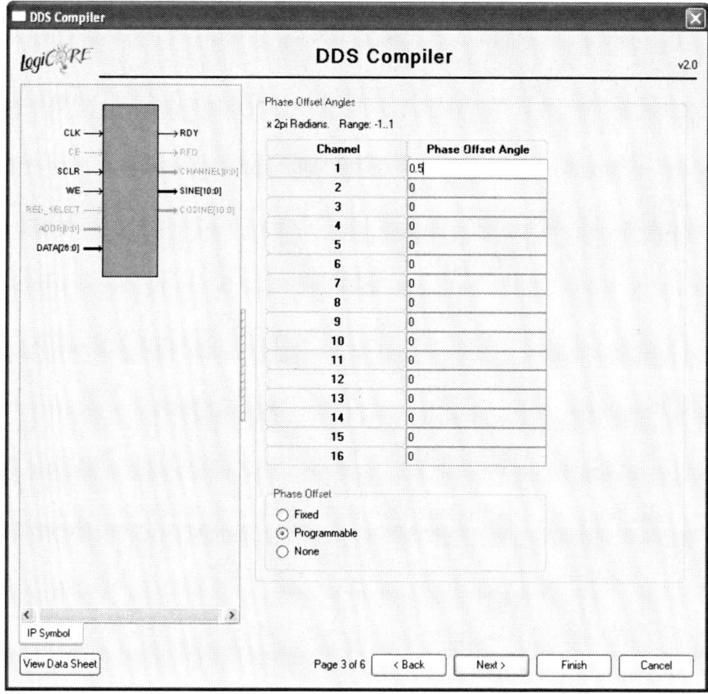


**Figure 4.28** Third design window of the LogiCORE DDS Compiler

As described in Chapter 2 Verilog Design Automation, the Design Utilization Summary for the module s3eddspsk.v shows the use of 205 slice flip-flops (2%), 118 4-input LUTs (1%), 162 occupied slices (3%), 1 block RAM (5%) and a total of 75 785 equivalent gates in the XC3S500E Spartan-3E FPGA synthesis.

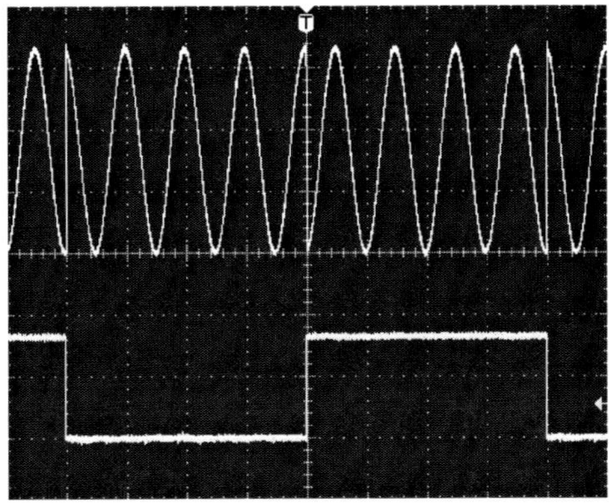

**Figure 4.29** PSK modulator output with 1 kHz at 180° for binary 1 and 1 kHz at 0° for binary 0 at 500 mV/div (top) and 250 b/sec data at 2 V/div (bottom) and 1 msec/div

## Quaternary Phase Shift Keying Modulator

A multilevel (M-ary) phase shift keying (PSK) signal *shifts* the phase angle of the carrier frequency to one of M discrete values during the *symbol time* $T_s$ for the representation of $N = \log_2 M$ binary logic signals for the transmission of information. The modulated sinusoidal carrier signal has an amplitude of $A$ V, a carrier frequency of $f_c$ Hz, and a 0° reference phase angle, as given by the *analytical expression* in Equation 4.37 [Silage06].

$$s_j(t) = \mathrm{A}\cos(2\pi\, f_c t + k_p m_j(t)) \quad (i-1)T_s \le t \le iT_s \quad j = 1, 2, ..., M \tag{4.37}$$

A convenient set of equally spaced phase angles $\varphi_j = k_p m_j(t)$ for 4-PSK, also known as quaternary (Q) PSK, is ± 45° and ± 135° for one symbol time $T_s$ where – 45° = 315° and –135° = 225°. QPSK is the terminology often used for 4-PSK and the signal then is given by Equation 4.38.

$$\begin{aligned}
s_1(t) &= A\cos(2\pi\, f_c t + 45°) && \text{di-bit } 10 \quad b_{2i}b_{2i+1} = 11\\
s_2(t) &= A\cos(2\pi\, f_c t + 135°) && \text{di-bit } 00 \quad b_{2i}b_{2i+1} = 00\\
s_3(t) &= A\cos(2\pi\, f_c t + 225°) && \text{di-bit } 01 \quad b_{2i}b_{2i+1} = 01\\
s_4(t) &= A\cos(2\pi\, f_c t + 315°) && \text{di-bit } 11 \quad b_{2i}b_{2i+1} = 10\\
& && (i-1)T_s \le t \le iT_s
\end{aligned} \tag{4.38}$$

The input information is straight-binary coded data ($b_{2i}$, $b_{2i+1}$), but is a Gray-coded *di-bit* so that if a symbol error is interpreted as an adjacent symbol it would now only produce one bit in error. The Gray code improves the BER performance of the M-ary digital communication system [Silage06]. Using the trigonometric identity $A\cos(u + v) = A\cos(u)\cos(v) - A\sin(u)\sin(v)$, Equation 4.38 for QPSK can be described in terms of an *in-phase* (I, cosine) and *quadrature* (Q, sine) components, as given by Equation 4.39.

$$\begin{aligned}
s_1(t) &= A_v\cos(2\pi\, f_c t) - A_v\sin(2\pi\, f_c t) && \text{di-bit } 10 \quad b_{2i}b_{2i+1} = 11\\
s_2(t) &= -A_v\cos(2\pi\, f_c t) - A_v\sin(2\pi\, f_c t) && \text{di-bit } 00 \quad b_{2i}b_{2i+1} = 00\\
s_3(t) &= -A_v\cos(2\pi\, f_c t) + A_v\sin(2\pi\, f_c t) && \text{di-bit } 01 \quad b_{2i}b_{2i+1} = 01\\
s_4(t) &= A_v\cos(2\pi\, f_c t) + A_v\sin(2\pi\, f_c t) && \text{di-bit } 11 \quad b_{2i}b_{2i+1} = 10\\
& && (i-1)T_s \le t \le iT_s
\end{aligned} \tag{4.39}$$

From Equation 4.38, the carrier amplitude $A_v = 0.707A$ in Equation 4.39. The QPSK signal can be decomposed into two binary PSK signals with same carrier frequency $f_c$, with one corresponding to a cosine carrier (I component) and the other corresponding to the sine carrier (Q component), as given by Equation 4.39. Because of the I-Q decomposition of the quaternary PSK signal, this bandpass modulation method is also often referred to as a quadrature PSK signal or, redundantly, as QPSK.

A QPSK modulator is implemented with the LogiCORE DSS Compiler and programmable phase deviations $\Delta\varphi$ or phase offsets in the Verilog top module s3eddsqpsk.v in Listing 4.14, which is in the *Chapter 4\dds\s3eddsqpsk* folder. The six Verilog modules operate in parallel and some independently in the top module. The file download procedure is described in the Appendix.

The generation of the DDS LogiCore block here specifies a fixed phase increment $\Delta\theta$ of 53Eh for an output sinusoidal carrier frequency of 1 kHz but a programmable phase offset in the third design window of the LogiCORE DDS Compiler, as shown in Figure 4.28. The phase offsets of a single channel DDS discrete cosine output are reprogrammed by the di-bit $b_{2i}b_{2i+1}$ to output the four sinusoidal signals given in Equation 4.38. The 26-bit phase deviation 800000h represents 45°, 1800000h as 135°, 2800000h as 225° and 3800000h as 315°.

The phase offset, in the range ± 1 which is multiplied by $2\pi$ radians, can be changed in the third design window of the LogiCORE DDS Compiler and the resultant phase offset can be read in the sixth design window without finishing the design to ascertain the value needed. Table 4.1 lists the four QPSK signals, the di-bits they represent, the cosine phases and DDS Compiler phase offsets.

**Table 4.1** Phase deviation $\Delta\varphi_j$ and DDS Compiler phase offset for a QPSK modulated signal

| Signal | Di-bit | Phase Deviation $\Delta\varphi_j$ | DDS Compiler Phase Offset |
|---|---|---|---|
| $s_1(t)$ | 11 | 45° | 800000h |
| $s_2(t)$ | 00 | 135° | 1800000h |
| $s_3(t)$ | 01 | 225° | 2800000h |
| $s_4(t)$ | 10 | 315° | 3800000h |

The QPSK modulator uses a data rate $r_b$ = 1 kb/sec, a resulting symbol rate $r_s$ = 500 symbols/sec since there are two bits per symbol, $f_c$ = 1 kHz and $\Delta\varphi_j$ = 45°, 135°, 225° and 315°. The LogiCORE DSS Compiler has $f_{clock}$ = 50 MHz, $B_\theta$ = 26 bits, $\Delta f$ = 1 Hz, SFDR = 66 dB, a phase increment $\Delta\theta$ of 53Eh for 1 kHz and an initial phase offset $\varphi_j$ of 0°. A clock module clock.v divides the 50 MHz crystal oscillator and provides the 1 kb/sec data rate to a linear feedback shift register (LFSR) module lfsrdibit.v [Sklar01].

The input variable lfsrd is a *flag* whose logic 1 value indicates that di-bit data is available. The output of the LFSR is a pseudorandom di-bit stream as test data to be inputted to the datapath module genddsqpsk.v and is described in this Chapter.

**Listing 4.14** QPSK modulator top module for the Spartan-3E Starter Board s3eddsqpsk.v

```
// Spartan-3E Starter Board
// Digital-to-Analog Converter s3eddspsk.v
// c 2008 Embedded Design using Programmable Gate Arrays  Dennis Silage

module s3eddsqpsk (input CCLK, output SPIMOSI, SPISCK, DACCS, DACCLR, SPISF,
                        output AMPCS, ADCON, SFCE, FPGAIB, JC1, JC2, JC3);
wire dacdav, davdac, dacsck, dacspid, clrdac, csdac, rdy, sclr, we, lfsrd, lfsrclk;
wire [11:0] dacdata;
wire [3:0] daccmd, dacaddr;
wire [25:0] data;
```

```
wire [10:0] cosine;
wire [1:0] lfsrout;

assign SPISCK = dacsck;        // SPI clock
assign SPIMOSI = dacspid;      // SPI data in

assign DACCS = csdac;          // disable DAC
assign DACCLR = clrdac;        // DAC clear
assign AMPCS = 1;              // disable prog amp
assign ADCON = 0;              // disable ADC

assign SPISF = 1;              // disable serial Flash
assign SFCE = 1;               // disable StrataFlash
assign FPGAIB = 1;             // disable Platform Flash

assign JC1 = dacdav;           // monitor DAC rate
assign JC2 = lfsrout[0];       // monitor LFSR data
assign JC3 = lfsrout[1];

s3edac M0 (CLKFX, dacdav, davdac, dacdata, dacaddr, daccmd, dacsck, dacspid, csdac, clrdac);
dds4 M1 (.clk(CLKOUT),.sclr(sclr),.we(we),.data(data),.rdy(rdy),.sine(sine));
dacs3edcm M2 (.CLKIN_IN(CCLK), .RST_IN(0), .CLKFX_OUT(CLKFX),
              .CLKIN_IBUFG_OUT(CLKINIBO), .CLK0_OUT(CLKOUT),
              .LOCKED_OUT(LOCK));
clock M3 (CLKOUT, 25000, lfsrclk);      // 1 kHz
lfsrdibit M4 (lfsrclk, lfsrd, lfsrout);
genddsqpsk M5 (CLKOUT, lfsrd, lfsrout, sclr, we, data, rdy, cosine, dacdav, davdac, dacdata, dacaddr,
               daccmd);

endmodule

module genddsqpsk (input genclk, lfsrd, input [1:0] lfsrout, output reg sclr = 0, output reg we = 0,
                   output reg [25:0] data = 26'h0, input rdy, input [10:0] cosine,
                   output reg dacdav = 0, input davdac, output reg [11:0] dacdata = 0,
                   output reg [3:0] dacaddr = 0, output reg [3:0] daccmd = 3);

reg [2:0] gstate = 0;  // state register
reg [1:0] cosd = 0;    // cosine phase offset

always@(posedge genclk)
    begin
        case (gstate)
            0:  begin
                    if (lfsrd == 1)          // di-bit available
                        begin
                            case (lfsrout)   // LFSR data
                                0:  cosd = 3;
                                1:  cosd = 5;
                                2:  cosd = 7;
                                3:  cosd = 1;
                            endcase
                        end
```

```
                    gstate = 1;
                end
            1:  begin
                    data[25:23] = cosd[2:0];
                    we = 1;                         // write DDS data
                    gstate = 2;
                end
            2:  gstate = 3;
            3:  begin
                    we = 0;
                    if (rdy == 1)
                        begin
                            dacdata[10:0] = cosine[10:0];
                            dacdata[10] = ~dacdata[10];
                            gstate = 4;
                        end
                end
            4:  begin
                    dacaddr = 0;
                    dacdav = 1;             // DAC conversion
                    if (davdac == 1)
                        begin
                            dacdav = 0;
                            gstate = 0;
                        end
                end
            default: gstate = 0;
        endcase
    end

endmodule
```

The 11-bit two's complement sinusoidal output cosine is converted to 12-bit straight binary data for two channels of the DAC of the Spartan-3E Starter Board by complementing bit 10 and setting the most significant bits (MSB) to 0. The output of the DAC is the QPSK modulator signal and is shown in Figure 4.29.

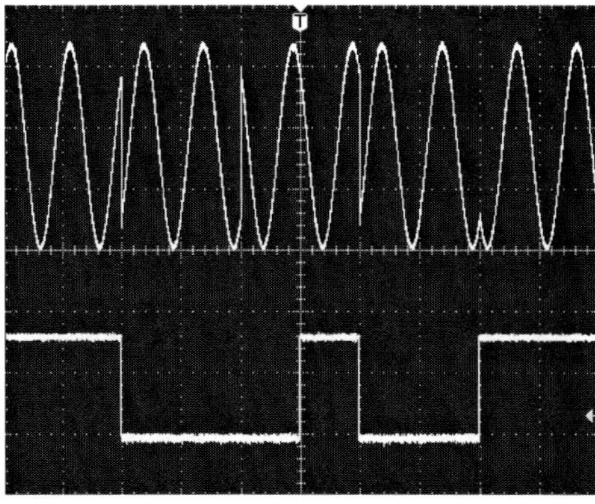

**Figure 4.29** QPSK modulator output at 500 mV/div, 500 symbols/sec (top) and 1 kb/sec data at 2 V/div (bottom) and 1 msec/div

A di-bit is semaphored with a 1 kHz cosine with a phase angle for a period of 2 msec by reprogramming the most significant three bits of the phase offset data to the DDS LogiCORE block at each symbol time $T_s = 1/r_s$.

As described in Chapter 2 Verilog Design Automation, the Design Utilization Summary for the module s3eddsqpsk.v shows the use of 178 slice flip-flops (1%), 125 4-input LUTs (1%), 145 occupied slices (3%), 1 block RAM (5%) and a total of 75 458 equivalent gates in the XC3S500E Spartan-3E FPGA synthesis.

## Linear Finite Shift Register

A binary sequence generator is a digital logic circuit consisting of a sequential logic *shift register* and register *taps* that feedback the output to a network of combination logic. The output of this *feedback shift register* (FSR) is pseudo-random since the output must repeat after finite number of clock cycles. Applications of the FSR include the generation of random numbers and binary data and in the generation of direct sequence spread spectrum (DSSS) digital modulation [Haykin01].

If the combinational logic of the FSR is an *exclusive or* (xor) gate, then it is called a *linear* FSR (LFSR) because the input bit feedback to the input of the shift register is a linear combination of the previous state, as shown in Figure 4.30. If the FSR consists of $M$ shift registers and the output repeats only after $2^M - 1$ clock inputs, the FSR is called a *maximal length sequence* or an $M$-sequence [Simon01].

The taps that are fedback from the output of the shift register to the xor combinational logic gates for an LFSR can be represented as a *characteristic polynomial*. The polynomial is modulus 2 and the coefficients are binary numbers. The binary sequence generator shown in Figure 4.30 is an $M$-sequence LFSR ($M = 8$). The characteristic polynomial is $x^8 + x^6 + x^5 + x^4 + 1$ which indicates that the register taps occur on the eighth, sixth, fifth and fourth registers. The last term is the input to the shift register ($x^0 = 1$). This type of binary sequence generator is also known as a *Fibonacci* implementation.

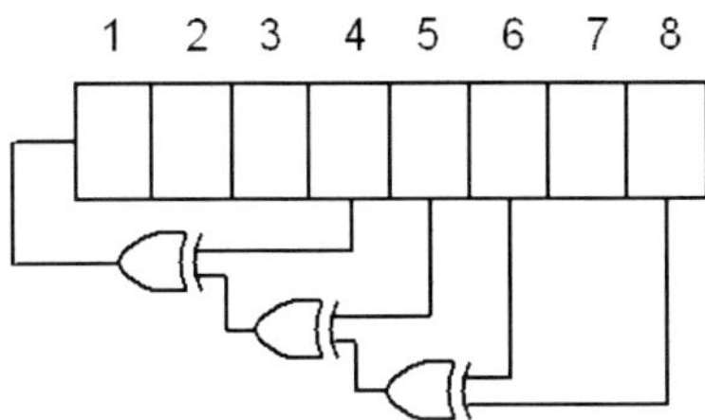


**Figure 4.30** Maximal length sequence linear feedback shift register (M = 8) with characteristic polynomial $x^8 + x^6 + x^5 + x^4 + 1$

An alternative construct for the binary sequence generator, the *Galois* implementation, uses exclusive or combinational logic gates between the shift registers and is not shown here. Table 4.2 lists some of the possible characteristic polynomials for $M$-sequence LFSRs and the non-repeating, maximal length [Simon01].

The LFSR provides *pseudorandom* binary data for test and measurement. The binary data for the FSK modulator in Listing 4.13 and the PSK modulator are provided by the $M$-sequence LFSR ($M$ = 4) Verilog module lfsr.v with the characteristic polynomial $x^4 + x^3 + 1$, as given in Listing 4.15. The di-bit data for the QPSK modulator is provided by the Verilog module lfsrdibit.v which uses the same characteristic polynomial and is given in Listing 4.16. Both of these LFSR modules are in the Xilinx ISE WebPACK project directories for the digital communication modulators as subfolders in the *Chapter 4 \dds* folder. The file download procedure is described in the Appendix.

**Table 4.2** Characteristic polynomials of some maximal length sequence linear feedback shift registers

| Shift Registers $M$ | Characteristic Polynomial | Non-Repeating Length $2^M - 1$ |
|---|---|---|
| 3 | $x^2 + x^1 + 1$ | 7 |
| 4 | $x^4 + x^3 + 1$ | 15 |
| 5 | $x^5 + x^3 + 1$ | 31 |
| 6 | $x^6 + x^5 + 1$ | 63 |
| 7 | $x^7 + x^6 + 1$ | 127 |
| 8 | $x^8 + x^6 + x^5 + x^4 + 1$ | 255 |
| 9 | $x^9 + x^5 + 1$ | 511 |
| 10 | $x^{10} + x^7 + 1$ | 1023 |
| 11 | $x^{11} + x^9 + 1$ | 2047 |
| 12 | $x^{12} + x^{11} + x^{10} + x^4 + 1$ | 4095 |
| 13 | $x^{13} + x^{12} + x^{11} + x^8 + 1$ | 8191 |
| 14 | $x^{14} + x^{13} + x^{12} + x^2 + 1$ | 16383 |
| 15 | $x^{15} + x^{14} + 1$ | 32767 |
| 16 | $x^{16} + x^{14} + x^{13} + x^{11} + 1$ | 65535 |
| 17 | $x^{17} + x^{14} + 1$ | 131071 |
| 18 | $x^{18} + x^{11} + 1$ | 262143 |
| 19 | $x^{19} + x^{18} + x^{17} + x^{14} + 1$ | 524287 |
| 20 | $x^{20} + x^{17} + 1$ | 1048575 |
| 21 | $x^{21} + x^{19} + 1$ | 2097151 |
| 22 | $x^{22} + x^{21} + 1$ | 4194303 |

The maximum sequence length of the LFSR can be increased by increasing the number $M$ of shift registers, exclusive or-ing the appropriate bits according to the characteristic polynomial as shown in Figure 4.30 and as given in Listing 4.15 or Listing 4.16 and concatenating the resulting shift register. Note that register lfsr[n–1] is shift register n of the characteristic polynomial.

**Listing 4.15** *M*-sequence LFSR ($M = 4$) module lfsr.v

```
// Spartan-3E Starter Board
// Linear Finite Shift Register lfsr.v
// c 2008 Embedded Design using Programmable Gate Arrays  Dennis Silage

module lfsr (input lfsrclk, output reg lfsrout);

reg [3:0] lfsr = 1;

always @(posedge lfsrclk)
    begin
        lfsr = {lfsr[2:0], lfsr[3] ^ lfsr[2]};      // x^4 + x^3 + 1
        lfsrout = lfsr[3];
    end

endmodule
```

The 4-bit shift register variable lfsr is event driven by the input clock signal lfsrclk and is *seeded* with 0001 binary to initiate the binary sequence generation. The seed is any number except 0000 binary. The di-bit LFSR Verilog module lfsrdibit.v outputs the most significant bit (MSB) of the register variable lfsr and the previous MSB. The output register variable lfsrd is the *flag* whose logic 1 value indicates that a new di-bit is available.

**Listing 4.16** Di-bit *M*-seqence LFSR (*M* = 4) module lfsrdibit.v

```
// Spartan-3E Starter Board
// Linear Finite Shift Register lfsrdibit.v
// c 2008 Embedded Design using Programmable Gate Arrays  Dennis Silage

module lfsrdibit (input lfsrclk, output reg lfsrd = 1, output reg [1:0] lfsrout = 0);

reg [3:0] lfsr = 1;

always @(posedge lfsrclk)
    begin
        lfsr = {lfsr[2:0], lfsr[3] ^ lfsr[2]};      // x^4 + x^3 + 1
        lfsrout[1] = lfsrout[0];
        lfsrout[0] = lfsr[3];
        lfsrd = ~lfsrd;
    end

endmodule
```

The Xilinx Linear Feedback Shift Register LogiCORE block is not available in the Xilinx ISE WebPACK. However, the Xilinx LSFR data sheet DS257 and application note XAPP210 (*www.xilinx.com*) provides a discussion of the tenets of binary sequence generation, implementations and maximum sequence characteristic polynomials through M = 168.

The Xilinx application note XAPP217 (*www.xilinx.com*) describes the Gold code binary sequence generator which demonstrates correlation properties that are ideally suited for Code Division Multiple Access (CDMA) applications [Silage06]. The Gold code generator is the exclusive or of the outputs of two preferred LFSRs seeded with a specific initial value [Simon01].

## Data Communication

Several serial data communication protocols with various degrees of complexity and application properties are used in embedded design. The serial peripheral interface (SPI) can connect local devices such as an analog-to-digital converter, digital-to-analog converter, or flash memory PROM to the programmable gate array in an embedded design, as described in Chapter 3 Programmable Gate Array Hardware.

The SPI is a *synchronous* bus and utilizes a SPI clock to coordinate bi-directional binary data transfers on separate output and input signals. The SPI bus is also described in the Xilinx application note XAPP348 (*www.xilinx.com*). The inter-integrated circuit ($I^2C$) bus is a serial data communication interface that uses only a clock and a single data line to transfer bi-directional data to multiple devices. The $I^2C$ bus is described in the Xilinx application note XAPP333.

More complicated serial data communication protocols include the universal serial bus (USB), FireWire (IEEE1394) and Ethernet (IEEE 802.3). These protocols require extensive hardware and software support and are not featured here. However, an established protocol for serial data communication with equipment external to an embedded system is the RS-232/RS-422 standard and data source coding is delineated with the Manchester encoder-decoder for serial data.

## RS-232 Standard

The RS-232 (Recommended Standard 232, now EIA/TIA-232F) standard is a binary serial data communication bus that is usually *asynchronous* and which connects data terminal equipment (DTE) to data circuit-terminating equipment (DCE) commonly used in computer ports. Although a synchronous clock is part of the RS-232 standard it is not usually employed. The synchronous binary serial data communication RS-422 standard (now EIA/TIA-422B) uses differential signaling for high speed and noise immunity.

Unlike the complexity of the protocol of USB which is replacing it in many computer systems, the RS-232 protocol is simple and easily implemented. The RS-232 standard remains commonly in use to connect legacy DTE and DCE devices and industrial equipment. However, there are some disparities and limitations in the RS-232 protocol.

Although the RS-232 standard requires the transmitter to use –12 V and +12 V for the binary data, the receiver must distinguish binary data with voltages as low as –3 V and +3 V. The large voltage required for the transmitter increases power consumption and complicates power supply design. As a result, so-called RS-232 compatible serial ports using –5 V and +5 V are often employed. The hardware flow control lines, such as request to send (RTS) and clear to send (CTS), are problematic and are often not used in simple data communication. Finally, the asymmetry of the DTE and DCE data protocol and connector induces problems in the correct equipment interface.

The frame format for RS-232 binary serial data transmission is shown in Figure 4.31. The transmitter *idles* at $-V_S$ volts until data are available. The start of transmission is asynchronous and begins with a *start bit* of one bit time ($T_b$) duration and $+V_S$ volts. The 8 bit data is sent least significant bit (LSB) first where logic 1 is $-V_S$ volts and at least one *stop bit* concludes the transmission.

Binary serial data communication using the RS-232 standard utilizes the universal asynchronous receiver transmitter (UART) hardware to provide serial-to-parallel and parallel-to-serial conversion, bit rate generation, control functions and buffering. The type 16550 UART (National Semiconductor, *www.national.com*) can be implemented in an embedded design as a *soft-core* peripheral using the on-chip peripheral bus (OPB) LogiCORE block (data sheet DS430, *www.xilinx.com*). However, the OPB 16550 UART LogiCORE block is not available in the Xilinx ISE WebPACK. A simple UART originally configured for a Xilinx complex programmable logic device (CPLD) is described in the Xilinx application note XAPP341 and is used with modifications here.

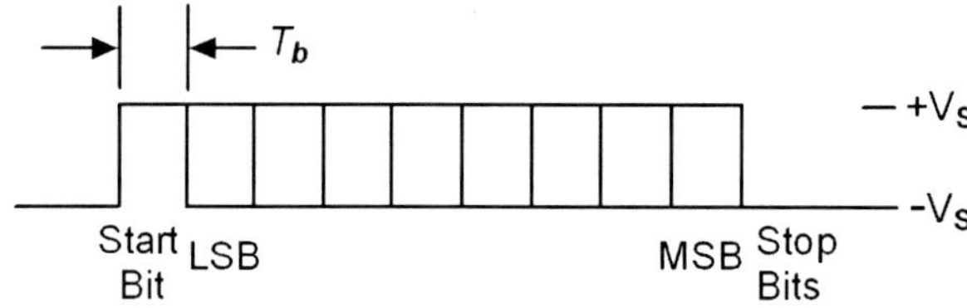


**Figure 4.31** RS-232 frame format for binary serial data transmission

The UART RS-232 standard transmitter Verilog module txmit.v is shown in Listing 4.17. The 8-bit data is inputted from the variable din to the 8-bit buffer register variable tbr with logic 1 on the *write strobe* wrn. Register variable tsr receives the data from tbr and is used to sequentially shift the data out as txd. The input variable clk16x is the reference clock at 16 times the nominal bit rate $r_b = 1/T_b$. The register variable tbre as logic 1 indicates that the input buffer tbr is empty.

An asynchronous input signal rst is used to reset the functionality of the UART transmitter. The txmit.v Verilog module is in the *Chapter 4\rs232\s3elcdkbduart* folder and utilizes multiple concurrent cyclical behavioral events to affect the data transfer with non-blocking assignments, as described in Chapter 1 Verilog Hardware Description Language. The file download procedure is described in the Appendix.

**Listing 4.17** UART transmitter module txmit.v

```
// Spartan-3E Starter Board
// UART Transmitter txmit.v
// c 2000 Xilinx, Inc. XAPP341
// c 2008 Embedded Design using Programmable Gate Arrays Dennis Silage

module txmit (input [7:0] din, output reg tbre, input clk16x, wrn, rst, output reg txd);

reg clk1x_enable, parity, wrn1, wrn2;
reg [7:0] tsr;
reg [7:0] tbr;
reg [3:0] clkdiv;
reg [3:0] no_bits_sent;
assign clk1x = clkdiv[3];

always @(posedge clk16x or posedge rst)
	begin
		if (rst)
			begin
				wrn1 <= 1;
				wrn2 <= 1;
			end
		else
			begin
				wrn1 <= wrn;
				wrn2 <= wrn1;
			end
	end

always @(posedge clk16x or posedge rst)
	begin
		if (rst)
			begin
				tbre <= 0;
				clk1x_enable <= 0;
			end
		else if (!wrn1 && wrn2)
			begin
				clk1x_enable <= 1;
				tbre <= 1;
			end
		else if (no_bits_sent == 2)
			tbre <= 1'b1 ;
		else if (no_bits_sent == 13)
			begin
				clk1x_enable <= 0;
				tbre <= 0;
			end
	end
```

```
always @(posedge wrn or posedge rst)
      begin
            if (rst)
                  tbr = 0;
            else
                  tbr = din;
end

always @(posedge clk16x or posedge rst)
      begin
            if (rst)
                  clkdiv = 0;
            else if (clk1x_enable)
                  clkdiv = clkdiv + 1;
      end

always @(negedge clk1x or posedge rst)
      begin
            if (rst)
                  begin
                        txd <= 1 ;
                        parity <= 1 ;
                        tsr <= 0 ;
                  end
            else
                  begin
                        if (no_bits_sent == 1)
                              tsr <= tbr;
                        else if (no_bits_sent == 2)
                              txd <= 0;
                        else if ((no_bits_sent >= 3) && (no_bits_sent <= 10))
                              begin
                                    txd <= tsr[0];          // transmitted data
                                    tsr[6:0] <= tsr[7:1];
                                    tsr[7] <= 0;
                                    parity <= parity ^ tsr[0];
                              end
                        else if (no_bits_sent == 11)
                              txd <= parity;
                        else if (no_bits_sent == 12)
                              txd <= 1;
                  end
      end

always @(posedge clk1x or posedge rst or negedge clk1x_enable)
      begin
            if (rst)
                  no_bits_sent = 0;
            else if (!clk1x_enable)
                  no_bits_sent = 0;
```

```
        else
            no_bits_sent = no_bits_sent + 1;
    end

endmodule
```

The UART RS-232 standard receiver Verilog module rcvr.v is shown in Listing 4.18. The binary serial data is inputted as rxd and the 8-bit data is outputted from the reg variable rbr. Register variables rxd1 and rxd2 are used to adjust the read timing for the receiver. The register variable data_ready indicates that received data is available with logic 1 on the *read strobe* rdn. Register variable rbr receives the data from rsr. The input variable clk16x is the reference clock at 16 times the nominal bit rate $r_b = 1/T_b$. An asynchronous input signal rst is used to reset the functionality of the UART receiver.

The rcvr.v Verilog module is in the *Chapter 4\rs232\s3elcdkbduart* folder and utilizes multiple concurrent cyclical behavioral events to affect the data transfer with non-blocking assignments, as described in Chapter 1 Verilog Hardware Description Language. The file download procedure is described in the Appendix.

**Listing 4.18** UART receiver module rcvr.v

```
// Spartan-3E Starter Board
// UART Receiver rcvr.v
// c 2000 Xilinx, Inc. XAPP341
// c 2008 Embedded Design using Programmable Gate Arrays Dennis Silage

module rcvr (output reg [7:0] rbr, output reg data_ready, input rxd, clk16x, rst, rdn);

reg rxd1, rxd2, parity, clk1x_enable;
reg [3:0] clkdiv;
reg [7:0] rsr;
reg [7:0] rbr;
reg [3:0] no_bits_rcvd;
assign clk1x = clkdiv[3];

always @(posedge clk16x or posedge rst)
    begin
        if (rst)
            begin
                rxd1 <= 1;
                rxd2 <= 1;
            end
        else
            begin
                rxd1 <= rxd;            // received data
                rxd2 <= rxd1;
            end
    end

always @(posedge clk16x or posedge rst)
    begin
        if (rst)
            clk1x_enable <= 0;
```

```
		else if (!rxd1 && rxd2)
			clk1x_enable <= 1;		// start bit
		else if (no_bits_rcvd == 12
			clk1x_enable <= 0;
	end

always @(posedge clk16x or posedge rst or negedge rdn)
	begin
		if (rst)
			data_ready <= 0;
		else if (!rdn)
			data_ready <= 0;
		else if (no_bits_rcvd == 11
			data_ready <= 1;
	end

always @(posedge clk16x or posedge rst)
	begin
		if (rst)
			clkdiv = 0;
		else if (clk1x_enable)
			clkdiv = clkdiv + 1;
	end

always @(negedge clk1x or posedge rst)
	begin
		if (rst)
			begin
				rsr <= 0;
				rbr <= 0;
				parity <= 1;
			end
		else
			begin
				if (no_bits_rcvd >= 0 && no_bits_rcvd <= 8)
					begin
						rsr[7] <= rxd2;
						rsr[6:0] <= rsr[7:1];
					end
				else if (no_bits_rcvd == 9)
					rbr <= rsr;
			end
	end

always @(posedge clk1x or posedge rst or negedge clk1x_enable)
	begin
		if (rst)
			no_bits_rcvd = 0;
		else if (!clk1x_enable)
			no_bits_rcvd = 0;
```

```
        else
            no_bits_rcvd = no_bits_rcvd+1;
    end

endmodule
```

The PS/2 keyboard produces a complex scan code rather than ASCII characters, as described in Chapter 3 Programmable Gate Array Hardware. The scan code received by the Spartan-3E Starter Board from the PS/2 keyboard must be converted to a 7-bit ASCII character for RS-232 serial data communication. The eighth bit of the RS-232 transmission is used for the odd or even parity bit but is set to 0 or no parity here. The scan code to ASCII conversion Verilog module scancodeASCII.v is shown in Listing 4.19.

The 8-bit input variable scancode is available with logic 1 on the input dav and converted to the register variable ascii. The output register variable asciidav is logic 1 when the conversion is complete. PS/2 keyboard left and right side shift keys actually have different scan codes, 12h = 18 and 59h = 89 respectively as listed in Table 3.3, and the application must keep track of the depression and release of the shift key with the register variable shiftreg for the upper case alphabetical characters and alternate symbols.

This module is a datapath utilizing a finite state machine (FSM) with the 4-bit state register ascstate and provides status signals to the controller. The complete Verilog module scancodeASCII.v is in the *Chapter 4\rs232 \s3elcdkbduart* folder and only a portion is listed here. The file download procedure is described in the Appendix.

**Listing 4.19** PS/2 keyboard scan code to ASCII character conversion module scancodeASCII.v

```
// Spartan-3E Starter Board
// PS/2 Keyboard to ASCII  scancodeASCII.v
// c 2008 Embedded Design using Programmable Gate Arrays  Dennis Silage

module scancodeASCII (input CCLK, dav, input [7:0] scancode, output reg asciidav = 0,
                        output reg [7:0] ascii);

reg shiftreg = 0;
reg [7:0] asciiw;
reg [3:0] ascstate = 0;

always@(posedge CCLK)
    begin
        case (ascstate)
            0:  begin
                    if (dav == 0)
                        ascstate = 1;
                end
            1:  begin
                    asciidav = 0;
                    if (dav == 1)
                        ascstate = 2;
                end
            2:  begin                   // repeat or left shift or right shift
                    if (scancode == 240 || scancode == 18 || scancode == 89)
                        ascstate = 4;
```

```
                    else
                        ascstate = 3;
                end
            3:  begin
                    case(scancode)
                        16'h1C:  ascii = 16'h41;       // A
                        16'h32:  ascii = 16'h42;       // B
                        16'h21:  ascii = 16'h43;       // C
{ASCII characters D through W are similarly decoded}
                        16'h22:  ascii = 16'h58;       // X
                        16'h35:  ascii = 16'h59;       // Y
                        16'h1A:  ascii = 16'h5A;       // Z
                        16'h0E:  begin
                                    if (shiftreg == 1)
                                        ascii = 16'h7E;  // ~
                                    else
                                        ascii = 16'h27;  // `
                                 end
{Keyboard shift and special characters are similarly decoded}
                        16'h52:  begin
                                    if (shiftreg == 1)
                                        ascii = 16'h22;  // "
                                    else
                                        ascii = 16'h60;  // '
                                 end
                        16'h45:  ascii = 16'h30;       // 0
                        16'h16:  ascii = 16'h31;       // 1
{Numeric characters 2 through 7 are similarly decoded}
                        16'h3E:  ascii = 16'h38;       // 8
                        16'h46:  ascii = 16'h39;       // 9
                        16'h29:  ascii = 16'h20;       // space
                        16'h5A:  ascii = 16'h0D;       // ENTER = CR
                        default: ascii = 16'h31;
                    endcase

                    if (shiftreg == 0 && ascii >= 65 && ascii <= 90)
                        ascii = ascii + 32;            // A a through Z z
                    if (shiftreg == 1 && ascii >= 48 && ascii <= 57)
                        begin                          // shift 0 through 9
                            case (ascii)
                                48:  asciiw = 41;      // )
                                49:  asciiw = 33;      // !
                                50:  asciiw = 64;      // @
                                51:  asciiw = 35;      // #
                                52:  asciiw = 36;      // $
                                53:  asciiw = 37;      // %
                                54:  asciiw = 94;      // ^
                                55:  asciiw = 38;      // &
                                56:  asciiw = 42;      // *
                                57:  asciiw = 40;      // (
                                default: asciiw = 0;
                            endcase
```

```
                        ascii = asciiw;
                    end
                asciidav = 1;
                ascstate = 0;
            end
        4:  begin
                if (scancode == 240)        // repeat
                    ascstate = 5;
                if (scancode == 18 || scancode == 89)       // left or right shift
                    ascstate=8;
            end
        5:  begin                           // break
                if (dav == 0)
                    ascstate = 6;
            end
        6:  begin                           // character break
                if (dav == 1)
                    ascstate = 7;
            end
        7:  begin                           // shift break?
                if (scancode == 18 || scancode == 89)
                    shiftreg = 0;
                ascstate = 0;
            end
        8:  begin                           // shift
                shiftreg = 1;
                ascstate = 0;
            end
        default: ascstate = 0;
    endcase
end

endmodule
```

A serial data communication system is implemented with the Verilog top module s3elcdkbduart.v in Listing 4.20, which is in the *Chapter 4\rs232\s3elcdkbduart* folder and only a portion is listed here. The file download procedure is described in the Appendix. The Xilinx ISE WebPACK project uses the UCF lcdkbduarts3esb.ucf which uncomments the signals CCLK, PS2CLK, PS2DAT, BTN0, LCDRS, LCDRW, LCDE, LCDDAT, DCETXD, DCERXD, JC1 and JC2 in the Spartan-3E Starter Board UCF of Listing 3.2. The seven Verilog modules operate in parallel and some independently in the top module.

A PS/2 keyboard is interfaced to the Spartan-3E Starter Board and a DB9 male-female cable is connected to the RS-232 DCE port (J9) and to an RS-232 DTE port of a PC executing a terminal emulator. The nominal output bit rate $r_b$ = 9600 b/sec and is set by the clock module clock.v. The UART requires a clock at 16 times the nominal bit rate or 153.6 kb/sec. The clock scale factor, as determined by Equation 3.1, is 162.76, rounded to 163 and the output bit rate is 9585.9 b/sec or 0.15% low. However, this bit rate tolerance is well within the EIA/TIA-232F standard of less than 4%.

Note that the Microsoft *Windows XP* terminal communications program, *HyperTerminal*, and some USB to RS-232 serial data converters have significant problems in this and other applications and produce dropped ASCII characters in the data transmission. The Spartan-3E Starter Board RS-232 port does not have hardware flow control (RTS/CTS). Other available PC terminal emulators are shown to perform correctly.

The characters typed on the PS/2 keyboard connected to the Spartan-3E Starter Board are sent to the PC terminal emulator and are echoed on the top line of the LCD. The characters typed on the PC keyboard of the terminal emulator are sent to the Spartan-3E Starter Board and echoed on the bottom line of the LCD.

Position wrap-around and character overwrite occurs on each line of the LCD independently. Register variables txpos and rxpos are the current location for the ASCII character on the LCD. Push button BTN0 resets the UART and LCD datapath modules and the controller module.

The genlcd.v module used here is the controller for the lcd.v datapath module and is similar to the genlcd.v module used for keyboard data for the Spartan-3E Starter Board, as given in Listing 3.20. The genlcd.v module is an FSM that provides control signals to and receives status signals from the LCD, keyboard and UART modules. The net variable rcvdatardy indicates with logic 1 to the genlcd.v controller module that received data is available.

The net variable asciidav from the datapath module scancode.v indicates with logic 1 that a scan code has been converted to an ASCII character. The register variable gcount in the controller module genlcd.v is a delay variable that sets the proper duration for the UART read and write strobe variables rdn and wrn.

The transmitted and received data variables txd and rxd of the UART are continuously assigned to the DCE port of the Spartan-3E Starter Board by the UCF lcdkbduarts3esb.ucf and are monitored at JC1 (J4-1) and JC2 (J4-2). The parity bit from the PS/2 keyboard module keyboard.v is not used here and the PC terminal emulator is set for 8 bits, 9600 b/sec, no parity and no hardware control.

**Listing 4.20** Serial data communication terminal emulator s3elcdkbduart.v

```
// Spartan-3E Starter Board
// PS/2 Keyboard LCD UART Test  s3elcdkbduart.v
// c 2008 Embedded Design using Programmable Gate Arrays  Dennis Silage
module s3elcdkbduart (input CCLK, PS2CLK, PS2DAT, BTN0, output LCDRS, LCDRW, LCDE,
                      output [3:0] LCDDAT, output DCETXD, input DCERXD, output JC1, JC2);

wire rslcd, rwlcd, elcd, dav, asciidav, tbre, rst, clk16x, wrn, txd, rxd, rcvdatardy, rdn;
wire [7:0] kbddata, ascii;
wire [3:0] lcdd;
wire [7:0] lcddatin;
wire [7:0] rcvdata;
wire [31:0] clkscale;

assign LCDDAT[3] = lcdd[3];
assign LCDDAT[2] = lcdd[2];
assign LCDDAT[1] = lcdd[1];
assign LCDDAT[0] = lcdd[0];

assign LCDRS = rslcd;
assign LCDRW = rwlcd;
assign LCDE = elcd;

assign rxd = DCERXD;
assign DCETXD = txd;

assign JC1 = rxd;            // monitor received data
assign JC2 = txd;            // monitor transmitted data
```

```
clock M0 (CCLK, 163, clk16x);       // 9600 b/sec
keyboard M1 (PS2CLK, PS2DAT, dav, kbddata, parity);
rcvr M2 (rcvdata, rcvdatardy, rxd, clk16x, rst, rdn);
txmit M3 (ascii, tbre, tsre, rst, clk16x, wrn, txd);
scancodeASCII M4 (CCLK, dav, kbddata, asciidav, ascii);
lcd M5 (CCLK, resetlcd, clearlcd, homelcd, datalcd, addrlcd, lcdreset, lcdclear, lcdhome, lcddata,
        lcdaddr, rslcd, rwlcd, elcd, lcdd, lcddatin, initlcd);
genlcd M6 (CCLK, BTN0, resetlcd, clearlcd, homelcd, datalcd, addrlcd, initlcd, lcdreset, lcdclear,
           lcdhome, lcddata, lcdaddr, lcddatin, dav, asciidav, ascii, rst, tbre, wrn, rcvdata,
           rcvdatardy, rdn);

endmodule

module genlcd (input CCLK, BTN0, output reg resetlcd, output reg clearlcd, output reg homelcd,
           output reg datalcd, output reg addrlcd, output reg initlcd, input lcdreset, lcdclear,
           input lcdhome, lcddata, lcdaddr, output reg [7:0] lcddatin, input dav,
           input asciidav, input [7:0] ascii, output reg rst, input tbre, output reg wrn,
           input [7:0] rcvdata, input rcvdatardy, output reg rdn);

reg [4:0] gstate = 0;      // state register
reg [15:0] gcount = 0;     // delay counter
reg [3:0] txpos;           // LCD position for transmitted data
reg [3:0] rxpos;           // LCD position for received data

always@(posedge CCLK)
    begin
        gcount = gcount + 1;
        if (BTN0 == 1)       // reset
            begin
                resetlcd = 0;
                clearlcd = 0;
                homelcd = 0;
                datalcd = 0;
                gcount = 0;
                gstate = 0;
            end

        case (gstate)
            0:  begin
                    txpos = 0;
                    rxpos = 0;
                    rst = 0;
                    wrn = 0;
                    rdn = 1;
                    initlcd = 1;
                    gstate = 1;
                end
            1:  begin
                    rst = 1;          // reset UART
                    initlcd = 0;
                    gstate = 2;
                end
```

```
        2:   begin
                  resetlcd = 1;      // reset LCD
                  if (lcdreset == 1)
                        begin
                          resetlcd = 0;
                              gstate = 3;
                        end
             end
{LCD initialization and clearing proceeds in states 3 through 7}
        8:   begin
                  if (asciidav == 1)
                        gstate = 9;
                  else if (rcvdatardy == 1)
                        begin
                              gcount = 0;
                              gstate = 16;
                        end
             end
        9:    begin
                  if (ascii == 0)            // null character
                        gstate = 15;
                  else
                        begin
                              lcddatin[7:4] = 0;            // LCD top line
                              lcddatin[3:0] = txpos[3:0];
                              addrlcd = 1;
                              if (lcdaddr == 1)
                                    begin
                                          addrlcd = 0;
                                          gstate = 10;
                                    end
                        end
             end
        10:  begin
                  txpos = txpos + 1;
                  initlcd = 1;
                  gstate = 11;
             end
        11:  begin
                  initlcd = 0;
                  gstate = 12;
             end
        12:  begin
                  lcddatin = ascii;        // transmit data output to LCD
                  datalcd = 1;
                  if (tbre == 0)
                        begin
                              gcount=0;
                              gstate=13;
                        end
             end
```

```
13:  begin
          wrn = 1;
          if (gcount == 1024)
               gstate = 14;
     end
14:  begin
          wrn = 0;
          if (lcddata == 1)
               begin
                    datalcd = 0;
                    gstate = 15;
               end
     end
15:  begin
          if (asciidav == 0)
               gstate = 6;
     end
16:  begin
          if (gcount == 4096)
               begin
                    gcount = 0;
                    gstate = 17;
               end
     end
17:  begin
          rdn = 0;
          if (gcount == 1024)
               gstate = 18;
     end
18:  begin
          rdn = 1;
          lcddatin[7:4] = 4'b0100;          // LCD bottom line
          lcddatin[3:0] = rxpos[3:0];
          addrlcd = 1;
          if (lcdaddr == 1)
               begin
                    addrlcd = 0;
                    gstate = 19;
               end
     end
19:  begin
          rxpos = rxpos + 1;
          initlcd = 1;
          gstate = 20;
     end
20:  begin
          initlcd = 0;
          gstate = 21;
     end
```

```
            21: begin
                    lcddatin = rcvdata;    // received data output to LCD
                    datalcd = 1;
                    gstate = 22;
                end
            22: begin
                    if (lcddata == 1)
                        begin
                            datalcd = 0;
                            gstate = 23;
                        end
                end
            23: begin
                    if (rcvdatardy == 0)
                        gstate = 6;
                end
            default: gstate=0;
        endcase
    end

endmodule
```

The LCD of the Spartan-3E Starter Board with RS-232 data communication is shown in Figure 4.32. The Verilog HDL top module s3elcdkbduart.v displays both ASCII data streams on the LCD and processes and converts scan code data from the keyboard. The UART transmit and receiver modules have a one ASCII character buffer.

**Figure 4.32** RS-232 serial data communication with transmission on the top line and reception on the bottom line of the LCD of the Spartan-3E Starter Board

Figure 4.33 shows RS-232 standard serial data reception at the programmable gate array (PGA) for ASCII *v* and transmission from the PGA for ASCII *B* at 9600 b/sec. The logic signals from the PGA are inverted by the RS-232 voltage interface integrated circuit (Maxim MAX3232, *www.maxim-ic.com*) on the Spartan-3E Starter Board, as shown in Figure 4.31. The bit time $T_b = 1/r_b =$ 104.66 μsec. The received bit pattern (LSB…MSB) after the start bit (logic 0) is 01101110 binary or by reversal 76h (ASCII *v*). The transmitted bit pattern is 01000010 binary or 42h (ASCII *B*).

As described in Chapter 2 Verilog Design Automation, the Design Utilization Summary for the module s3elcdkbduart.v shows the use of 245 slice flip-flops (2%), 751 4-input LUTs (8%), 475 occupied slices (10%) and a total of 7267 equivalent gates in the XC3S500E Spartan-3E FPGA synthesis.

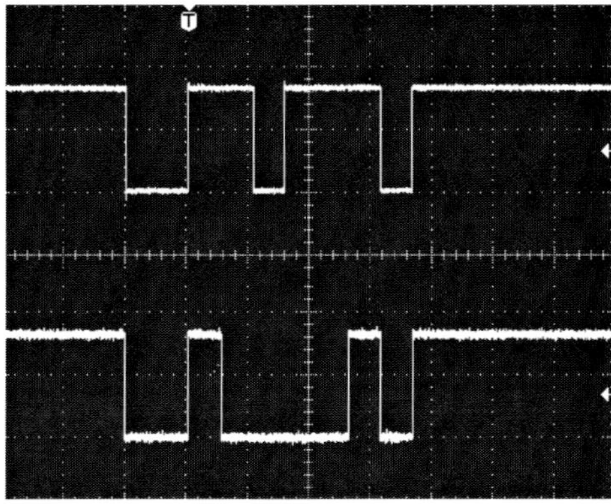

**Figure 4.33** RS-232 9600 b/sec serial data reception for ASCII *v* (top) and transmission for ASCII *B* (bottom) at 2V/div and 200 μsec/div

## Manchester Encoder-Decoder

The binary data source code represented by the RS-232 serial data communication is termed a unipolar *non-return to zero* (NRZ) line code and utilizes a rectangular pulse p($t$) of +V volts for one bit time $T_b$, as shown in Figure 4.33 [Haykin01]. Alternatively, a unipolar *return-to*-zero (RZ) line code uses a rectangular pulse p($t$) of +V volts for half a bit time $T_b/2$ and then returns to 0 V for the remaning half a bit time $T_b/2$. The pulse p($t$) of +V volts can represent either a logic 1 or logic 0 which is set by the *protocol* (specification) of the line code.

Some desirable properties of a line code include a power spectral density (PSD) which is 0 (a null) at a frequency of 0 Hz (DC) because many data communication systems utilize AC coupling and magnetic transformers and *transparency* in which there are no long strings of binary 1s or 0s regardless of the data source information [Sklar01].

The magnitude | $P(f)$ | of the Fourier transform of a single unipolar rectangular NRZ pulse $p(t)$ is determined by Equation 4.40. The normalized PSD of the unipolar NRZ line code is then determined by Equation 4.41 [Silage06]. A normalized PSD has a load $R_L = 1\ \Omega$.

$$\left| P_{rectangular}(f) \right| = A\ T_b \operatorname{sinc}(2\pi (T_b/2) f) \qquad (4.40)$$

$$\text{PSD}_{\text{unipolar NRZ}}(f) = A^2\, T_b \operatorname{sinc}^2(2\pi (T_b/2) f) \left[ \frac{1}{4} + \frac{1}{4T_b} \sum_{n=-\infty}^{\infty} \delta(f - n/T_b) \right] \qquad (4.41)$$

The term $\delta(f)$ is the impulse at a frequency $f$ and sinc $x$ is the function (sin $x$)/$x$. Although Equation 4.41 indicates that the normalized PSD of the unipolar rectangular NRZ line code has components at multiples of the bit rate $r_b = 1/T_b$, the sinc$^2$ term is zero at the same frequency $n/T_b$ Hz, $n \neq 0$. As a result the unipolar rectangular NRZ line code does not enable the extraction of bit timing and synchronization information. The resulting normalized PSD is then actually determined by Equation 4.42.

$$\text{PSD}_{\text{unipolar NRZ}}(f) = A^2\, T_b \operatorname{sinc}^2(2\pi (T_b/2) f) \left[ \frac{1}{4} + \frac{1}{4T_b} \delta(f) \right] \qquad (4.42)$$

The Manchester or split-phase NRZ line code uses a rectangular pulse $p(t)$ which is a positive constant amplitude for the first half and an equal negative constant amplitude for the second half of the bit time $T_b$ [Haykin01]. The magnitude | $P(f)$ | of the Fourier transform of a single Manchester NRZ pulse $p(t)$ is determined by Equation 4.43. The normalized PSD of the Manchester NRZ line code is determined by Equation 4.44 [Xiong00].

$$|P_{Manchester}(f)| = A T_b \operatorname{sinc}(2\pi(T_b/2)f)\sin(2\pi(T_b/2)f) \qquad (4.43)$$

$$\text{PSD}_{Manchester}(f) = A^2 T_b \operatorname{sinc}^2(2\pi(T_b/4)f)\ \sin^2(2\pi(T_b/4)f) \qquad (4.44)$$

The PSD in decibels referenced to 1 milliwatt (dBm) verses frequency of the unipolar NRZ and the Manchester NRZ line code is shown in Figure 4.34. The PSD is generated using pseudorandom binary at a data rate $r_b$ = 1 kb/sec or a bit time $T_b$ = 1 msec using the Agilent Technologies *SystemVue* digital communication system simulation environment [Silage06].

The normalized PSD of the Manchester NRZ line code has no component at $f$ = 0 Hz (DC), unlike that of the unipolar rectangular NRZ line code which displays an impulse frequency component in Figure 4.34. However, the $\sin^2$ term in Equation 4.44 implies that the Manchester NRZ line code has a wide transmission bandwidth with the first null occurring at $f = 2/T_b = 2r_b$ = 2 kHz, which is twice that of the unipolar NRZ line code as shown in Figure 4.34.

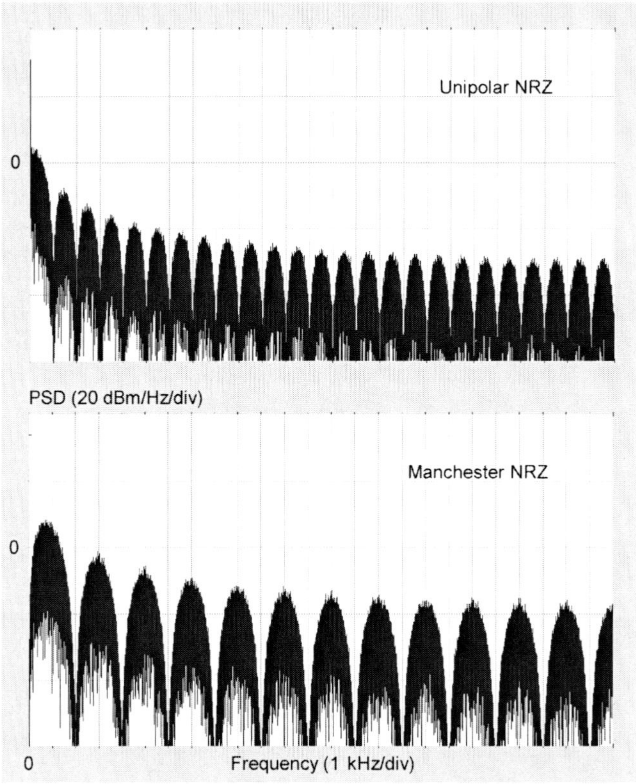


**Figure 4.34** Normalized power spectral density dBm/Hz/div of the unipolar NRZ and Manchester NRZ lines codes [Silage06]

Finally, unlike the unipolar NRZ line code, the Manchester NRZ line code is fully transparent, since a long string of binary 1s or binary 0s display voltage transitions which implies no loss of timing and synchronization information [Lathi98]. The Manchester NRZ line code is used in the 10 Mb/sec Ethernet standard. A general mBnN code maps m bits into n bits and the split phase line code is termed a 1B2B code [Sklar01].

Binary serial data communication with a Manchester NRZ line code can also utilize a UART to provide serial-to-parallel and parallel-to-serial conversion, bit rate generation, control functions and buffering. A Manchester NRZ line code encoder-decoder originally configured for a Xilinx complex programmable logic device (CPLD) is described in the Xilinx application note XAPP339 and is used with modifications here.

The Manchester NRZ encoder Verilog module encoder.v is shown in Listing 4.21. The module is comparable to the RS-232 standard transmitter module txmit.v in Listing 4.17 and also utilizes a UART configuration. The 8-bit data is inputted from the variable din to the 8-bit buffer register variable tbr with a logic 1 on the write strobe wrn.

Register variable tsr receives the data from tbr and is used to sequentially shift the data out as mdo. The input variable clk16x is the reference clock at 16 times the nominal bit rate $r_b = 1/T_b$. The register variable tbre as logic 1 indicates that the input buffer tbr is empty. An asynchronous input signal rst is used to reset the functionality of the Manchester NRZ encoder.

The encoder.v Verilog module is in the *Chapter 4\Manchester\s3emanchester* folder and utilizes multiple concurrent cyclical behavioral events to affect the data transfer with non-blocking assignments, as described in Chapter 1 Verilog Hardware Description Language. The file download procedure is described in the Appendix.

**Listing 4.21** Manchester NRZ line code encoder module encoder.v

```
// Spartan-3E Starter Board
// Manchester NRZ Encoder encoder.v
// c 2000 Xilinx, Inc. XAPP339
// c 2008 Embedded Design using Programmable Gate Arrays  Dennis Silage

module encoder (input [7:0] din, output reg tbre, input rst, clk16x, wrn, output mdo);

reg clk1x_enable, parity, wrn1, wrn2;
reg [7:0] tsr;
reg [7:0] tbr;
reg [3:0] clkdiv;
reg [3:0] no_bits_sent;

assign clk1x = clkdiv[3];
assign mdo = tsr[0] ^ clk1x;            // transmitted data

always @(posedge clk16x or posedge rst)
    begin
        if (rst)
            begin
                wrn2 <= 1;
                wrn1 <= 1;
            end
        else
            begin
                wrn2 <= wrn1;
                wrn1 <= wrn;
            end
    end
```

```
always @(posedge clk16x or posedge rst)
    begin
        if (rst)
            clk1x_enable <= 0;
        else if (wrn1 && !wrn2)
            clk1x_enable <= 1;
        else if (no_bits_sent == 15)
            clk1x_enable <= 0;
    end

always @(posedge clk16x or posedge rst)
    begin
        if (rst)
            tbre <= 1;
        else if (wrn1 && !wrn2)
            tbre <= 0;
        else if (no_bits_sent == 10)
            tbre <= 1;
        else
            tbre <= 0;
    end

always @(posedge clk16x or posedge rst)
    begin
        if (rst)
            tbr = 0;
        else if (wrn1 && !wrn2)
            tbr = din;
    end

always @(posedge clk16x or posedge rst)
    begin
        if (rst)
            clkdiv = 0;
        else if (clk1x_enable)
            clkdiv = clkdiv+1;
    end

always @(negedge clk1x or posedge rst)
    begin
        if (rst)
            tsr <= 0 ;
        else if (no_bits_sent == 1)
            tsr<=tbr;
        else if ((no_bits_sent >= 2) && (no_bits_sent < 10))
            begin
                tsr[6:0] <= tsr[7:1];
                tsr[7] <= 0;
            end
    end
```

```
always @(negedge clk1x or posedge rst)          // parity (not used)
    begin
        if (rst)
            parity <= 0;
        else
            parity <= parity ^ tsr[0];
    end

always @(posedge clk1x or posedge rst)
    begin
        if (rst)
            no_bits_sent = 0;
        else if (clk1x_enable)
            no_bits_sent = no_bits_sent + 1;
        else if (!clk1x_enable)
            no_bits_sent = 0;
    end

endmodule
```

The Manchester NRZ decoder Verilog module rcvr.v is shown in Listing 4.22. The binary serial data is inputted as mdi and the 8-bit data is outputted from the reg variable rbr. Register variables mdi1 and mdi2 are used to adjust the read timing for the receiver. The register variable data_ready indicates that received data is available with logic 1 on the *read strobe* rdn. Register variable rbr receives the data from rsr.

The input variable clk16x is the reference clock at 16 times the nominal bit rate $r_b = 1/T_b$. An asynchronous input signal rst is used to reset the functionality of the Manchester NRZ decoder. The continuous assignment variable *sample* determines when the NRZ decoding occurs outputting the register variable nrz.

The decoder.v Verilog module is in the *Chapter 4\Manchester\s3emanchester* folder and utilizes multiple concurrent cyclical behavioral events to affect the data transfer with non-blocking assignments, as described in Chapter 1 Verilog Hardware Description Language. The file download procedure is described in the Appendix.

**Listing 4.22** Manchester NRZ line code decoder module decoder.v

```
// Spartan-3E Starter Board
// Manchester NRZ Decoder decoder.v
// c 2000 Xilinx, Inc. XAPP339
// c 2008 Embedded Design using Programmable Gate Arrays  Dennis Silage

module decoder (output reg [7:0] rbr, output reg data_ready, input mdi, input clk16x, rst, rdn);

reg mdi1, mdi2, nrz, clk1x_enable;
reg [3:0] clkdiv;
reg [7:0] rsr;
reg [3:0] no_bits_rcvd;

assign clk1x = clkdiv[3];
assign sample = (!clkdiv[3] && !clkdiv[2] && clkdiv[1] && clkdiv[0]) ||
                        (clkdiv[3] && clkdiv[2] && !clkdiv[1] && !clkdiv[0]);
```

```
always @(posedge clk16x or posedge rst)
    begin
        if (rst)
            begin
                mdi1 <= 0;
                mdi2 <= 0;
            end
        else
            begin
                mdi2 < = mdi1;
                mdi1 <= mdi;            // received data
            end
    end

always @(posedge clk16x or posedge rst)
    begin
        if (rst)
            clk1x_enable <= 0;
        else if (!mdi1 && mdi2)
            clk1x_enable <= 1;          // start bit
        else if (!mdi1 && !mdi2 && (no_bits_rcvd == 8))
            clk1x_enable <= 0;
    end

always @(posedge clk16x or posedge rst)
    begin
        if (rst)
            nrz = 0;
        else if ((no_bits_rcvd > 0) && sample)
            nrz = mdi2 ^ clk1x;
    end

always @(posedge clk16x or posedge rst)
    begin
        if (rst)
            clkdiv = 0;
        else if (clk1x_enable)
            clkdiv = clkdiv + 1;
    end

always @(negedge clk1x or posedge rst)
    begin
        if (rst)
            rsr <= 0;
        else
            begin
                rsr[7] <= nrz;
                rsr[6:0] <= rsr[7:1];
            end
    end
```

```
always @(negedge clk1x or posedge rst)
    begin
        if (rst)
            rbr <= 0;
        else
            rbr <= rsr;
    end

always @(posedge clk1x or posedge rst or negedge clk1x_enable)
    begin
        if (rst)
            no_bits_rcvd = 0;
        else if (!clk1x_enable)
            no_bits_rcvd = 0;
        else
            no_bits_rcvd = no_bits_rcvd + 1;
    end

always @(negedge clk1x_enable or posedge rst)
    begin
        if (rst)
            data_ready <= 0;
        else if (!rdn)
            data_ready <= 0;
        else
            data_ready <= 1;
    end

endmodule
```

A Manchester NRZ encoder-decoder serial data communication system is implemented with the Verilog top module s3emanchester.v which is in the *Chapter 4\Manchester\s3emanchester* folder. The file download procedure is described in the Appendix.

The top module s3emanchester.v is similar to the top module s3elcdkbduart.v in Listing 4.20 and not given here. The ASCII character output of the scan code module scancode.v, which decodes the PS/2 keyboard input, is inputted to the encoder and displayed on the top line of the LCD. Since the serial data communication uses the Manchester NRZ line code a PC terminal emulator, which employs the RS-232 standard unipolar NRZ line code, cannot be utilized to display the data transmission. Instead, the output of the Manchester NRZ encoder is ported through the decoder.

The ASCII character output of the Manchester NRZ decoder is then displayed on the bottom line of the LCD, similar to that shown in Figure 4.32. Push button BTN0 resets the encoder, decoder and LCD datapath modules and the controller module.

As described in Chapter 2 Verilog Design Automation, the Design Utilization Summary for the module s3emanchester.v shows the use of 263 slice flip-flops (2%), 782 4-input LUTs (8%), 427 occupied slices (9%) and a total of 6584 equivalent gates in the XC3S500E Spartan-3E FPGA synthesis.

One protocol for the unipolar Manchester NRZ or split phase line code uses +V for $T_b/2$ then 0 for $T_b/2$ as logic 1 and 0 for $T_b/2$ then +V for $T_b/2$ as logic 0. Figure 4.35 shows the Manchester NRZ bit pattern with this protocol from the PGA (LSB…MSB) at $r_b$ = 9600 b/sec, $T_b$ = 52.1 μsec, which after the start bit (logic 1) is 01101110 binary or by reversal 76h (ASCII *v*).

Other line codes are often employed which feature desirable properties, in addition to a spectral null at 0 Hz and transparency, such the detection and preferably the correction of errors in the data transmission and transmitter power efficiency [Lathi98]. These lines codes can be conveniently encoded and decoded in an embedded design with Verilog HDL datapaths similar to the modules described here for the RS-232 standard unipolar NRZ and the Manchester NRZ line code.

The alternate mark inversion (AMI) line code has the spectral null at 0 Hz and is partially transparent but also has the properties of error detection and power efficiency [Sklar01]. AMI is used in high speed digital communication systems and was the original line code for T1 (1.544 Mb/sec) data transmission. The polar AMI NRZ line code uses +V for $T_b$ as a logic 1, *alternates* and uses by –V for $T_b$ for the next logic 1 and 0 as logic 0. The polar AMI RZ line code uses +V for $T_b/2$ then 0 for $T_b/2$ as a logic 1, *alternates* and uses by –V for $T_b/2$ then 0 for $T_b/2$ for the next logic 1 and 0 as logic 0. Figure 4.36 compares the polar AMI NRZ and RZ, the unipolar NRZ and Manchester NRZ line codes.

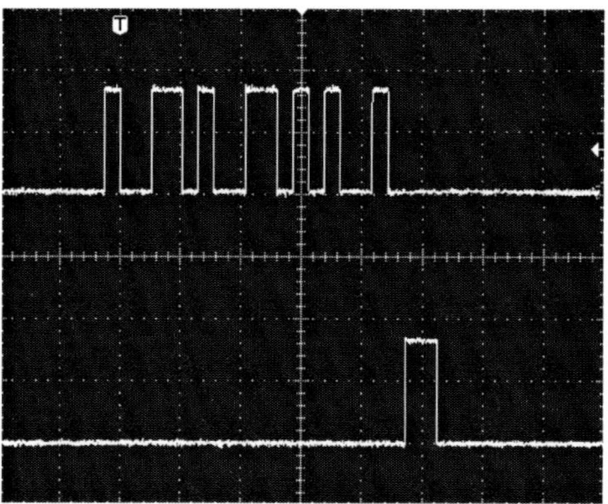

**Figure 4.35** Manchester NRZ 9600 b/sec serial data transmission for ASCII *v* (top) and transmit data buffer empty signal *tbre* (bottom) at 2V/div and 200 μsec/div

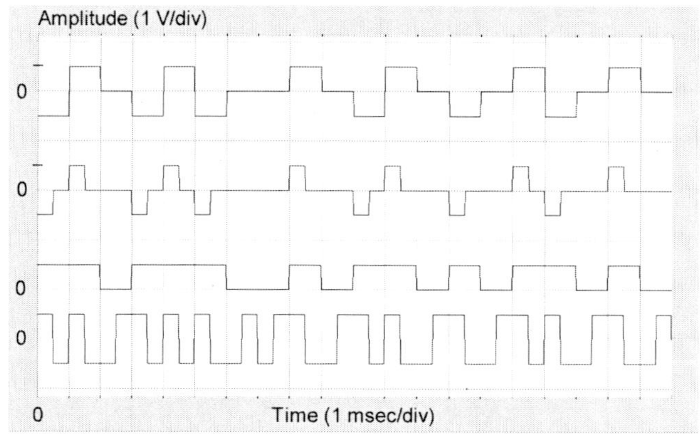


**Figure 4.36** AMI NRZ (top), AMI RZ, unipolar NRZ and Manchester NRZ (bottom) line codes at 1 kb/sec [Silage06]

The available Xilinx LogiCORE communication blocks have several *source* encoders and decoders which provide data transmission error detection and correction capabilities, as shown in Table 2.2. The convolutional, Turbo, Reed-Solomon and Viterbi LogiCORE blocks are not line code encoders and decoders but extend the bit error rate (BER) performance of digital communication systems [Sklar01]. A *source encoder* determines the appropriate parity bits to be appended to a group of data bits before data transmission. As a result, the overall bit rate for the encoded data increases and the resulting baseband or bandpass *bandwidth* also increases [Haykin01].

However, the advantageous increase in BER performance with encoded data offsets the disadvantage of the increase in bandwidth. The *source decoder* utilizes the parity bits for error detection and correction of the data bits. Both source encoding and decoding are required to be accomplished in real-time for data transmission and are suitable for the controller-datapath construct of the PGA in an embedded system.

## Digital Control

The digital control of a process can be provided by an embedded design with various degrees of complexity. Sensors are used to convert physical properties, such as temperature, pressure, electromagnetic field intensity, electrical current, voltage or power, mechanical position, velocity, and acceleration, as the input to the system usually to an analog electrical signal. The analog signal is then sampled and quantized by an analog-to-digital converter (ADC) or a threshold logic interface to a discrete signal as described in this Chapter.

The discrete signal can be processed by a programmable gate array (PGA) using the controller and datapath construct and the basic tenants of digital signal processing and digital control. Finally, the processed signals are used to affect or control the output of the system using a digital-to-analog converter (DAC) or digital logic with actuators, servomotors, valves and electronic devices [Kuo91].

An *open-loop controller* has no direct connection between the actual output of a system or response and the desired or reference input, as shown in Figure 4.37. Thus the open-loop controller does not use *feedback* to determine if the input has produced the desired response. However, an open-loop controller is reasonable for well-defined systems not subject to unpredictable responses and is often used in simple control processes.

**Figure 4.37** An open-loop controller

A *closed-loop controller* monitors the actual output of the system through a reference sensor and *feeds back* a signal to the input. The reference feedback signal is subtracted from the desired input producing an error signal which is then inputted to the controller, as shown in Figure 4.38. A closed-loop controller cancels errors and obviates the effects of changes in the parameters of operation.

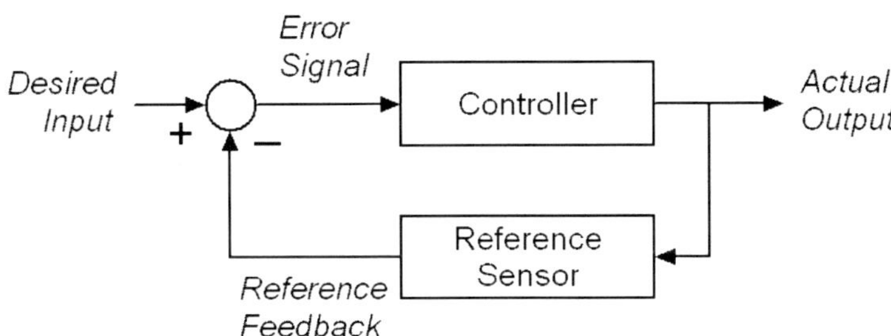


**Figure 4.38** A closed-loop controller with feedback

A common type of closed loop control with feedback for a process utilizes the proportion-integral-derivative (PID) controller in which the error signal is processed by three distinct algorithms operating in parallel, as determined by Equation 4.45. The term e(t) is the error signal, sp(t) is the desired *set point* for the process, o(t) is the output from and i(t) is the input to the process. The constants $K_p$, $K_i$, and $K_d$ in Equation 4.45 described the PID controller algorithm.

$$e(t) = sp(t) - o(t)$$
$$i(t) = K_p\, e(t) + K_i \int e(\tau)\, d\tau + K_d \frac{de(t)}{dt} \qquad (4.45)$$

The constants $K_p$, $K_i$, and $K_d$ of the PID controller have appropriate magnitudes and the proper conversion units to provide the input i(t) to the process. Proportional control processes the weighted error, integral control processes the weighted accumulated error and derivative control processes the weighted rate of change of the error [Kuo97].

The processed weighted error signals are used to affect the output of the system. Separation of the controller into three distinct algorithms facilitates the adjustment of the output response to various desired inputs. However, the PID controller does not assure that *optimal control* actually occurs in the system and more complex *state space control* is often required [Ogata02].

Some applications may not require all three algorithms of the PID controller and proportional (P), proportion-integral (PI) or proportion-derivative (PD) controllers are used in embedded design. The PID, P, PI and PD controllers have associated gains for the error signal, however these control adjustments can lead to instability, oscillation or overshoot in the output [Kuo91].

The PI controller is often utilized since the derivative process is very sensitive to noise on the desired input. However, integral control responds to the accumulated error signal of the past it can cause the instability and overshoot in the response. Finally, although proportional control can also lead to instability and usually produces a steady-state offset in the desired output, it is the foremost contributor to the actual output.

## Pulse Width Modulation

A prevalent application in open-loop control is setting the speed of a *DC servomotor* in industrial processes and robotics using pulse width modulation (PWM) in an embedded design. In PWM a rectangular pulse is modulated by an input signal which increases or decreases the pulse width $T_w$ for a fixed period $T_p$, as shown in Figure 4.39. The ratio $T_w/T_p$ is the *duty cycle* of the PWM signal. In one protocol a unipolar input signal $V_{input}$ is sampled at a rate $f_p = 1/T_p$ and the magnitude of the discrete sample determines the pulse width $T_w$. Other protocols include pulse position modulation (PPM), in which the location of a fixed pulse within the period $T_p$ is varied and pulse amplitude amplitude modulation (PAM) in which the amplitude of a fixed pulse within the period $T_p$ is varied.

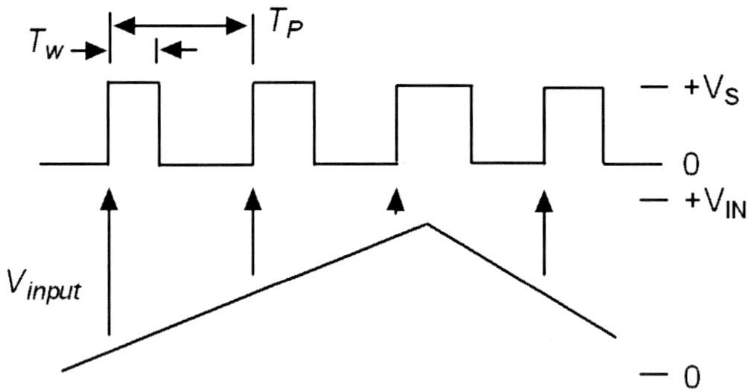


**Figure 4.39** Pulse width modulation

PWM can be implemented with a programmable gate array counter in an embedded design. The counter is used to set the pulse width $T_w$ with the pulse on by counting a clock signal with a rate $f_s$ where $f_s >> f_p$. A high sampling rate $f_s$ increases the resolution of the pulse width $T_w$. The counter then sets the fixed period $T_p$ with the pulse off before the PWM cycle repeats.

## DC Servomotor Speed Control

The output of the pulse width modulator (PWM) is connected to a DC servomotor by an *H-bridge* circuit, a portion of which is shown in Figure 4.40. The H-bridge circuit provides voltage isolation for the programmable gate array (PGA) to the higher operating voltage of the DC servomotor. It also isolates deleterious voltage spikes that can occur during operation of the DC servomotor. An H-bridge system has additional combinational logic which uses control inputs from the PGA, commonly direction and enable control signals, to affect the motion of the DC servomotor.

If the MOSFET transistors Q1 and Q4 are turned on by the control logic, the DC servomotor turns in one direction since the voltage supply $V_{supply}$ is connected to the left terminal of the servomotor and the right terminal is grounded. If Q2 and Q3 are turned on, then the DC servomotor turns in the opposite direction since the voltage supply $V_{supply}$ is connected to the right terminal and the left terminal is grounded.

If the pulse width $T_w$ equals the period $T_p$, from Figure 4.39, then the maximum amount of power is delivered to DC servomotor resulting in maximal torque and speed. If $T_w = 0$ then the delivered power is zero and the friction of the load of the DC servomotor eventually will stop its rotation. For critical embedded design applications the direction control signal can be reversed for a short period of time.

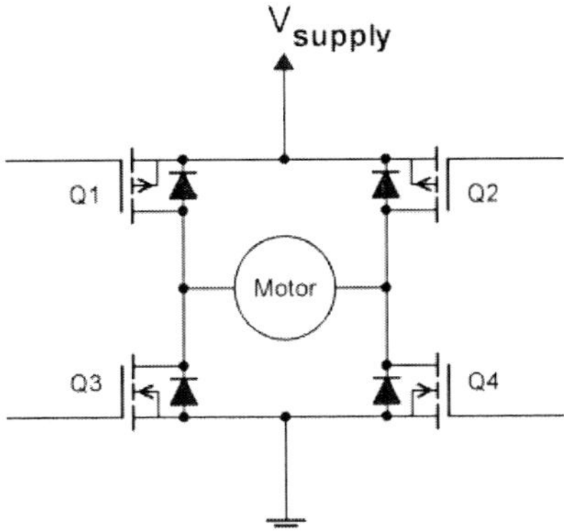


**Figure 4.40** H-bridge DC servomotor circuit

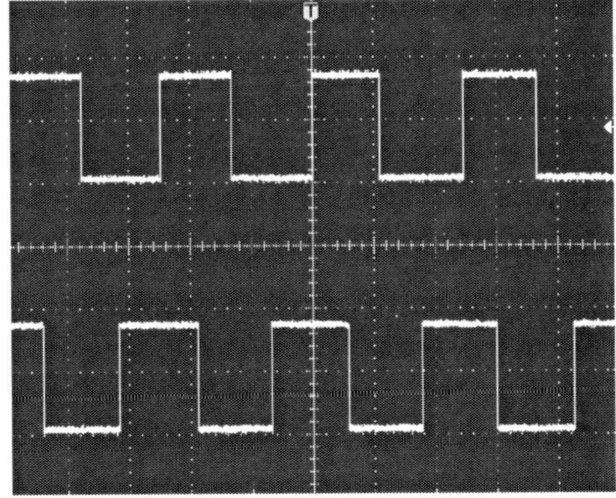

**Figure 4.41** DC servomotor quadrature encoder signals A (top) and B (bottom) at 2 V/div and 2 msec/div

A DC servomotor provides feedback of its rotational speed and direction by quadrature encoder logic signals (A and B) from Hall effect magnetic sensors, as shown in Figure 4.41. The quadrature encoder signals have a period which is proportional to the rotational speed of the DC servomotor and are 90° out of phase.

The period of the encoder signal in Figure 4.41 is approximately 4.9 msec. The direction of rotation of the DC servomotor is sensed by which quadrature encoder signal (A or B) leads in phase. In Figure 4.41 the positive pulse of quadrature encoder signal B leads that of A by 90°.

Closed-loop speed control of a DC servomotor using a proportional feedback controller is implemented with the Verilog top module s3eservomotor.v in Listing 4.23, which is in the *Chapter 4 \rs232\s3eservocontrol* folder. The file download procedure is described in the Appendix. The Xilinx ISE WebPACK project uses the UCF servocontrols3esb.ucf which uncomments the signals CCLK, SW0, BTN0, ROTA, ROTB, ROTCTR, JA1, JA2, JA3, JA4, JC1, JC2 and JC3 in the Spartan-3E Starter Board UCF of Listing 3.2. The five Verilog modules operate in parallel and some independently in the top module.

The clock module clock.v and rotary shaft encoder module rotary.v are in the *Chapter 3 \peripherals* folder. The clock.v module provides a 1 MHz clock signal mclk by dividing the 50 MHz crystal oscillator and the rotary.v module provides debounced shaft encoder rotational signals rotAreg and rotBreg and a debounced shaft pushbutton signal rotCTRreg, as described in Chapter 3 Programmable gate Array Hardware.

**Listing 4.23** DC servomotor closed-loop proportional speed control s3eservocontrol.v

```
// Spartan-3E Starter Board
// Servomotor Control s3eservocontrol.v
// c 2008 Embedded Design using Programmable Gate Arrays  Dennis Silage

module s3eservocontrol (input CCLK, SW0, BTN0, ROTA, ROTB, ROTCTR, output JA1, JA2,
                        input JA3, JA4, output JC1, JC2, JC3);

wire mclk, rotAreg, rotBreg, rotCTRreg, spddav, flagpwm, dir, enable;
wire [14:0] speed;
wire [14:0] dspeed;
wire [8:0] pwm;

assign JA1=dir;
assign JA2=enable;

assign sa=JA3;
assign sb=JA4;
assign JC1=sa;
assign JC2=sb;

assign JC3=enable;

clock M0 (CCLK, 25, mclk);          // 1 MHz
servospeed M1 (mclk, sa, spddav, speed);
servomotor M2 (mclk, speed, spddav, rotAreg, rotBreg, rotCTRreg, pwm);
servocontrol M3 (mclk, BTN0, SW0, pwm, dir, enable);
rotary M4 (CCLK, ROTA, ROTB, ROTCTR, rotAreg, rotBreg, rotCTRreg);

endmodule
```

```
module servomotor (input smmclk, input [17:0] speed, input spddav, rota, rotb, rotctr,
                   output reg [8:0] pwm);

reg [17:0] dspeed = 5000;       // desired speed
reg [8:0] tpwm = 200;           // temporary PWM on count
reg davspd = 0;

always@(posedge smmclk)
    begin
        if (rota)                       // increment speed
            begin
                dspeed = dspeed + 50;
                if (dspeed > 50000)
                    dspeed = 50000;
            end
        if (rotb)                       // decrease speed
            begin
                dspeed = dspeed – 50;
                if (dspeed == 0)
                    dspeed = 1000;
            end
        if (rotctr)                     // reset speed and PWM on
            begin
                dspeed = 5000;
                pwm = 200;      // 50% duty cycle
            end

        if (spddav && !davspd)
            begin
                pwm = tpwm;
                davspd = 1;
                if (dspeed > speed)
                    begin
                        tpwm = tpwm – 4;
                        if (tpwm <= 50)
                            tpwm = 50;
                    end
                if (dspeed <= speed)
                    begin
                        tpwm = tpwm + 4;
                        if (tpwm <= 350);
                            tpwm = 350;
                    end
            end

        if (!spddav)
            davspd = 0;
    end

endmodule
```

```
module servospeed (input ssmclk, sa, output reg spddav = 0, output reg [14:0] speed = 0);

reg [17:0] cspeed = 0;          // calculated speed
reg flagsa = 0;

always@(posedge ssmclk)
    begin
        cspeed = cspeed + 1;

        if (sa && !flagsa)
            begin
                speed = cspeed;
                spddav = 1;
                cspeed = 0;
                flagsa = 1;
            end

        if (!sa)
            begin
                spddav = 0;
                flagsa = 0;
            end
    end

endmodule

module servocontrol (input scsclk, input encmd, dircmd, input [8:0] pwm, output reg dir,
                    output reg en = 0);

reg [8:0] pwmon;            // PWM on
reg [8:0] pwmcount;         // PWM count
reg flagpwm;

always@(posedge scsclk)        // 1 MHz
    begin
        dir = dircmd;

        if (flagpwm == 1)
            begin
                pwmon = pwm;
                flagpwm = 0;
                pwmcount = 0;
            end

        if (flagpwm == 0)
            begin
                pwmcount = pwmcount + 1;
                if (pwmcount <= pwmon)
                    en = encmd;
                else
                    en = 0;
```

```
            if (pwmcount > 400)
                flagpwm = 1;
        end
end

endmodule
```

The complete ISE WebPACK project utilizes three other modules which demonstrate the capability of the PGA to process signals in parallel. The servospeed.v module uses the clock signal mclk to measure the count period as the output register variable speed of the quadrature encoder signal A as the net variable sa. The period is recorded as a count of the 1 MHz clock signal with a resolution of 1 μsec.

The register variable flagsa is use to *flag* when the net variable sa has completed a complete period. The output register variable spddav as logic 1 indicates to the servomotor.v module that an updated DC servomotor speed measurement is available.

The servomotor.v module increments or decrements the desired speed register variable dspeed with the rotation of the shaft encoder. The desired speed is initialed or reset with the debounced shaft pushbutton signal rotCTRreg to a count of 5000. If the desired speed count is greater than the actual speed count of the DC servomotor then the PWM pulse width $T_w$ as the output register pwm is decreased. If the desired speed count is less than the actual speed count, the output register pwm is increased.

The recommended PWM pulse period $T_p$ for the DC servomotor is 0.4 msec or a count of 0.4 msec/1 μsec = 400. The output register pwm as the PWM pulse width $T_w$ is limited to a count range of 50 to 350, which is equivalent to width of 0.05 to 0.35 msec and a resulting duty cycle of 12.5% to 87.5%.

Finally, the servocontrol.v module generates the two command signals for the DC servomotor. The direction control signal is set from slide switch SW0. The enable command signal of the DC servomotor is set by the pushbutton BTN0. The direction of the DC servomotor should not be changed if the enable control signal set is logic 1 which can damage or destroy the H-bridge MOSFET transistors, as shown in Figure 4.40.

The register variable pwmon, determined by the net variable pwm from the servomotor.v module, is the PWM pulse width $T_w$ count in the range from 50 to 350. The register variable pwmcount determines the interval during which the DC servomotor enable control signal is determined either by the enable command signal net variable encmd or set to 0 to disable.

As described in Chapter 2 Verilog Design Automation, the Design Utilization Summary for the module s3eservocontrol.v shows the use of 127 slice flip-flops (1%), 211 4-input LUTs (2%), 175 occupied slices (3%) and a total of 3309 equivalent gates in the XC3S500E Spartan-3E FPGA synthesis.

The modules in the ISE WebPACK project for the closed-loop proportional digital controller for the speed of a DC servomotor are a connection of interrelated modules and not an evident controller and datapath. The servomotor.v and servocontrol.v modules together represent a near datapath construct with an input command signal spddav but no status signal in return.

The module servospeed.v is a processing module which determines the speed count as a data input to the near datapath construct. Using several smaller modules in Verilog HDL facilitates the efficient development and hardware synthesis of a project.

The stability and response of the closed-loop proportional digital controller for the speed of a DC servomotor is determined by the electromechanical characteristics of the servomotor and its mechanical load, the sampling rate $f_s$ and the choice of the proportionality constant $K_p$ [Ogata02]. The proportionality constant $K_p$ relates the difference in the desired and actual speed count as the error signal to the incrementation or decrementation of PWM signal. The process of *tuning* of this parameter and the selection of the sampling rate $f_s$ is an iterative process but could be facilitated by inputting them as variables from an external peripheral.

The DC servomotor is connected to Pmod HB5 H-bridge hardware module peripheral (*www.digilentinc.com*), as shown in Figure 4.42. The DC servomotor operates with an external +6 V DC power supply.

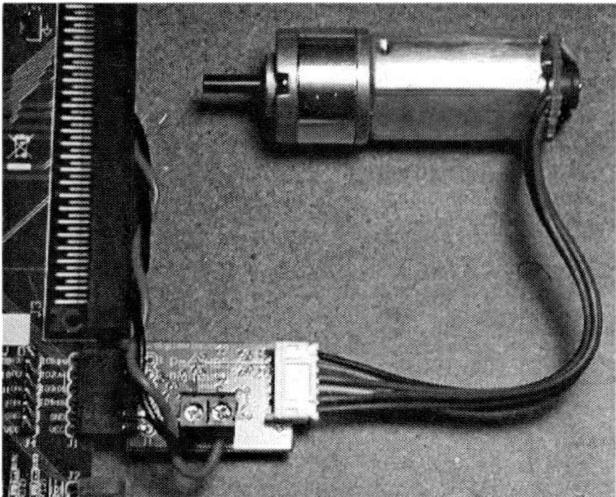

**Figure 4.42** Pmod HB5 H-bridge module and DC servomotor connected to the JA (J1) 6-pin peripheral connector of the Spartan-3E Starter Board

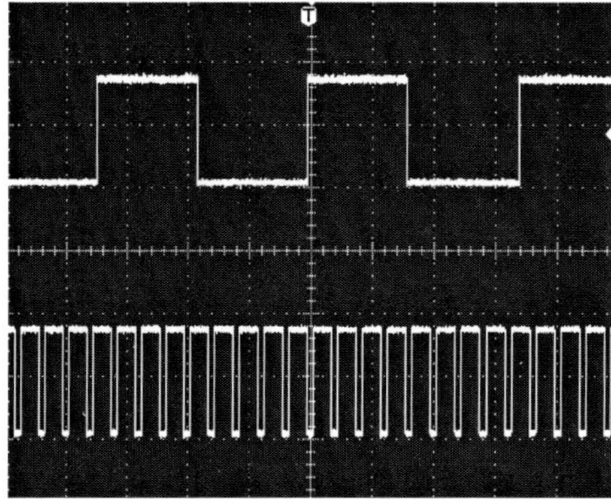

**Figure 4.43** DC servomotor quadrature encoder signal A (top) and enable control signal (bottom) at 2 V/div and 1 msec/div

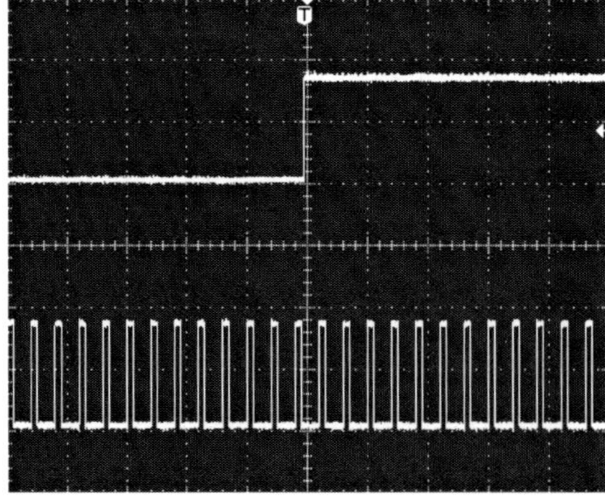

**Figure 4.44** DC servomotor quadrature encoder signal A (top) and enable control signal (bottom) at 2 V/div and 1 msec/div

The Pmod HB5 H-bridge hardware module outputs the quadrature encoder signals to and receives the direction and enable control signals for the DC servomotor from the PGA. The Pmod HB5 H-bridge hardware module is connected to the J1 (JA) 6-pin peripheral connector of the Spartan-3E Starter Board in this embedded design servomotor application.

Figure 4.43 shows DC servomotor the quadrature encoder signal A and the enable control signal. The speed count from the quadrature encoder signal A is approximately 3.4 msec/1 μsec = 3400. The duty cycle of the enable control signal is approximately 75% here and the DC servomotor is rotating at a relatively high rate. Figure 4.44 shows a lower rotation rate where the speed count from the quadrature encoder signal A is approximately 10 msec/1 μsec = 10 000 and the duty cycle of the enable control signal is approximately 25%.

## Addtional Embedded Design Projects

The use of the Xilinx LogiCORE blocks and the controller-datapath construct in Verilog HDL facilitates the development of projects in embedded design. Other LogiCORE blocks, such as the Floating Point processor, Memory Interface Generator, Fast Fourier Transform, CORDIC algorithm, Reed-Solomon coder and the Ethernet MAC, as described in Chapter 2 Verilog Design Automation, can be used for embedded design projects. Other peripherals of the Spartan-3E Starter Board, such as the USB, Ethernet and VGA ports, the DDR SDRAM and the flash PROM devices provide the capability for additional embedded design projects.

Although these embedded design projects target the Spartan-3E Starter Board some are capability of being ported to the Basys Board and its peripherals, as described in Chapter 3 Programmable Gate Array Hardware. Other embedded design projects not described in the Chapters may be available as further development continues, as discussed in the Appendix.

## Summary

In this Chapter several embedded design projects in digital signal processing, digital communications and digital control are presented using the Xilinx ISE WebPACK for the Spartan-3E Starter Board. The controller-datapath construct is utilized for these embedded projects that require multiple parallel processes to execute and the resulting real-time performance. However, the soft core processor is efficient when the constraints of real-time are not evident. Chapter 5 Embedded Soft Core Processors describes the auxiliary electronic design automation (EDA) software for the Xilinx PicoBlaze 8-bit soft-core processor and its comparison to the Verilog controller and datapath construct and Xilinx LogiCORE blocks in the Xilinx ISE WebPACK.

# References

[Cavicchi00] Cavicchi, Thomas J., *Digital Signal Processing*. Wiley, 2000.

[Chen01] Chen, Chi-Tsong, *Digital Signal Processing: Spectral Computation and Filter Design*. Oxford, 2001.

[Haykin01] Haykin, Simon, *Communication Systems*. Wiley, 2001.

[Ifeachor02] Ifeachor, Emmanuel C. and Barrie W. Jervis, *Digital Signal Processing: A Practical Approach*. Prentice Hall, 2002.

[Kuo91] Kuo, Benjamin, *Automatic Control Systems*. Prentice Hall, 1991.

[Kuo97] Kuo, Benjamin, *Digital Control Systems*. Oxford, 1997.

[Lathi98] Lathi, B.P., *Modern Digital and Analog Communication*. Oxford, 1998.

[Mano07] Mano, M. Morris and Michael D. Cilletti, *Digital Design*. Prentice Hall, 2007.

[Ogata02] Ogata, Katsuhiko, *Modern Control Engineering*, Prentice Hall, 2002.

[Mitra06] Mitra, Sanjit K., *Digital Signal Processing: A Computer Based Approach*. McGraw-Hill, 2006.

[Proakis07] Proakis, John G. and Dimitris G. Manolakis, *Digital Signal Procesing*. Prentice Hall, 2007.

[Silage06] Silage, Dennis, *Digital Communication Systems using SystemVue*. Thomson Delmar, 2006.

[Simon01] Simon, Marvin et al., *Spread Spectrum Communications Handbook*, Mc-Graw-Hill, 2001.

[Sklar01] Sklar, Bernard, *Digital Communications*, Prentice-Hall, 2001.

[Xiong00] Xiong, Fuqin, *Digital Modulation Techniques*, Artech, 2000.

# 5

# Embedded Soft Core Processors

Embedded design utilizing the Verilog hardware description language (HDL), electronic design automation (EDA) tools and the programmable gate array (PGA) can supplement finite state machines (FSM) and controller and datapath constructs by the implementation of a soft core processor. Although the constraints of real-time processing in digital signal processing, communications and control often require the rapid execution and parallel processing capabilities of the Verilog HDL controller and datapath construct, other embedded projects are well suited to a resource efficient sequential processor or microprocessor.

This Chapter presents the 8-bit PicoBlaze soft core processor implemented with auxiliary EDA software in the Xilinx ISE WebPACK and its integration with Verilog controller-datapath modules and Xilinx LogiCORE blocks. The discussion of the PicoBlaze processor here is not meant to be comprehensive but is intended to be comparison to the concepts of the controller and datapath construct and the finite state machine (FSM). The available Xilinx PicoBlaze reference projects described in this Chapter illustrate not only the use of soft core peripherals, ports and external hardware devices in an embedded design but the inherent versatility of the soft core processor.

## Programmable Gate Array Processors

The programmable gate array (PGA) has provided the opportunity for the design and implementation of a soft core processor in embedded design. Although the early *coarse grained* complex programmable logic device (CPLD) had limited capabilities and resources, soft core processors could still be implemented within their restricted structure [Lee06]. The coarse gained CPLD is an assembly of macrocells and a programmable interconnection system without the benefit of the random access memory (RAM) of the recent *fine grained* PGA devices, as described in Chapter 1 Verilog Hardware Description Language.

The GNOME soft core processor (1994) is implemented on an Intel NFX780 CPLD with only 5000 logic gates (XESS, *www.xess.com*) [Bout94]. GNOME has a 4-bit arithmetic logic unit (ALU) and accumulator, 16 4-bit general purpose registers, 128 locations by 10-bit program memory and 14 instructions. A variety of 8-bit and 16-bit microprocessors have also been rendered as soft core processors on a PGA (*www.opencores.com*).

## Xilinx PicoBlaze Development Tools

The optional Xilinx Embedded Development Kit (EDK) supports the 32-bit MicroBlaze reduced instruction set computer (RISC) soft core processor. The Xilinx EDK includes the Xilinx Platform Studio (XPS) and Software Development Kit (SDK) which have an integrated development environment (IDE). GNU C/C++ compilers, debuggers and software utilities are provided for the MicroBlaze processor in the SDK.

Support for the 32-bit MicroBlaze processor in the Xilinx EDK is extended to on-chip busing and operating systems (OS). The Xilinx Microkernel for the MicroBlaze processor is a modular library of nucleus system calls for OS services. The processor local bus (PLB), on-chip peripheral bus (OPB) and soft core peripherals facilitate the concept of the *system on a chip* (SOC) in embedded design.

Although the Xilinx ISE WebPACK can be configured to provide an electronic design automation (EDA) environment for the 8-bit PicoBlaze soft core processor, a separate download and installation procedure is required. The PicoBlaze processor EDA software tools as a partial EDK for the Spartan-3E field programmable gate array (FPGA) are downloaded as a ZIP archive file

*KCPSM3.zip* (*www.xilinx.com/products/ipcenter/picoblaze*) and extracted to a working directory. The subfolders and files in the extracted directory are shown in Figure 5.1.

| Name | Size | Type |
|---|---|---|
| Assembler | | File Folder |
| DATA2MEM_assistance | | File Folder |
| JTAG_loader | | File Folder |
| Verilog | | File Folder |
| VHDL | | File Folder |
| kcpsm3.ngc | 50 KB | NGC File |
| KCPSM3_Manual.pdf | 609 KB | Adobe Acrobat 7.0 Document |
| read_me.txt | 22 KB | Text Document |
| UART_Manual.pdf | 111 KB | Adobe Acrobat 7.0 Document |
| UART_real_time_clock.pdf | 316 KB | Adobe Acrobat 7.0 Document |

**Figure 5.1** Subfolders and files extracted from the file *KCPSM3.zip*

Originally termed the Constant (k) Coded Programmable State Machine (KCPSM) (or more appropriately Ken Chapman's Programmable State Machine after the Xilinx system designer who first implemented it) and now called the Xilinx PicoBlaze, the subfolders include an Assembler, Verilog files, a JTAG design file loader and a manual KCPSM3_Manual.pdf. The PicoBlaze processor is also described in a User Guide (UG129, *www.xilinx.com*).

The PicoBlaze processor is provided as a synthesizable Verilog hardware description language (HDL) module kcpsm3.v in the *Verilog* subfolder in Figure 5.1, as shown in Figure 5.2. The Verilog module embedded_kcpsm3.v instantiates the PicoBlaze processor in the Xilinx ISE WebPACK project and connects it to a Spartan-3E FPGA block RAM as program memory.

| Name | Size | Type |
|---|---|---|
| bbfifo_16x8.v | 13 KB | V File |
| embedded_kcpsm3.v | 5 KB | V File |
| kcpsm3.v | 93 KB | V File |
| kcpsm3_int_test.v | 5 KB | V File |
| kcuart_rx.v | 18 KB | V File |
| kcuart_tx.v | 13 KB | V File |
| testbench.v | 4 KB | V File |
| uart_clock.v | 11 KB | V File |
| uart_rx.v | 5 KB | V File |
| uart_tx.v | 5 KB | V File |

**Figure 5.2** Files in the KCPSM3 *Verilog* subfolder

The KCPSM Assembler for the PicoBlaze processor executes in a DOS based environment in a Command Prompt window in Microsoft *Windows XP* and *Vista*. The Command Prompt window is available as an *All Programs* file under *Accessories*, as shown in Figure 5.3. Note that DOS *command line* functions are used, such as *dir* for a directory listing, and names greater than eight characters or are truncated as in *PICOBL~1* for *PICOBLAZE* in the *directory tree*.

The files in the *Assembler* folder are shown in Figure 5.4 and include the KCPSM3.EXE assembler. The assembled PicoBlaze processor program is stored in the block random access memory (RAM) of the Spartan-3E FPGA configured as read-only memory (ROM). The extension for a PicoBlaze assembly language source file is *.psm* and the output is the *instruction code* Verilog module that replaces the ROM_form.v template which configures the block RAM.

There is no source text editor for the KCPSM Assembler or an IDE for the PicoBlaze processor in the Xilinx ISE WebPACK EDA. Although Microsoft Windows *Wordpad* can be used to edit the assembly language program, DOS commandsin the Command Prompt window are required for the assembly operation and file download. However, DOS *batch command* files (*.bat*) can be implemented to simplify the operations.

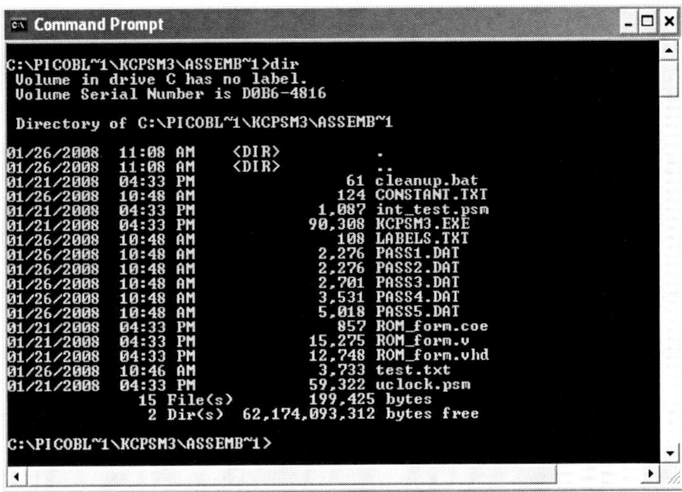


**Figure 5.3** DOS Command Prompt window

| Name | Size | Type |
|---|---|---|
| cleanup.bat | 1 KB | MS-DOS Batch File |
| int_test.psm | 2 KB | PSM File |
| KCPSM3.EXE | 89 KB | Application |
| ROM_form.coe | 1 KB | COE File |
| ROM_form.v | 15 KB | V File |
| ROM_form.vhd | 13 KB | jGRASP VHDL file |
| uclock.psm | 58 KB | PSM File |

**Figure 5.4** Files in the KCPSM3 *Assembler* subfolder

The available Mediatronix pBlazIDE EDA software tool (*www.mediatronix.com*) for the PicoBlaze processor provides a IDE for the Xilinx PicoBlaze processor that can edit, assemble and debug the source code, however it does not provide Verilog HDL formatted output files.

Although the Mediatronix pBlazIDE software tool provides an instruction set simulator with breakpoints, register and memory display which facilitates software debugging, it also utilizes a slightly different set of instruction mnemonics and assembler directives. A *File…Import* function converts the KCPSM instruction, such as *RETURN* rendered as *RET* and *ADDRESS* as *ORG* for the pBlazIDE EDA software tool.

The files KCPSM3.EXE, ROM_form.v and ROM_form.coe are copied into the Xilinx ISE WebPACK project directory. Without the use of the pBlazIDE EDA software tool, a Command Prompt or *DOS box* window can be opened and the PicoBlaze source file name.psm assembled with the command: *kcpsm3 name.psm.*

If assembly errors are detected, the process halts and a message is displayed with the line that caused the error. Since the Command Prompt is somewhat inconvenient for text display, the output of the assembly can be redirected by a DOS command to a text file by the command: *kcpsm3 name.psm > textname.txt*.

The KCPSM3 assembler utilizes the name.psm, ROM_form.v and ROM_form.coe input files and produces instruction code files name.v, name.coe and name.m which initialize the block RAM as

ROM by setting the *write enable* input to logic 0. The KCPSM3 assembler provides report files name.log, constant.txt and labels.txt and a formatted version of the source file name.fmt. The *.log* file shows the memory address and the 18-bit instruction in hexadecimal for the assembled source file. The constant.txt provides a list of constant and their values as defined by the *constant* assembler directive. The labels.txt gives a list of line labels and their associated memory addresses.

The KCPSM3 assembler also outputs files for the intermediate stages of the process *PASSn.DAT* which typically can be ignored. However, these files may aid in identifying how the assembler has interpreted program file syntax. If an error in assembly occurs, the intermediate files will only be completed to that stage. Finally, the assembled PicoBlaze instruction code formatted for the download utilities are outputted as name.hex and name.dec.

After the initial Verilog HDL synthesis and the iMPACT programming tool download of the bit file to Spartan-3E FPGA in the Xilinx ISE WebPACK project, as described in Chapter 2 Verilog Design Automation, a PicoBlaze program can be easily reconfigured by the auxiliary EDA software JTAG loader. The files in the *JTAG_loader* subfolder are shown in Figure 5.5 and include the hex2svfsetup.exe, hex2svf.exe and svf2xsvf.exe, where SVF is the serial vector format protocol. This procedure is described in the JTAG_loader_quick_guide.pdf manual in the *JTAG_loader* subfolder, as shown in Figure 5.6.

| Name | Size | Type |
|---|---|---|
| hex2svf | 0 KB | SpeedDial |
| hex2svf.exe | 92 KB | Application |
| hex2svfsetup.exe | 94 KB | Application |
| jtag_loader.bat | 1 KB | MS-DOS Batch File |
| JTAG_loader_quick_guide.pdf | 318 KB | Adobe Acrobat 7.0 Document |
| JTAG_Loader_ROM_form.v | 19 KB | V File |
| JTAG_Loader_ROM_form.vhd | 18 KB | jGRASP VHDL file |
| Normal_ROM_form.v | 15 KB | V File |
| Normal_ROM_form.vhd | 13 KB | jGRASP VHDL file |
| playxsvf.exe | 60 KB | Application |
| svf2xsvf.exe | 92 KB | Application |

**Figure 5.5** Files in the KCPSM3 *JTAG_loader* subfolder

The hex2svfsetup.exe utility, which is used to describe the JTAG chain on the target hardware board is executed in a DOS Command Prompt window once, as given from a *screen capture* in Listing 5.1. The JTAG chain for the Spartan-3E Starter Board from the Xilinx iMPACT programming tool is shown in Figure 5.6. The input hex to SVF setup utility is the configuration of JTAG chain and the instruction length of the other devices in the chain. The instruction length can also be obtained from the IEEE 1149.1 standard boundary-scan description language (BSDL) file of the FPGA, CPLD or flash programmable read-only memory (PROM) device (*www.xilinx.com*).

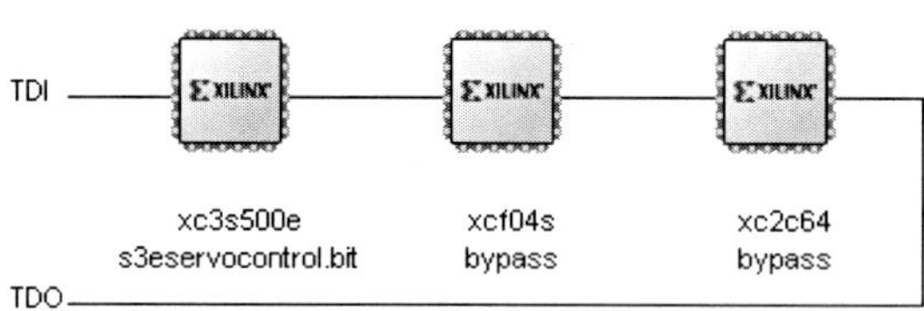


**Figure 5.6** JTAG chain of the Spartan-3E Starter Board

The new PicoBlaze source, with an eight character limit, newname.psm, is then processed by the KCPSM3 assembler. The hexadecimal (hex) output newname.hex is processed by the hex to SVF utility hex2svfsetup.exe by the DOS command: *hex2svf newname.hex newname.svf.*

A translation to a Xilinx SVF format is then performed by the utility svf2xsvf.exe by the DOS command: *svf2xsvf newname.svf newname.xsvf.* Finally, the transfer utility playxsvf.exe is used to download the new PicoBlaze design to the Spartan-3E FPGA by the DOS command: *playxsvf name.xsvf.* The Xilinx iMPACT programming tool must be closed before issuing the *playxsvf* command to avoid a conflict with the JTAG port of the Spartan-3E Starter Board.

**Listing 5.1** Xilinx hexadecimal to SVF utility setup hex2svfsetup.exe for the Spartan-3E Starter Board

```
Hex2svf Setup Utility
=====================
By Kristian Chaplin 2003
Build date Aug 26 2004 - Version 0.4
Bugs/questions: kristian.chaplin@xilinx.com

Are you programming:
(1) A Virtex-II device
(2) A Virtex-II Pro device (2vp2 - 2vp7)
(3) A Virtex-II Pro device (2vp20 - 2vp100)
(4) A Spartan-3 device
4
How many JTAG devices are before the PicoBlaze FPGA in your chain ? >0
How many JTAG devices are after the PicoBlaze FPGA in your chain ? >2
For instruction lengths, use the following guide,
or the BSDL file for the device
Spartan / SpartanXL                        - 3
Spartan2 / Spartan2e / Virtex / VirtexE    - 5
Spartan-3                                  - 6
Coolrunner (XCR3xxx XCR5xxx)               - 4
Virtex2                                    - 6
18vXX / 9500 / 9500XL / 9500XV             - 8
XCFXXS                                     - 8
XCFXXP                                     - 16
Virtex2P 2  to 7                           - 10
Virtex2P 20 to 100                         - 14

*** Device number 1 is the FPGA with PicoBlaze II ***
What is the instruction length of device #2 ?8
What is the instruction length of device #3 ?4

Results:
========

TDR 0 TDI() SMASK ();
TIR 0 TDI() SMASK ();
HDR 2 TDI(0) SMASK (3);
HIR 12 TDI(FFF) SMASK (FFF);
SIR 6 TDI (02) SMASK (3f);
Done - You are now setup to run HEX2SVF!
```

**Listing 5.2** Xilinx JTAG loader batch file JTAG_loader.bat

```
if exist .\hex2svf.cnf goto one
  echo Need to set up jtag chain first
  hex2svfsetup.exe
  echo jtag chain set up
  pause
:one
hex2svf %1.hex %1.svf
svf2xsvf -d -i %1.svf -o %1.xsvf
playxsvf %1.xsvf
```

A DOS batch command file JTAG_loader.bat facilitates the sequence of steps required to download an assembled PicoBlaze source, as given in Listing 5.2. DOS batch file processing is executed by the DOS command: *JTAG_loader name*, where the argument *name* on the DOS command line replaces the *%1* (as the first argument) in the batch file.

## Xilinx PicoBlaze Processor Architecture

The Xilinx PicoBlaze is an 8-bit reduced instruction set computer (RISC) soft core processor optimized for efficiency and very low use of the available resources of the Xilinx field programmable gate arrays (FPGA). The PicoBlaze processor does not require any external resources, such as random access memory (RAM), other than peripherals not available on the Xilinx FPGA, such as an analog-to-digital converter and uses only a minuscule 96 occupied slices and a single block RAM for program memory.

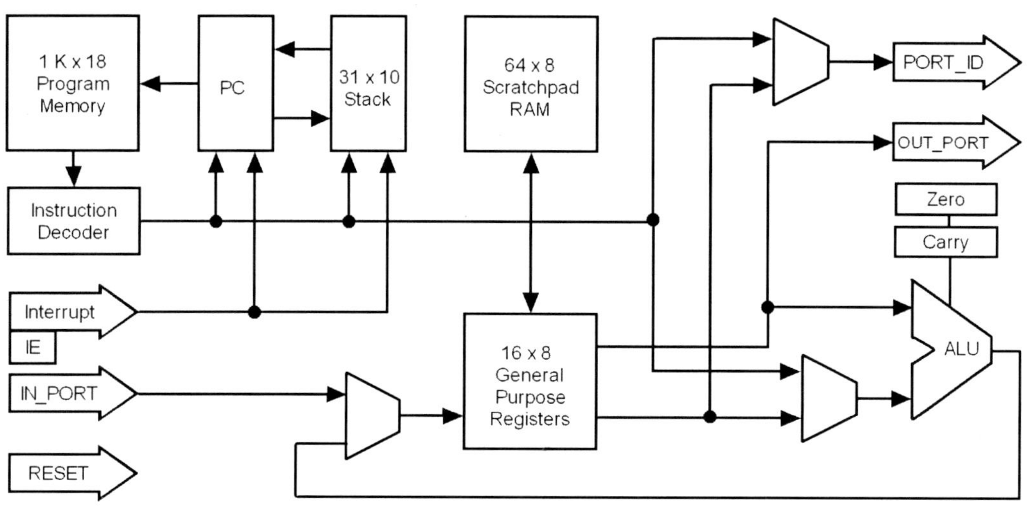


**Figure 5.7** Xilinx PicoBlaze soft core processor architecture

As described in Chapter 2 Verilog Design Automation, the Design Utilization Summary for the PicoBlaze processor shows the use of 76 slice flip-flops (1%), 109 4-input LUTs (2%), 96 occupied slices (2%) and a total of 74 814 equivalent gates in the XC3S500E Spartan-3E FPGA synthesis. The PicoBlaze processor features 16 8-bit general data registers, 1024 (1 K) 18-bit program locations typically using a single block RAM, an 8-bit arithmetic logic unit (ALU), a 64 byte internal *scratchpad* RAM, 256 input and 256 output ports and a 31 location stack for subroutine call and return, as shown in Figure 5.7. The 10-bit program counter (PC) addresses the 1 K program memory and indirect addressing is not supported.

Unlike several traditional microprocessor architectures, the Xilinx PicoBlaze processor does not have a specific accumulator and any of the general purpose registers can be utilized [Mano07]. The ALU provides addition, subtraction, arithmetic and bitwise compare and test and shift and rotate operations. ALU operations affect the zero and carry flags and interrupts can be enabled (IE). The reset input (RESET) of the PicoBlaze processor sets the PC to 0, clears the zero and carry flags, disables interrupts and the sets the stack pointer to the top of the stack.

The input-output (IO) interface signals to and from the PicoBlaze processor demonstrates the usefulness of this architecture as a controller, as shown in Figure 5.8. The 8-bit input data port IN_PORT provides data on the rising edge of the clock with the INPUT instruction.

Data appears on the 8-bit output port OUT_PORT for two clock cycles during the OUTPUT instruction. The input or output port address appears on the 8-bit PORT_ID during the INPUT or OUTPUT instruction. The INPUT and OUTPUT instructions can be either directly address as an 8-bit immediate constant or indirectly addressed as the contents of any of the general purpose registers [Mano07].

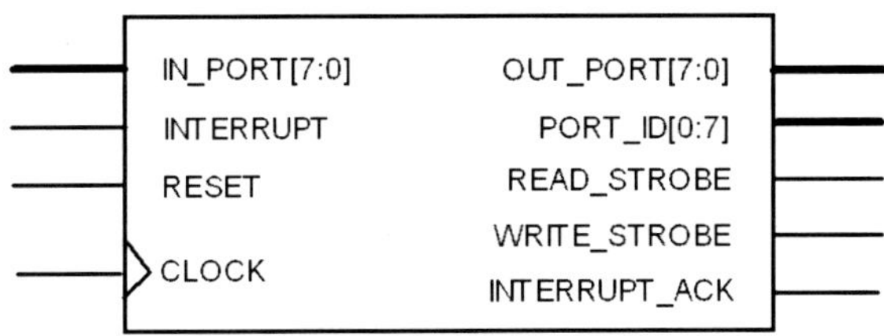


**Figure 5.8** Xilinx PicoBlaze soft core processor interface signals

The READ_STROBE output when logic 1 indicates that the input data was captured and the WRITE_STROKE output when logic 1 indicates validates the output data. These operations of the PicoBlaze processor are convenient to use as a controller or finite state machine (FSM) for low speed peripherals in an embedded design, such as the liquid crystal display (LCD) or auxiliary PS/2 mouse and keyboard of the Spartan-3E Starter Board.

The PicoBlaze processor operates at the maximum clock frequency of the Xilinx FPGA, which for the Spartan-3E (-4 speed grade) FPGA is 88 MHz and utilizes two clock cycles per instruction. When used as a controller or FSM for low speed peripherals and processes in an embedded design the PicoBlaze can operate at a lower clock frequency which reduces idle clock cycles and lowers DC power consumption. The PicoBlaze processor is a fully static logic design and can actually operate down to DC.

The Xilinx PicoBlaze processor instruction set is more expansive that merely the INPUT and OUTPUT instructions. The instruction set is standard and furthermore includes add and subtract with and without the carry (borrow) flag for multiple byte arithmetic, unconditional and condition subroutine call and comparison and test of unaffected registers using the zero and carry flags, bitwise combination logic operations (AND, OR and XOR), store and fetch data to the scratchpad RAM, rotate and shift the general purpose registers and disabling or enabling and processing asynchronous interrupts [Wakerly00]. The extensive PicoBlaze processor User Guide (UG129, *www.xilinx.com*) provides complete details of the architecture, interface signals, performance, assembler directives and EDA software tools.

The performance of an embedded design using a soft core processor can be maximized if multiple PicoBlaze processors partition the IO tasks and coordinate their operation using *semaphores* implemented with the input and output ports. The reduced number of IO ports on a single PicoBlaze processor then simplifies the input data multiplexing and decoding of the 8-bit IO address. If only one or two IO ports are needed, then no multiplexing or decoding is necessary. If eight or less IO ports are used, then a single input multiplexer for binary encoding of IN_PORT and an efficient *one-hot* encoding of OUT_PORT is required.

An innovative application in embedded design is to use the PicoBlaze processor as an *on-chip* test and debugging processor. The rapid download of new PicoBlaze soft-core processor code using the JTAG loader without resynthesizing the FPGA hardware provides a means of inputting a new *test vector* to verify the performance of the logic.

## Xilinx PicoBlaze Reference Projects

The Xilinx PicoBlaze soft core processor reference projects for the Xilinx Spartan-3E Starter Board are instantiated using the alternative VHDL (Very high speed integrated circuit Hardware Description Language) templates rather than the Verilog hardware description language (HDL)

templates which are also available. This use of VHDL does not preclude their demonstration here and discussion of the performance of the Xilinx PicoBlaze soft core processor, but can complicate any further modification of the reference projects.

The PicoBlaze reference projects for the Xilinx Spartan-3E Starter Board are available (*www.xilinx.com/products/boards/s3estarter/reference_designs.htm*) and include a scrolling message on the LCD, an analog-to-digital converter (ADC) voltage display, digital-to-analog converter (DCA) output, frequency generator, frequency counter and pulse width modulation (PWM) for control. These PicoBlaze reference projects are similar to the controller and datapath constructs in the Xilinx ISE WebPACK projects for the Spartan-3E Starter Board in Chapter 3 Programmable Gate Array Hardware and Chapter 4 Digital Signal Processing, Communications and Control and are comparatively described here.

## Initial Design

The initial design Xilinx PicoBlaze reference project is installed by default on the Spartan-3E Starter Board but is also available as a ZIP archive file (*www.xilinx.com/products/boards/s3estarter/files/s3esk_startup.zip*). The reference design is extracted to the *Initial Design* folder, as shown in Figure 5.9. The DOS batch file install_s3esk_startup.bat opens a DOS Command Prompt window and runs the Xilinx ISE iMPACT programming tool to download the project.

| Name | Size | Type |
|---|---|---|
| s3esk_startup.zip | 406 KB | WinZip File |
| control.psm | 37 KB | PSM File |
| control.vhd | 24 KB | jGRASP VHDL file |
| install_s3esk_startup.bat | 2 KB | MS-DOS Batch File |
| s3esk_startup.bit | 278 KB | BIT File |
| s3esk_startup.ucf | 3 KB | UCF File |
| s3esk_startup.vhd | 14 KB | jGRASP VHDL file |
| s3esk_startup_rev2.mcs | 780 KB | MCS File |
| s3esk_startup_rev2.pdf | 395 KB | Adobe Acrobat 7.0 Document |

**Figure 5.9** Files in the *Initial Design* folder

The initial design Xilinx PicoBlaze project is described in s3esk_startup_rev2.pdf and initializes the LCD and scrolls a fixed message, as shown in Figure 5.10. The somewhat equivalent Verilog HDL project is the top module lcdtest.v in Chapter 3 Programmable Gate Array Hardware which initializes the LCD, displays a fixed message *hello world* but also incorporates a complete LCD datapath module lcd.v.

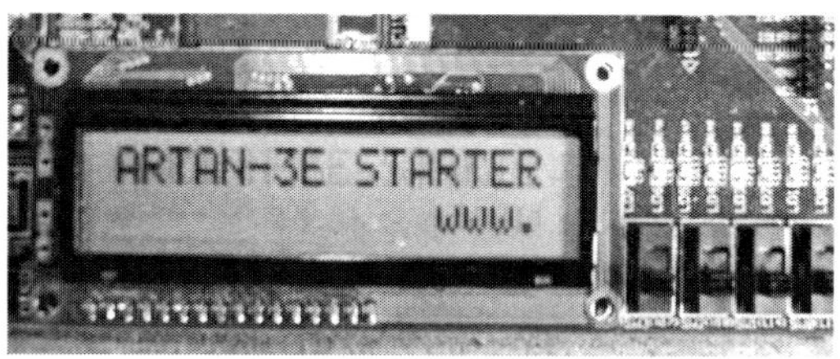


**Figure 5.10** Initial design Xilinx PicoBlaze project

As described in Chapter 2 Verilog Design Automation, the Design Utilization Summary for the initial design Xilinx PicoBlaze project shows the use of 113 occupied slices (2%), 1 block RAM and a total of 75 945 equivalent gates in the XC3S500E Spartan-3E FPGA synthesis. Similarly, the Verilog HDL top module lcdtest.v shows the use of 417 occupied slices (8%) but only a total of 6383 equivalent gates which is less than 10% of that of the initial design project because the block RAM is not utilized.

However, the control.log file of the assembly of the source file control.psm shows that only 289 of 1024 (28%) of the 18-bit locations in the block RAM of the XC3S500E Spartan-3E FPGA are used here. The remaining instructions in the block RAM can be utilized for additional features for the soft core processor project.

## Programmable Amplifier and Analog-to-Digital Converter

The programmable amplifier (PA) and analog-to-digital converter (ADC) Xilinx PicoBlaze reference project for the Spartan-3E Starter Board is available as a ZIP archive file which is extracted to the *PA-ADC* folder (*www.xilinx.com/products/boards/s3estarter/files/s3esk_picoblaze_amplifier_and_adc_control.zip*), as shown in Figure 5.11. The DOS batch file install_picoblaze_amp_ adc_ control.bat opens a DOS Command Prompt window and runs the Xilinx ISE iMPACT programming tool to download the project.

| Name | Size | Type |
|---|---|---|
| s3esk_picoblaze_amplifier_and_adc_control.zip | 990 KB | WinZip File |
| adc_ctrl.psm | 73 KB | PSM File |
| adc_ctrl.vhd | 24 KB | jGRASP VHDL file |
| install_picoblaze_amp_adc_control.bat | 2 KB | MS-DOS Batch File |
| picoblaze_amp_adc_control.bit | 278 KB | BIT File |
| picoblaze_amp_adc_control.ucf | 4 KB | UCF File |
| picoblaze_amp_adc_control.vhd | 13 KB | jGRASP VHDL file |
| PicoBlaze_Amplifier_and_ADC_control_rev2.pdf | 1,055... | Adobe Acrobat 7.0 Document |

**Figure 5.11** Files in the *PA-ADC* folder

The PA-ADC Xilinx PicoBlaze project initializes the PA and the LCD, initiates an analog signal conversion to digital data by the ADC, reads the data using the serial peripheral interface (SPI) bus and displays the results on the LCD, as shown in Figure 5.12. The PA-ADC PicoBlaze project is described in PicoBlaze_Amplifier_and_ADC_control_rev2.pdf. The somewhat equivalent Verilog HDL project is the top module s3eadctest.v in Chapter 3 Programmable Gate Array Hardware which performs approximately the same operations except that the binary data output from the ADC is converted to binary coded decimal (BCD).

**Figure 5.12** PA-ADC Xilinx PicoBlaze project

As described in Chapter 2 Verilog Design Automation, the Design Utilization Summary for the initial design Xilinx PicoBlaze project shows the use of 128 occupied slices (2%), 1 block RAM (5%) and a total of 76 260 equivalent gates in the XC3S500E Spartan-3E FPGA synthesis. Similarly, the Verilog HDL top module s3eadctest.v shows the use of 852 occupied slices (18%) and only a total of 13 184 equivalent gates which is less than 20% of that of the PA-ADC project because the block RAM is not utilized. The adc_ctrl.log file of the assembly of the source file adc_ctrl.psm shows that

546 of 1024 (53%) of the 18-bit locations in the block RAM of the XC3S500E Spartan-3E FPGA are used here.

## Digital-to-Analog Converter

The digital-to-analog converter (DAC) Xilinx PicoBlaze reference project for the Spartan-3E Starter Board is available as a ZIP archive file (*www.xilinx.com/products/boards/s3estarter/files /s3esk_picoblaze_dac_control.zip*) which is extracted to the *DAC* folder, as shown in Figure 5.13. The DOS batch file install_picoblaze_dac_control.bat opens a DOS Command Prompt window and runs the Xilinx ISE iMPACT programming tool to download the project.

The DAC Xilinx PicoBlaze project initiates a four channel digital data transfer using the SPI bus to the DAC, conversion and outputting the four analog signals by the DAC, as shown in Figure 5.14 and Figure 5.15 and described in PicoBlaze_DAC_control_rev2.pdf. The somewhat equivalent Verilog HDL project is the top module s3edactest.v in Chapter 3 Programmable Gate Array Hardware which outputs a linear ramp from DAC A at a data rate $R_{DAC} \approx 1.517$ Msamples/sec, which is nearly the maximum throughput data rate of the DAC of 1.666 Msamples/sec.

| Name | Size | Type |
|---|---|---|
| s3esk_picoblaze_dac_control.zip | 469 KB | WinZip File |
| dac_ctrl.psm | 36 KB | PSM File |
| dac_ctrl.vhd | 24 KB | jGRASP VHDL file |
| install_picoblaze_dac_control.bat | 2 KB | MS-DOS Batch File |
| picoblaze_dac_control.bit | 278 KB | BIT File |
| picoblaze_dac_control.ucf | 4 KB | UCF File |
| picoblaze_dac_control.vhd | 12 KB | jGRASP VHDL file |
| PicoBlaze_DAC_control_rev2.pdf | 531 KB | Adobe Acrobat 7.0 Document |

**Figure 5.13** Files in the *DAC* folder

The DAC Xilinx PicoBlaze project outputs a 2 kHz square wave on DAC A, 200 Hz triangle wave on DAC B, another 2 kHz square wave on DAC C and an approximate 770 Hz sine wave on DAC D. The PicoBlaze processor is interrupted to generate the four analog signal outputs every 125 μsec or an 8 kHz sampling rate, as shown in Figure 5.14 and Figure 5.15.

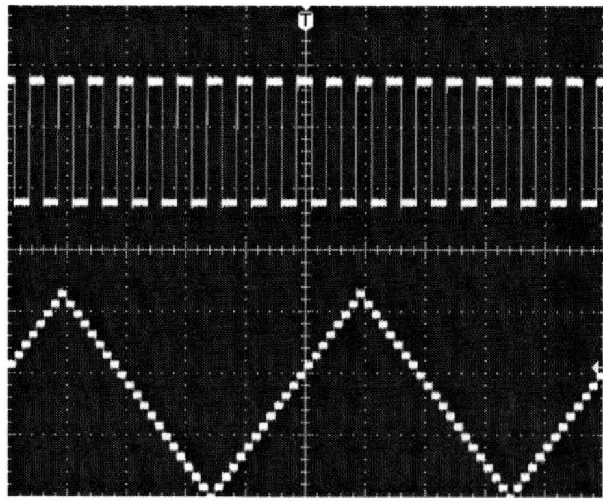

**Figure 5.14** DAC Xilinx PicoBlaze project DAC A (top) and DAC B output (bottom) at 1 V/div and 1 msec/div

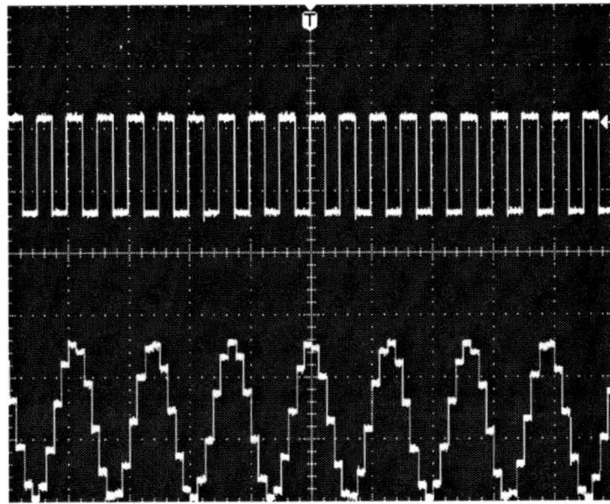

**Figure 5.15** DAC Xilinx PicoBlaze project DAC C (top) and DAC D output (bottom) at 1 V/div and 1 msec/div

As described in Chapter 2 Verilog Design Automation, the Design Utilization Summary for the DAC Xilinx PicoBlaze project shows the use of 116 occupied slices (2%), 1 block RAM (5%) and a total of 75 911 equivalent gates in the XC3S500E Spartan-3E FPGA synthesis. Similarly, the Verilog HDL top module s3edactest.v shows the use of 55 occupied slices (1%) and a total of 8194 equivalent gates which is less than 11% of that of the DAC project because the block RAM is not utilized. However, the dac_ctrl.log file of the assembly of the source file dac_crtl.psm shows that only 210 of 1024 (21%) of the 18-bit locations in the block RAM of the Spartan-3E FPGA are used here.

Another somewhat equivalent Verilog HDL project is the top module s3esincosdtmf.v in Chapter 4 Digital Signal Processing, Communication and Control which generates a dual-tone multiple frequency (DTMF) audio signal using the Xilinx CORE Sine-Cosine Look-Up Table LogiCORE block. In that project the rate of the sinusoidal *tone* generation is at the design maximum of the DAC of the Spartan-3E Starter Board of 1.47 Msamples /sec, as shown in Figure 4.24.

## Frequency Generator

The frequency generator Xilinx PicoBlaze reference project for the Spartan-3E Starter Board is available as a ZIP archive file which is extracted to the *Frequency Generator* folder (*www.xilinx.com/products/boards/s3estarter/files /s3esk_frequency_generator.zip*), as shown in Figure 5.16. The DOS batch file install_frequency_generator.bat opens a DOS Command Prompt window and runs the Xilinx ISE iMPACT programming tool to download the project.

| Name | Size | Type |
|---|---|---|
| s3esk_frequency_generator.zip | 555 KB | WinZip File |
| fg_ctrl.psm | 91 KB | PSM File |
| fg_ctrl.vhd | 24 KB | jGRASP VHDL file |
| frequency_generator.bit | 278 KB | BIT File |
| frequency_generator.ucf | 4 KB | UCF File |
| frequency_generator.vhd | 27 KB | jGRASP VHDL file |
| frequency_generator_v100.pdf | 614 KB | Adobe Acrobat 7.0 Document |
| install_frequency_generator.bat | 2 KB | MS-DOS Batch File |

**Figure 5.16** Files in the *Frequency Generator* folder

The frequency generator Xilinx PicoBlaze project, as shown in Figure 5.17, is described in frequency_generator_v100.pdf. The somewhat equivalent Verilog HDL project is the top module s3eddsfreqgen.v in Chapter 4 Digital Signal Processing, Communication and Control which generates a nominal 25 kHz sinusoidal audio signal using the Xilinx CORE Generator Direct Digital Synthesis (DDS) LogiCORE block and the DAC of the of the Spartan-3E Starter Board. The data throughput rate of the Verilog top module s3eddsfreqgen.v is 1.429 Msamples/sec and reduced from the design maximum of 1.47 Msamples /sec by the latency of the DDS and the datapath module genddsfreq.v.

The frequency generator Xilinx PicoBlaze reference project uses a configured DDS circuit, rather than the Xilinx CORE Generator DDS LogiCORE block, to output a square wave at a frequency of 1 Hz to approximately 100 MHz on the *SMA* connector (J17) and J4-IO12 (JC-4) of the Spartan-3E Starter Board. The formulation of the DDS circuit in this project is instructive and utilizes two digital clock managers (DCM) and a 32-bit register as a phase accumulator.

The frequency generator Xilinx PicoBlaze reference project utilizes the rotary shaft encoder and the LEDs to provide an editor that can modify the individual output frequency digit values, as shown in Figure 5.17. This features demonstrates the utility of the embedded soft core PicoBlaze processor to implement an intricate function.

The Verilog HDL top module s3eddsfreqgen.v sets the sinusoidal audio frequency output using the rotary shaft encoder. The push buttons are used to reset the nominal frequency to 25 kHz and adjust the frequency by a coarse amount and the LCD is not used here.

As described in Chapter 2 Verilog Design Automation, the Design Utilization Summary for the frequency generator Xilinx PicoBlaze project shows the use of 172 occupied slices (3%), 1 block RAM (5%), 2 DCMs (50%) and a total of 91 537 equivalent gates in the XC3S500E Spartan-3E FPGA synthesis. Similarly, the Verilog HDL top module s3eddsfreqgen.v shows the use of 155 occupied slices (3%), 1 block RAM (5%), 1 DCM (25%) and a total of 75 895 equivalent gates which is nearly the same as that of the frequency generator PicoBlaze project.

**Figure 5.17** Frequency generator Xilinx PicoBlaze project frequency display

This near resource equivalency demonstrates the efficiency of the embedded soft core PicoBlaze processor to implement a complex control function. Finally, the fg_ctrl.log file of the assembly of the source file fg_crtl.psm shows that only 574 of 1024 (56%) of the 18-bit locations in the block RAM of the Spartan-3E FPGA are used here even with the functionality of the project.

## Frequency Counter

The frequency counter Xilinx PicoBlaze reference project for the Spartan-3E Starter Board is described in frequency_counter_v100.pdf and is available as a ZIP archive file (*www.xilinx.com /products/boards/s3estarter/files/s3esk_frequency_counter.zip*) which is extracted to the *Frequency Counter* folder, as shown in Figure 5.18. The DOS batch file install_frequency_generator.bat opens a DOS Command Prompt window and runs the Xilinx ISE iMPACT programming tool to download the project.

The Xilinx PicoBlaze project initial display selects one of four input signals used to the frequency counter, as shown in Figure 5.19. The reference clock oscillator for the frequency counter is the nominal 50 MHz crystal clock oscillator on the Spartan-3E Starter Board.

| Name | Size | Type |
|---|---|---|
| dcm_fixed_osc.vhd | 5 KB | jGRASP VHDL file |
| fc_ctrl.psm | 60 KB | PSM File |
| fc_ctrl.vhd | 24 KB | jGRASP VHDL file |
| frequency_counter.bit | 278 KB | BIT File |
| frequency_counter.ucf | 3 KB | UCF File |
| frequency_counter.vhd | 19 KB | jGRASP VHDL file |
| frequency_counter_v100.pdf | 550 KB | Adobe Acrobat 7.0 Document |
| install_frequency_counter.bat | 2 KB | MS-DOS Batch File |
| ring_osc.vhd | 6 KB | jGRASP VHDL file |
| s3esk_frequency_counter.zip | 525 KB | WinZip File |

**Figure 5.18** Files in the *Frequency Counter* folder

**Figure 5.19** Frequency counter Xilinx PicoBlaze project initial display

The first signal that can be inputted to the frequency counter is a *Ring* oscillator implemented on the Spartan-3E FPGA. A ring oscillator employs the inherent delay of the logic elements and feedback to provide an unstable clock source that is sensitivity to the device fabrication but also the temperature and applied voltage [Lee06]. The *DCM* input signal is a non-standard, oscillating configuration of the digital clock manager (DCM) of the Spartan-3E FPGA not normally utilizes in an embedded design using the FPGA.

Since the *50M* input signal is the nominal 50 MHz crystal clock oscillator, the frequency counter is measuring the reference clock oscillator and reports an exact result, as shown in Figure 5.20. Finally, the *SMA* input signal is an external signal inputted at the SMA connector (J17) and J4-IO12 (JC-4) of the Spartan-3E Starter Board. The maximum unipolar voltage that can be inputted at the SMA connector is +3.3 V DC, as described in the Spartan-3E Starter Board User's Guide (UG230.pdf, *www.xilinx.com* or *www.digilentinc.com*).

**Figure 5.20** Frequency counter Xilinx PicoBlaze project counter display

As described in Chapter 2 Verilog Design Automation, the Design Utilization Summary for the frequency counter Xilinx PicoBlaze project shows the use of 190 occupied slices (4%), 1 block RAM (5%), 1 digital clock managers (DCM) (50%) and a total of 85 032 equivalent gates in the XC3S500E Spartan-3E FPGA synthesis. The fc_ctrl.log file of the assembly of the source file fc_crtl.psm shows that only 616 of 1024 (60%) of the 18-bit locations in the block RAM of the Spartan-3E FPGA are used here even with the complexity and utility of the project.

## Pulse Width Modulation and Control

The pulse width modulation (PWM) and control Xilinx PicoBlaze reference project for the Spartan-3E Starter Board is described in PicoBlaze_PWM_control_rev1.pdf and is available as a ZIP archive file which is extracted to the *PWM* folder (*www.xilinx.com/products/boards/s3estarter/files/s3esk_picoblaze_pwm_control.zip*), as shown in Figure 5.21. The DOS batch file install_ picoblaze_pwm_control.bat opens a DOS Command Prompt window and runs the Xilinx ISE iMPACT programming tool to download the project.

The PWM Xilinx PicoBlaze project implements a twelve channel system, eight channels of which control the relative intensity of the LEDs and four channels are outputted to the J4 (JC) of the Spartan-3E Starter Board. The PWM has a pulse frequency $f_p$ = 1 kHz or a pulse period $T_p$ = 1 msec and a pulse width $T_w$ with a resolution of 256 increments (8-bit) or 3.90625 μsec. The PWM Xilinx PicoBlaze project provides an RS-232 standard universal aynchronous receiver transmitter (UART) communication port that is used to set the duty cycle $T_w/T_p$ of the eight LEDs or the four output signals independently.

Note that the Microsoft *Windows XP* terminal communications program, *HyperTerminal*, and some USB to RS-232 serial data converters have significant problems in this application and produce dropped ASCII characters in the data transmission. The Spartan-3E Starter Board RS-232 port does not have hardware flow control (RTS/CTS). Other available PC terminal emulators are shown to perform correctly. The PicoBlaze implementation of the RS-232 standard UART communication port is innovative and can be used as a part of an *on-chip* test and debugging soft core processor for other embedded design projects.

The PWM Xilinx PicoBlaze project also illustrates the use of the JTAG_loader by the DOS batch file update_ picoblaze_auto_pwm.bat which opens a DOS Command Prompt window and runs the Xilinx ISE iMPACT programming tool to download the assembled PicoBlaze source to the project.

| Name | Size | Type |
|---|---|---|
| s3esk_picoblaze_pwm_control.zip | 425 KB | WinZip File |
| auto_pwm.psm | 32 KB | PSM File |
| auto_pwm.xsvf | 71 KB | XSVF File |
| install_picoblaze_pwm_control.bat | 2 KB | MS-DOS Batch File |
| picoblaze_pwm_control.bit | 278 KB | BIT File |
| picoblaze_pwm_control.ucf | 2 KB | UCF File |
| picoblaze_pwm_control.vhd | 14 KB | jGRASP VHDL file |
| PicoBlaze_PWM_control_rev1.pdf | 501 KB | Adobe Acrobat 7.0 Document |
| pwm_ctrl.psm | 43 KB | PSM File |
| pwm_ctrl.vhd | 24 KB | jGRASP VHDL file |
| update_picoblaze_auto_pwm.bat | 1 KB | MS-DOS Batch File |

**Figure 5.21** Files in the *PWM* folder

The first two channels of the eight channel output of the PWM Xilinx PicoBlaze project is shown in Figure 5.22. The pulse period is 1 msec and the approximate duty cycle for channel 1 is 10% and that of channel 2 is 75%.

The PWM Xilinx PicoBlaze project has a data throughput rate that is a function of the number of channels, the pulse width resolution and the pulse frequency or $12 \times 256 \times 10^3 = 3.072 \times 10^6$. Although the PicoBlaze can utilize an 88 MHz clock, the highest clock frequency available for the XC3S500E (-4 speed) Spartan-3E FPGA, the soft core processor uses the 50 MHz crystal oscillator of the Spartan-3E Starter Board.

The PicoBlaze soft core processor requires two clock cycles per instruction or an instruction period of 40 nsec. Therefore in the 3.90625 μsec available between a PWM step only about 97 PicoBlaze processor instructions can be executed. Although the PWM process here requires only approximately half of the available instructions, if either the channels, resolution or pulse frequency is doubled, the number of instructions available is halved and the PicoBlaze soft core processor could not perform the real-time task.

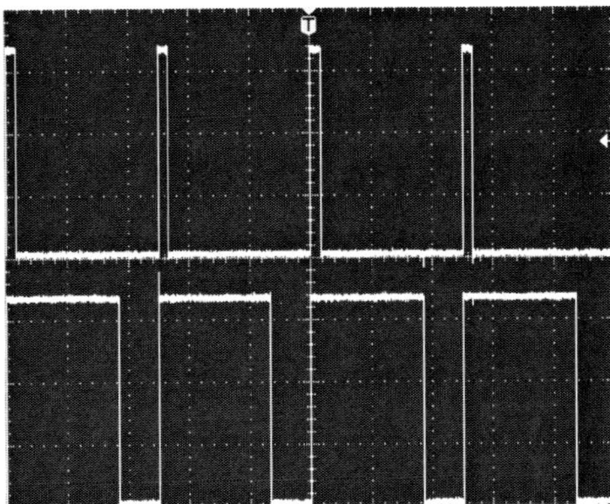

**Figure 5.22** PWM Xilinx PicoBlaze project channel 1 PWM output (top) and channel 2 PWM output (bottom) at 1 V/div and 400 μsec/div

This is certainly problematic for all sequential processors such as the PicoBlaze. Alternatively, each PWM channel can be tasked to a separate Verilog HDL module, all operating in parallel, to meet the requisite PWM performance. Each of the Verilog HDL modules then can even execute at one clock cycle per event, if warranted, as in the DC servomotor closed-loop proportional speed control top module s3eservocontrol.v in Chapter 4 Digital Signal Processing, Communications and Control.

As described in Chapter 2 Verilog Design Automation, the Design Utilization Summary for the PWM Xilinx PicoBlaze project shows the use of 165 occupied slices (3%), 1 block RAM (5%) and a total of 78 962 equivalent gates in the XC3S500E Spartan-3E FPGA synthesis. The auto_pwm.log file of the assembly of the source file auto_pwm.psm shows that only 305 of 1024 (30%) of the 18-bit locations in the block RAM of the Spartan-3E FPGA are used here even with the functionality of the project.

## Soft Core Processors and Peripherals

The availability of the soft core processor within the field programmable gate array (FPGA) has significantly altered the ambiance of embedded design. Although *hard core processors* are available for some FPGAs, notably the Xilinx Virtex-II Pro with the IBM 32-bit reduced instruction set computer (RISC) PowerPC, the soft core processor seemingly has more flexibility including the capability of utilizing several within the embedded design. Since the soft core processor can be modified easily, new and specialized architectures can also be readily developed.

The implementation of a soft core processor though could be hindered by the electronic design automation (EDA) software tools available. Although the Xilinx PicoBlaze soft core processor is programmed in its EDA by assembly language, well supported software compilers in conversation languages can possibly be developed. Nevertheless, the Verilog hardware description language (HDL), as an alternative EDA software tool for embedded design, is both conversational and easily deciphered when compared to assembly language.

Both the Xilinx 8-bit PicoBlaze soft core processor and the Verilog HDL controller and datapath construct can utilize *soft core peripherals* to augment their performance and functionality in an embedded design. The Xilinx LogiCORE blocks available in the Xilinx ISE WebPACK EDA software tool provide block and distributed memory, multipliers, dividers and floating point operations, as described in Chapter 2 Verilog Design Automation.

However, these useful Xilinx LogiCORE blocks often require more than 8-bits of data. External latches and logic would be required to interface the PicoBlaze soft core processor, as shown in Figure 5.8, to these Xilinx LogiCORE blocks. The Verilog HDL controller and datapath construct is readily extensible and interfacing to a LogiCORE block is therefore not so problematic, as described in the projects in Chapter 4 Digital Signal Processing, Communications and Control.

Alternatively, the Xilinx MicroBlaze 32-bit RISC soft core processor can conveniently interface to the Xilinx LogiCORE blocks and Xilinx and *third-party* Output Peripheral Bus (OPB) soft core peripheral devices. However, although the Xilinx MicroBlaze soft core processor includes a conversational C language compiler and a real-time operating system (RTOS), the Xilinx Embedded Design Kit (EDK) and third-party EDA software tools are required.

Other soft core peripherals nonetheless are quite comparable with the Xilinx 8-bit PicoBlaze soft core processor. Notable the soft core universal asynchronous receiver transmitter (UART) in the pulse width modulation (PWM) and control Xilinx PicoBlaze reference project can be easily extracted and used to provide RS-232 standard serial data communication in an embedded design.

The Xilinx PicoBlaze soft core processor and the Verilog HDL controller and datapath construct can also interact synergistically through the interface signals shown in Figure 5.8. An embedded design then would consist of processing elements appropriate for the task.

The Xilinx PicoBlaze soft core processor is quite resource efficient and is suitable for debouncing and reading slide switches and push buttons and peripherals with a low data throughput, such as the LCD of the Spartan-3E Starter Board. Nevertheless, the Verilog HDL controller and datapath construct is seeminbgly more suited to high throughput internal soft peripherals and external hard peripherals, such as a digital-to-analog converter (DAC) and an analog-to-digital converter (ADC) in a *wavefront* digital signal processing (DSP) systems, as described in Chapter 4 Digital Signal Processing, Communications and Control.

## Summary

In this Chapter the electronic design automation (EDA) software tools and the architecture of the Xilinx soft core 8-bit PicoBlaze processor is presented. A scrolling message on the LCD, an analog-to-digital converter (ADC) voltage display, digital-to-analog converter (DCA) output, frequency generator, frequency counter and pulse width modulation (PWM) for control as available PicoBlaze reference projects for the Xilinx Spartan-3E Starter Board are described. These PicoBlaze reference projects are compared to similar Verilog HDL projects in Chapter 3 Programmable Gate Array Hardware and Chapter 4 Digital Signal Processing, Communications and Control for their relative efficiency and functionality.

## References

[Bout94] David Van den Bout, *FPGA Workout*, XESS, 1994.

[Lee06] Lee, Sunggu., *Advanced Digital Logic Design*. Thomson, 2006.

[Mano07] Mano, M. Morris and Michael D. Cilletti, *Digital Design*. Prentice Hall, 2007.

[Wakerly00] Wakerly, John F., *Digital Design Principles and Practice*. Prentice Hall, 2000.

# Appendix

## Project File Download

The Xilinx ISE WebPACK projects and Verilog hardware description language (HDL) modules described in the Chapters are available for download in a ZIP file archive format from the website *astro.temple.edu/~silage/embeddes*. An alternative website for the ZIP file archive format is *www.dennis-silage.com/embeddes*. Separate ZIP file archives are available for specific projects for the Digilent Spartan-3E Basys Board and the Xilinx Spartan-3E Starter Board. The ZIP archive file download is password protected with *TUECEXUP08*.

These copyrighted materials are provided in support of this text and no other rights or license is implied by their availability for non-commercial use. Comments on the Xilinx ISE WebPACK projects and this file download procedure can be directed to the author via email at *silage@temple.edu*.

Although most of these embedded design projects target the Spartan-3E Starter Board some are capability of being easily ported to the Basys Board and its peripherals, as described in Chapter 3 Programmable Gate Array Hardware, or other Xilinx FPGA evaluation boards. Additional embedded design projects not described in the Chapters are to be available on the website as further development continues. The ZIP file archive on the website contains the most recent and verified Xilinx ISE WebPACK projects and other information.

Install the folders and subfolders in the directory *C:\EDPGA*. The Verilog modules (.v), user control files (.ucf), Xilinx Architecture Wizard files (.xaw) and Xilinx project files (.ise) are organized into folders and subfolders by Chapters, for example *C:\EDPGA\Chapter 2\elapsedtime*. Project files beginning with *s3e* are for the Spartan-3E Starter Board and those with *ba* are for the Digilent Basys Board. Project files listed in parentheses are in the folder *C:\EDPGA\Chapter 3\peripherals*.

Modifications to existing projects, hardware synthesis using other Xilinx programmable gate array devices or evaluation boards and the programming of the volatile bit map file (.bit) require the Xilinx ISE WebPACK electronic design automation environment (*www.xilinx.com*).

**\Chapter 2 Verilog Hardware Automation**

\elapsedtime
- s3eelapsedtime.ise
- s3eelapsedtime.v
  - (clock.v)
  - (lcd.v)
  - etlcd.v
  - elapsedtimes3e.v
- elapsedtimes3esb.ucf

\binBCDdivider
- binBCDdivider.ise
- binbcddivider.v
  - binbcd.v
  - (lcd.v)
  - genlcd.v
- binBCDs3esb.ucf

- \binBCDiterative
  - binBCDiterative.ise
  - binBCDiter.v
    - iterative.v
    - (lcd.v)
    - geniterlcd.v
  - binBCDs3esb.ucf

- \binBCDshiftadd3
  - binBCDshiftadd3.ise
  - binBCDshiftadd3.v
    - (lcd.v)
    - shiftadd3.v
    - genshiftaddlcd.v
  - binBCDs3esb.ucf

**\Chapter 3 Programmable Gate Array Hardware**

- \ucf
  - basys.ucf
  - s3esb.ucf

- \peripherals
  - ad1adc.v
  - bargradph.v
  - clock.v
  - da1dac.v
  - da2dac.v
  - da2dac2.v
  - dacs3edcm.xaw
  - keyboard.v
  - lcd.v
  - mouse.v
  - pbdebounce.v
  - rotary.v
  - s3eadc.v
  - s3edac.c
  - s3eprogamp.v
  - sevensegment.v

- \clocktest
  - \s3eclocktest
    - s3eclocktest.ise
    - clocktest.v
      - (clock.v)
    - clocktests3eucf.ucf

```
\bargraphtest
    \s3ebargraphtest
        s3ebargraph.ise
        bargraphtest.v
            (clock.v)
            (bargraph.v)
            gendata.v
        bargraphtests3esb.ucf

\pbsswtest
    \s3epbsswtest
        s3epbsswtest.ise
        pbsswtest.v
        pbsswtests3esb.ucf

\pbdebouncetest
    \s3epbdebounce
        s3epbdebounce.ise
        pbdebouncetest.v
            (pbdebounce.v)
            (clock.v)
            ledtest.v
        pbdebouncetests3esb.ucf

\rotarytest
    \s3erotarytest
        s3erotarytest.ise
        rotarytest.v
            (rotary.v)
            (clock.v)
            ledtest.v
        rotarytests3esb.ucf

\sevensegtest
    \basevensegtest
        basevensegtest.ise
        sevensegtest.v
        (clock.v)
        (sevensegment.v)
        elapsedtime.v

\lcdtest
    \s3elcdtest
        s3elcdtest.ise
        lcdtest.v
            (pbdebounce.v)
            (clock.v)
            (lcd.v)
            genlcd.v
        lcdtests3esb.ucf
```

```
\elapsedtime
    \s3eelapsedtime
        s3eelapsedtime.ise
        s3eelapsedtime.v
            (clock.v)
            (lcd.v)
            etlcd.v
            elapsedtimes3e.v
        elapsedtimes3esb.ucf

\keyboardtest
    \bakeyboardtest
        bakeyboardtest.ise
        bakeyboardtest.v
            (clock.v)
            (keyboard.v)
            (sevensegment.v)
            checkkbd.v
        keyboardtestbasys.ucf

    \s3ekeyboardtest
        s3ekeyboardtest.ise
        s3ekeyboardtest.v
            (keyboard.v)
            (lcd.v)
            genlcd.v
        keyboardtests3esb.ucf

\mousetest
    \bamousetest
        bamousetest.ise
        bamousetest.v
            (mouse.v)
            (sevensegment.v)
            checkmouse.v
        mousetestbasys.ucf
    \s3emousetest
        s3emousetest.ise
        s3emousetest.v
            (mouse.v)
            (lcd.v)
            checkmouse.v
            genlcd.v
        mousetests3esb.ucf
```

```
\dactest
    \s3edactest
        s3edactest.ise
        s3edactest.v
            (s3edac.v)
            dacs3edcm.v
            gendac.v
        dactests3esb.ucf

\da1test
    \bada1test
        bada1test.ise
        da1test.v
            (da1dac.v)
            gendac.v
        da1testbasys.ucf
    \s3eda1test
        s3eda1test.ise
        da1test.v
            (da1dac.v)
            gendac.v
        da1tests3esb.ucf

\da2test
    \bada2test
        bada2test.ise
        da2test.v
            (da2dac.v)
            gendac.v
        da2testbasys.ucf
    \s3eda2test
        s3eda2test.ise
        da2test.v
            (da2dac.v)
            gendac.v
        da2tests3esb.ucf

\adctest
    \s3eadctest
        s3eadctest.ise
        s3eadctest.v
            (s3eadc.v)
            (s3eprogamp.v)
            (lcd.v)
            adclcd.v
            genampadc.v
        adctests3esb.ucf
```

```
\ad1test
    \s3ead1test
        s3ead1test.ise
        ad1test.v
            (ad1adc.v)
            (lcd.v)
            adclcd.v
            genad1adc.v
        ad1tests3esb.ucf
    \baad1test
        baad1test.ise
        ad1test.v
            (ad1adc.v)
            (sevensegment.v)
            adcsevenseg.v
            genad1adc.v
        ad1testbasys.ucf
```

**\Chapter 4 Digital Signal Processing, Communications and Control**

```
\adcdac
    \s3eadcdac
        s3eadcdac.ise
        s3eadcdac.v
            (s3eadc.v)
            (s3edac.v)
            (s3eprogamp.v)
            (dacs3edcm.v)
            genadcdac.v
        adcdacs3esb.ucf
    \s3ead1da2
        s3ead1da2.ise
        s3ead1da2.v
            (ad12adc.v)
            (da2dac.v)
        genad1da2.v
        ad1da2s3esb.ucf
    \baad1da2
        baad1da2.ise
        baad1da2.v
            (ad12adc.v)
            (da2dac.v)
        genad1da2.v
        ad1da2basys.ucf
```

```
\s3eadclpfdac
    s3eadclpfdac.ise
    s3eadclpfdac.v
        (s3eadc.v)
        (s3eprogamp.v)
        (s3edac.v)
        (clock.v)
        (dacs3edcm.v)
        genadclpfdac.v
    adclpfdacs3esb.ucf
\s3eadcdtdac
    s3eadcdtdac.ise
    s3eadcdtdac.v
        (s3eadc.v)
        (s3eprogamp.v)
        (clock.v)
        (dacs3edcm.v)
    genadcdtdac.v
    adcdtdacs3esb.ucf
\s3eadcfirdac
    s3eadcfirdac.ise
    s3eadcfirdac.v
        (s3eadc.v)
        (s3eprogamp.v)
        (s3edac.v)
        (dacs3edcm.v)
        (fir1.v)
        genadcfirdac.v
    adcfirdacs3esb.ucf
\s3ead1firda2
    s3ead1firda2.ise
    s3ead1firda2.v
        (ad1adc.v)
        (da2dac.v)
        (fir1.v)
        genad1firda2.v
    ad1firda2s3esb.ucf
\s3eadchbdac
    s3eadchbdac.ise
    s3eadchbdac.v
        (s3eadc.v)
        (s3eprogamp.v)
        (s3edac.v)
        (dacs3edcm.v)
        (hb1.v)
        genadchbdac.v
    adchbdacs3esb.ucf
```

```
\sincoslut
    \s3esincosdtmf
        s3esincosdtmf.ise
        s3esincosdtmf.v
            (s3edac.v)
            (dacs3edcm.v)
            (clock.v)
            sincos1.v
            gendtmf.v
        sincosdtmfs3esb.ucf

\dds
    \s3eddsfreqgen
        s3eddsfreqgen.ise
        s3eddsfreqgen.v
            (s3edac.v)
            (rotary.v)
            (dacs3edcm.v)
            dds1.v
            genddsfreq.v
        ddsfreqgens3esb.ucf
    \s3eddsfsk
        s3eddsfsk.ise
        s3eddsfsk.v
            (s3edac.v)
            (dacs3edcm.v)
            (clock.v)
            dds2.v
            lfsr.v
            genddsfsk.v
        ddsfsks3esb.ucf
    \s3eddspsk
        s3eddspsk.ise
        s3eddspsk.v
            (s3edac.v)
            (dacs3edcm.v)
            (clock.v)
            dds3.v
            lfsr.v
            genddspsk.v
        ddspsks3esb.ucf
    \s3eddsqpsk
        s3eddsqpsk.ise
        s3eddsqpsk.v
            (s3edac.v)
            (dacs3edcm.v)
            (clock.v)
            dds4.v
            lfsrdibit.v
            genddsqpsk.v
        ddsqpsks3esb.ucf
```

```
\rs232
    \s3elcdkbduart
        s3elcdkbduart.ise
        s3elcdkbduart.v
            (clock.v)
            (keyboard.v)
            (lcd.v)
            scancode.v
            rcvr.v
            txmit.v
            genlcd.v
        lcdkbduarts3esb.ucf

\Manchester
    \s3emanchester
        s3emanchester.ise
        s3emanchester.v
            (clock.v)
            (keyboard.v)
            (lcd.v)
            scancode.v
            decoder.v
            encoder.v
            genlcd.v
        manchesters3esb.ucf

\servocontrol
    \s3eservocontrol
        s3eservocontrol.ise
        s3eservocontrol.v
            (clock.v)
            (rotary.v)
            servospeed.v
            servomotor.v
            servocontrol.v
        servocontrols3esb.ucf
```

# About the Author

Dennis Silage is a Professor in the Department of Electrical and Computer Engineering at Temple University. He has a PhD in Electrical Engineering from the University of Pennsylvania. He is a Senior Member of IEEE and Director of the System Chip Design Center *www.temple.edu/scdc* of Temple University which researches the application of programmable gate arrays in digital signal processing and digital communication. He is the author of *Digital Communication Systems using SystemVue* (Thomson Delmar 2006). He has published over 80 articles on digital signal processing, digital image processing and digital communication implementations. His email address is *silage@temple.edu* and academic website is *astro.temple.edu/~silage*.